WATER DIPLOMACY

Water is the resource that will determine the wealth, welfare, and stability of many countries in the twenty-first century. This book offers a new approach to managing water that will overcome the conflicts that emerge when the interactions among natural, societal, and political forces are overlooked. At the heart of these conflicts are complex water networks. In managing them science alone is not sufficient, but neither is policy-making that doesn't take science into account. Solutions will only emerge if a negotiated or diplomatic approach—that blends science, policy, and politics—is used to manage water networks. The authors show how open and constantly changing water networks can be managed successfully using collaborative adaptive techniques to build informed agreements among disciplinary experts, water users with conflicting interests, and governmental bodies with countervailing claims. Shafiqul Islam is an engineer with over twenty-five years of practical experience in addressing water issues. Lawrence Susskind is founder of MIT's Environmental Policy and Planning Program and a leader of the Program on Negotiation at Harvard Law School. Together they have developed a text that is relevant for students and experienced professionals working in a variety of engineering, science, and applied social science fields. They show how new thinking about water conflict can replace the zero-sum battles that pit experts, politicians, and stakeholders against each other in counterproductive ways. Their volume not only presents the key elements of a theory of water diplomacy, it includes excerpts and commentary from more than two dozen seminal readings as well as practice exercises that challenge readers to apply what they have learned.

Shafiqul Islam is the first Bernard M. Gordon Senior Faculty Fellow in Engineering and Professor of Water Diplomacy at the Fletcher School of Law and Diplomacy at Tufts University. He is the Director of the Water Diplomacy Initiative. His research group—a diverse network of national and international partners—integrates theory and practice to create actionable water knowledge. He has published over 100 refereed journal and other publications.

Lawrence E. Susskind is Ford Professor of Urban and Environmental Planning at the Massachusetts Institute of Technology. He has served on the faculty for 40 years. He is also Vice-Chair for Instruction at the Program on Negotiation at Harvard Law School, which he helped found in 1982, and where he heads the MIT–Harvard Public Disputes Program, and teaches advanced negotiation courses. In 1993, Susskind created the Consensus Building Institute.

WATER DIPLOMACY

A Negotiated Approach to Managing Complex Water Networks

Shafiqul Islam
and
Lawrence E. Susskind

RFF PRESS
RESOURCES FOR THE FUTURE

First published 2013
by RFF Press
Routledge, 711 Third Avenue, New York, NY 10017

Simultaneously published in the UK and ROW
by RFF Press
Routledge, 2 Park Square, Milton Park, Abingdon, Oxon OX14 4RN

RFF Press is an imprint of the Taylor & Francis Group, an informa business

Library of Congress Cataloging-in-Publication Data

Islam, Shafiqul, 1960-
Water diplomacy : a negotiated approach to managing complex water
networks / Shafiqul Islam, Lawrence Susskind, and associates.
    p. cm.
    Includes bibliographical references and index.
    1. Water-supply—Management. 2. Diplomacy. 3. Water-supply—
Political aspects. 4. Water-supply—Government policy.
    5. Water-supply—Environmental aspects. 6. Water resources
development. I. Susskind, Lawrence. II. Title.
    TD345.I84 2012
    333.91—dc23   2011053475

ISBN: 978-1-61726-102-2 (hbk)
ISBN: 978-1-61726-103-9 (pbk)

Typeset in Bembo
by Cenveo Publisher Services

ABOUT RESOURCES FOR THE FUTURE *AND* RFF PRESS

Resources for the Future (RFF) improves environmental and natural resource policymaking worldwide through independent social science research of the highest caliber. Founded in 1952, RFF pioneered the application of economics as a tool for developing more effective policy about the use and conservation of natural resources. Its scholars continue to employ social science methods to analyze critical issues concerning pollution control, energy policy, land and water use, hazardous waste, climate change, biodiversity, and the environmental challenges of developing countries.

RFF Press supports the mission of RFF by publishing book-length works that present a broad range of approaches to the study of natural resources and the environment. Its authors and editors include RFF staff, researchers from the larger academic and policy communities, and journalists. Audiences for publications by RFF Press include all of the participants in the policymaking process—scholars, the media, advocacy groups, NGOs, professionals in business and government, and the public.

CONTENTS

CONTRIBUTORS

Shafiqul ("Shafik") Islam is the Director of the Water Diplomacy Initiative at Tufts University. He is also the first Bernard M. Gordon Senior Faculty Fellow in Engineering and Professor of Water Diplomacy at the Fletcher School of Law and Diplomacy at Tufts. His research group, a diverse network of national and international partners, integrates theory and practice to create actionable water knowledge. Dr. Islam maintains an active national and international consulting and training practice ranging from flood forecasting in India and national water planning in Bangladesh, to water policy planning for ExxonMobil and advising the South Asia Consortium for Interdisciplinary Water Resources Studies. He has published over 100 refereed journal and other publications.

Lawrence ("Larry") E. Susskind is Ford Professor of Urban and Environmental Planning at the Massachusetts Institute of Technology and Director of the MIT Science Impact Collaborative. He was one of the founders of the inter-university Program on Negotiation at Harvard Law School, as well as the founder of the Consensus Building Institute, a not-for-profit provider of mediation services in complicated public disputes around the world. Professor Susskind is the author or co-author of more than 20 books, including *Breaking the Impasse*, *Environmental Diplomacy*, and *The Consensus Building Handbook*.

Catherine ("Cat") M. Ashcraft is a Visiting Assistant Professor in the Environmental Studies Program at Middlebury College. She has also served as a Visiting Assistant Professor at Pratt Institute, as a Visiting Instructor in the Government Department and Environmental Studies Program at Colby College, and as a Senior Consultant with the Consensus Building Institute. Catherine did her PhD research at the Massachusetts Institute of Technology focused on the adaptive governance of two international river basins, the Danube and the Nile.

Paola Cecchi-Dimeglio is an attorney-mediator and currently a post-doctoral researcher at the Program on Negotiation at Harvard Law School. She received the JAMS Foundation Weinstein Fellowship for her research on Alternative Dispute Resolution (ADR). She is Co-chair of the American Bar Association's IC Sub-committee on the Future of ADR and has served as an expert advisor for several EU projects on ADR. In addition to her academic career, she serves as a consultant for organizations in international partnerships.

Peter Kamminga is an Associate Professor of Law at VU University Amsterdam and postdoctoral researcher at the Program on Negotiation at Harvard Law School. He is a trained mediator and is currently consulting for organizations involved in infrastructure development. Prior to his academic career he practiced law at one of the Netherlands' largest law firms. He has published several articles and co-authored books on the subjects of dispute resolution and cooperation in complex multi-party situations.

Elizabeth ("Betsy") Fierman is an associate at the Consensus Building Institute. She holds a Master of Arts in Law and Diplomacy from the Fletcher School at Tufts University, and a Bachelor of Arts from Haverford College. A native of the Boston area, she has lived and studied in Chile.

Maia Majumder is a dual-degree student at Tufts University's School of Engineering and School of Medicine. Her graduate studies are concentrated in engineering science, epidemiology and biostatistics. She is the co-founder of The Village Zero Project: an initiative that synthesizes epidemiology and engineering, using mobile health technologies to spatially and temporally track the origination and propagation of endemic infectious diseases to better inform cost-effective prevention strategies.

PREFACE

To address the emerging realities of our globalized world, we can no longer rely on the popular twentieth-century paradigm to which we have become so accustomed: scientists innovate; politicians make policy; and people respond, especially when they are unhappy. We offer a twenty-first-century approach to water management that acknowledges the complexity and uncertainty of natural and societal systems, accepts the increasing interconnectivity and consequences of important decisions, and rejects the unquestioned authority of hierarchical governance structures.

Our views have been shaped by a number of important books—*The Consolation of Philosophy* (Boethius, 525AD), *The Reflective Practitioner* (Schon, 1983), *Managing the Unknowable* (Stacey, 1992), *At Home in the Universe* (Kauffman, 1995), *The End of Certainty* (Prigogine, 1996), *The Science of the Artificial* (Simon, 1996), *The Third Side* (Ury, 1999), *The Black Swan* (Taleb, 2007), *Thinking in Systems* (Meadows, 2008), *Working Together* (Poteete, Janssen, and Ostrom, 2010), *Practical Wisdom* (Schwartz and Shapiro, 2010) and *Water Wisdom* (Tal and Rabbo, 2010).

Our approach to water diplomacy starts with a question: How can we ensure effective management of water as a common pool resource given that we can neither predict nor control many of the forces involved in its allocation and use? We think of diplomacy as the process of defining and resolving water issues at every level—from the design of a small-scale sanitation system in a village, to the development of a contested hydroelectric facility in one region of a country, to formal treaty negotiations among different nations.

Water problems are shaped by many natural, societal, and political interactions that create complex water networks. As population growth, economic development and climate change put increasing pressure on water resources, the management of these networks becomes increasingly important. Science cannot provide all the answers. Policy-makers must take what scientists have to say into account, but beyond that, they also need to empower the relevant stakeholders to help formulate and implement solutions. To do this, we believe it will help to think of water as a flexible, even an expandable resource.

In our assessment, the most vexing water management problems are neither simple nor complicated. Simple problems are easily understood and manageable. Complicated problems, while not simple, involve interactions that are still knowable and predictable. Complex problems—and that is what most water management problems are—involve interactions that are both unknowable and unpredictable. Complex problems like these are not easily controlled. They involve too many variables, too many interactions and too much feedback.

For centuries we have taken nature apart and analyzed its components in ever-increasing detail. Now we realize that such "reductionism" can only provide limited insight. Water systems are more than the sum of their parts. "Systems engineering," which water managers have relied on for years, does not work well when natural, societal, and political boundaries are mismatched and cause–effect relationships are ambiguous.

We view water networks as an interconnected set of nodes representing natural, societal, and political variables. The flow of information among these nodes is what enables them to evolve and adjust. Our challenge is how best to manage the flow of information to formulate and achieve desired outcomes. It is in this context that we propose a new Water Diplomacy Framework (WDF) rooted in ideas from complexity theory and non-zero-sum negotiation. Water users and managers can use this Framework to link scientific objectivity and contextual understanding.

Throughout the development of this book, Shafik Islam has had the help of an extraordinary set of mentors, students, and friends. Several deserve special mention including A. Akanda, R. Bras, A. Chassot-Repella, E. Choudhury, Y. Gao, A. Jutla, P. Mollinga, I. Rodriguez-Iturbe, W. Moomaw, K. Portney, M. Reed, D. Small, and R. Vogel. Shafik also wants to acknowledge the love, support, and encouragement of his parents, his wife (Naaz), and their two wonderful daughters (Maia and Myisha). Without their unyielding support and their wise and diligent criticism during never-ending dinner-table conversations, this work would not exist.

Larry Susskind wants to thank Sossi Aroyan, Carri Hulet, Peter Kamminga, Paola Cecci-Dimeglio, Elizabeth Fierman, Todd Schenk, Noah Susskind and Nina Tamburello for their unstinting assistance in preparing this manuscript.

Shafik Islam
Larry Susskind
March 12, 2012

References

Boethius, A.M.S. (575 AD). *The Consolation of Philosophy*, English Translation by H.R. James, Signature Press Edition.

Kauffman, S. 1995. *At Home in the Universe: The Search for the Laws of Self Organization and Complexity*. Oxford: Oxford University Press.

Meadows, D.H. 2008. *Thinking in Systems: A Primer*, D. Wright (ed.), Chelsea Green Publishing.

Poteete, A.R., Janssen, M.A., and Ostrom, E. 2010. *Working Together: Collective Action, the Commons, and Multiple Methods in Practice*. Princeton, NJ: Princeton University Press.

Prigogine, I. 1996. *The End of Certainty: Time, Chaos, and the New Laws of Nature*. New York: The Free Press.

Schon, D.A. 1983. *The Reflective Practitioner: How Professionals Think in Action*. New York: Basic Books.

Schwartz, B. and Shapiro, K. 2010. *Practical Wisdom: The Right Way to do the Right Thing*. New York: Riverhead Books.

Simon, H.A. 1996. *The Science of the Artificial*. Cambridge, MA: The MIT Press.

Stacey, R.D. 1992. *Managing the Unknowable: Strategic Boundaries Between Order and Chaos in Organizations*. San Francisco, CA: The Jossey-Bass Management Series.

Tal, A. and Rabbo, A.A. 2010. *Water Wisdom: Preparing the Groundwork for Cooperation and Sustainable Water Management in the Middle East*. New Brunswick, NJ: Rutgers University Press.

Taleb, N.N. 2007. *The Black Swan: The Impact of the Highly Improbable*. New York: Random House.

Ury, W. 1999. *The Third Side: Why We Fight and How We Can Stop*. New York: Penguin Books.

ACKNOWLEDGEMENTS

Excerpts from the following materials have been reprinted in various chapters with the permission of the copyright holders. We would like to express our gratitude to those who have made it possible for us to include these works in this volume.

Allen, P. 2001. What is complexity science? Knowledge of the limits of knowledge, *Emergence, 3*(1), 24–42. Reprinted by permission of the publisher.

Ashcraft, C.M. (2011). *Indopotamia: Negotiating Boundary-Crossing Water Conflicts.* Program on Negotiation, Harvard Law School. Reprinted by permission.

Berkes, F. 2006. From community-based resource management to complex systems. *Ecology and Society, 11*(1): 45. Reprinted by permission of the publisher.

Bloschl, G. and Sivapalan, M. 1995. Scale issues in hydrological modeling: A review, *Hydrological Processes, 9*: 251–290. Reprinted by permission of the publisher.

Brooks, D. & Trottier, J. 2010. Confronting Water in an Israeli–Palestinian Peace Agreement, *Journal of Hydrology, 382*(1-4): 103–114. Reprinted by permission of the publisher.

Burkhard, R., Deletic, A., and Craig, A. 2000. Techniques for Water and Wastewater Management: A Review of Techniques and Their Integration in Planning Review, *Urban Water, 2*(3); 197–221. Reprinted by permission of the publisher.

Cash, D.W., Adger, N., Berkes, F., Garden, P., Lebel, L., Olsson, P., Pritchard, L., and Young, O. 2006. Scale and cross-scale dynamics: governance and information in a multilevel world, *Ecology and Society 11*(2): 8. Reprinted by permission of the publisher.

Cohen, A. and Davidson, S. 2011. The watershed approach: Challenges, antecedents, and the transition from technical tool to governance unit, *Water Alternatives, 4*(1): 1–14. Reprinted by permission of the publisher.

Coman, K. 1911. Some unsettled problems of irrigation, *American Economic Review, 1*(1): 1–19. [Reprinted in *American Economic Review, 101* (February 2011): 36-48]. Reprinted by permission of the publisher.

El-Sadek, A. 2010. Water desalination: An Imperative Measure for Water Security in Egypt, *Desalination, 250*(3): 876–884. Reprinted by permission of the publisher.

Fuller, B. 2006. "Trading zones: cooperating for water resource and ecosystem management when stakeholders have apparently irreconcilable differences." *Dissertation, Massachusetts*

Institute of Technology, Department of Urban Studies and Planning. Reprinted by permission of the author.

Fuller, B. 2009. Surprising cooperation despite apparently irreconcilable differences: agricultural water use efficiency and CALFED, *Environmental Science and Policy, 12*(6): 663–673. Reprinted by permission of the publisher.

Gibson, C.C., Ostrom, E. and Ahn, T.K. 2000. The concept of scale and the human dimensions of global change: a survey, *Ecological Economics 32*: 217–239. Reprinted by permission of the publisher.

Guan, D. and Hubacek, K. 2007. Assessment of regional trade and virtual water flows in China, *Ecological Economics, 61*: 159–170. Reprinted by permission of the publisher.

Kallis, G., Kiparsky, M., and Norgaard, R.B. 2009. Adaptive governance and collaborative water policy: California's CALFED Bay-Delta Program, *Environmental Science and Policy, 12*(6): 631–643. Reprinted by permission of the publisher.

Kiang, J.E., Olsen, J.R., and Waskom, R.M. 2011. Introduction to the featured collection on "Nonstationarity, Hydrologic Frequency Analysis, and Water Management." *Journal of the American Water Resources Association, 47*(3):433–435. Reprinted by permission of the publisher.

Liu, J., Dietz, T., Carpenter, S.R., Alberti, M., Folke, C., Moran, E., Pell, A.N., Deadman, P., Kratz, T., Lubchenco, J., Ostrom, E., Ouyang, Z., Provencher, W., Redman, C.L., Schneider, S.H., and Taylor, W.W. 2007. Complexity of coupled human and natural systems, *Science, 317*: 1513–1516. Reprinted by permission of the publisher.

Luijendijk, J. and Arriëns, W.L. 2007. *Water Knowledge Networking: Partnering for Better Results*. The Netherlands: UNESCO-IHE. Reprinted by permission of the publisher.

Mitleton-Kelly, E. 2003. Ten principles of complexity and enabling infrastructures, in E. Mitleton-Kelly (ed.) *Complex Systems and Evolutionary Perspectives on Organizations: The Applications of Complexity Theory of Organizations* (pp. 23–50). Oxford: Elsevier. Reprinted by permission of the publisher.

Narayanan, N.C. and Venot, J.P. 2009. Drivers of change in fragile environments: Challenges to governance in Indian wetlands, *Natural Resources Forum, 33*: 320–333. Reprinted by permission of the publisher.

Odeh, N. 2009. "Towards improved partnerships in the water sector in the Middle East: A case study of partnerships in Jordan's water sector." Dissertation, Department of Urban Studies and Planning, Massachusetts Institute of Technology. Reprinted by permission of the author.

Ostrom, K. 2011. Reflections on "Some Unsettled Problems of Irrigation," *American Economic Review, 101*: 49–63. Reprinted by permission of the publisher.

Pahl-Wostl, C., Craps, M., Dewulf, A., Mostert, E., Tabara, D., and Taillieu, T. 2007. Social learning and water resources management, *Ecology and Society, 12*(2): 5. Reprinted by permission of the publisher.

Pollard, S., du Toit, D., and Biggs, H. 2011. River management under transformation: The emergence of strategic adaptive management of river systems in the Kruger National Park, *Koedoe, 53*(2). No permission required to reprint.

Radosevich, G. 2010. *Mekong River Basin, Agreement & Commission.* IUCN Water Program Negotiate Toolkit: Case Studies. Reprinted by permission of the publisher.

Rittel, H.W. and Webber, M.M. 1973. Dilemmas in a general theory of planning, *Policy Sciences, 4*: 155–169. Reprinted by permission of the publisher.

Saravanan, V.S. 2008. A systems approach to unravel complex water management institutions, *Ecological Complexity, 5*(3): 202–215. Reprinted by permission of the publisher.

Sgobbi, A. and Carraro, C. 2011. A stochastic multiple players multi-issues bargaining model for the Piave river basin, *Strategic Behavior and the Environment, 1*(2): 119–150. Reprinted by permission of the publisher.

Sivapalan, M., Thompson, S.E., Harman, C.J., Basu, N.B., and Kumar, P. 2011. Water cycle dynamics in a changing environment: Improving predictability through synthesis, *Water Resources Research, 47*. Reprinted by permission of the publisher.

Tapela, B.N. 2006. Stakeholder participation in the transboundary management of the Pungwe river basin, in A. Earle and D. Malzbender (eds.) *Stakeholder Participation in Transboundary Water Management* (pp. 10–34), Cape Town: African Centre for Water Research. Reprinted by permission of the author.

Velázquez, E. 2007. Water trade in Andalusia, virtual water: An alternative way to manage water use, *Ecological Economics, 63*(1): 201–208. Reprinted by permission of the publisher.

Werick, B. 2007. Changing the rules for regulating Lake Ontario levels, in *Computer Aided Dispute Resolution, Proceedings from the CADRe Workshop* (pp. 119–128). Institute for Water Resources. Reprinted by permission of the publisher.

LIST OF ACRONYMS

ACF	Apalachicola-Chattahoochee-Flint
BATNA	Best Alternative to a Negotiated Agreement
CALFED	CALFED Bay-Delta Program (CA and Federal agreement)
CAM	Collaborative Adaptive Management
GWP	Global Water Program
IWRM	integrated water resources management
JFF	joint fact-finding
MRC	Mekong River Commission
PON	Program on Negotiation at Harvard Law School
RCN	research coordination network
USACE	U.S. Army Corps of Engineers
WDN	Water Diplomacy Network
WDF	Water Diplomacy Framework
WDW	Water Diplomacy Workshop
ZOPA	Zone of Possible Agreement

1

A WATER MANAGEMENT FABLE FOR ALL TIME

(with Maia Majumder)

Once upon a time, many millennia ago, three hunter-gatherer tribes settled in different parts of a river water basin known as Indopotamia. People were few and resources were plenty. As the population grew, the tribes realized that they could no longer depend entirely on wild foods. Slowly, they learned to cultivate rice. The best land for agriculture, however, became increasingly scarce. Land was transformed into a symbol of power. To protect themselves, and to assert their authority, the tribes established geographic boundaries and organized governments. These eventually became the modern day states of Alpha, Beta, and Gamma.

Flash forward to the present. For centuries, there have been tensions among the three countries. Alpha is the largest of the three, and is economically and politically dominant. It has long monopolized access to the river, insisting that only after its water needs have been met will Beta and Gamma be allowed more water. Continued population growth and increasing crop productivity in Alpha, however, seem never to leave enough water for the smaller upstream countries. As it has continued to devote all of its resources to its own political and economic development, Alpha has earned a reputation as self-serving, cruel, and uncompromising.

Periodic droughts and forest fires plague Gamma. Lack of stored water undermines its ability to deal with these problems, and has slowed its economic development. Also, much of Gamma's groundwater is contaminated with arsenic, forcing it to rely almost entirely on the river for its drinking water. Beta, on the other hand, has lots of farmable land, but it does not have adequate labor supplies. In recent years, it has tried to shift to less labor-intensive energy-powered agriculture. This move has been difficult because Beta can't afford the required gasoline or coal. Unless Beta can import cheap labor or generate hydropower from the river, its crop development and economic growth will continue to suffer.

In desperation, Beta and Gamma have decided to join forces and build a dam. They are seeking funding from the Regional Lending Agency (RLA). They won't qualify for loans and grants, though, unless they can reach a water usage agreement with Alpha.

Alpha is quite unhappy about the idea of the dam. But because the RLA will only give money to countries that have good relations with their neighbors—and because Alpha wants money from the RLA for its own purposes—Alpha has signed a statement saying it supports the dam. Simultaneously, Alpha has threatened Beta and Gamma through indirect channels, indicating that it won't be able to control local insurgents who may destroy the dam if it is built. This action has only increased the strain on Alpha's already poor relationships with Beta and Gamma.

Alpha has reason for alarm. If the dam is built, it won't have the water it needs to grow enough rice—the staple food of its population and economy. This concern has become increasingly acute in recent decades, as glacial melting and associated sea-level rise have caused saltwater intrusion in the Indopotamia basin. As a result, Alpha's coastal rice paddies are too salty to produce rice. Alpha used to be able to grow whatever it needed. It didn't have to rely on any of its neighbors, so friendly relationships with Beta or Gamma were not important. With the loss of coastal farmland, Alpha now needs help.

Alpha has such strained relationships with Beta and Gamma that it cannot ask for assistance. Instead, over the past few decades, Alpha has pumped more and more water from the basin to expand rice production. Beta and Gamma have not had the economic or military clout to stop this.

Alpha uses the additional water to irrigate dry land in Mu, its most populous state. Mu sits north of the coast at a relatively high elevation. It is one of the few parts of the country that has not been affected by saltwater intrusion. Its soil, though dry, is nutritious and can be made fertile for rice growing. But, as Alpha's population has continued to expand, irrigation of Mu's rice paddies has required many millions of gallons of water every season. Over the past few decades, Alpha's active pumping has resulted in the draining of lakes, ponds and streams in Beta and Gamma.

The residents of Mu feel that they have been robbed of their rights by the central government of Alpha. Wages have been cut drastically to pay for the ever-increasing energy costs associated with the active pumping scheme. Residential areas have been cleared to create more farmland. The central government has made it clear to the citizens of Mu that their purpose is to feed the country.

One faction in Mu has had enough. It is planning a rebellion, intending to cut off the nation's rice supply. Plans are afoot for organized military resistance. Mu is on the border with Beta, and has opened covert lines of communication with Beta's national government. Mu has offered to trade subsidized labor for arms and military support.

During diplomatic discussions about the dam, the relationship between Beta and Mu was revealed. The discovery undermined trust between Beta and Gamma. Gamma worries that Alpha's attempts to settle its dispute with Mu will make Beta's need for additional workers clear. Gamma believes that when Alpha realizes this, it will offer Beta the subsidized labor it needs in exchange for discontinuing its support for the dam project. Because the RLA intends to split the money for the dam between the two countries, Gamma won't be able to finish construction of the dam if Beta abandons the project.

Beta could, in fact, get by without the dam. It wants the dam primarily to support water-powered energy production as it attempts to convert its rice paddy fields to machine-operated farms. In the long term, these will be much more economically efficient. If, however, Beta can find additional workers whose wages are paid by Alpha, the conversion to energy-powered farming will not be necessary. By contrast, the only way for Gamma to deal with its recurring droughts, forest fires, and widespread dehydration is to increase its access to the river. Otherwise, Gamma will remain agriculturally and socioeconomically backward, falling further behind its neighbors.

In an ideal world, Alpha would allow Beta and Gamma to build the dam—free of insurgent attacks—in exchange for rice. This would allow Gamma to deal with dehydration, drought and wildfires. It would also allow Beta to pursue its conversion to machine-operated farming. By exporting rice to Alpha, Beta and Gamma would gain diplomatic clout, creating a stronger web of interdependencies among the three countries that would be more durable in the face of external adversity. Alpha's relationship with Mu, which would no longer be responsible for producing food for the entire country, could heal. Finally, with peace in the region, the RLA and other international players would be more inclined to increase their investments.

Is Alpha too proud to agree to this solution? Can Gamma trust Beta to follow through on the dam construction? Or, will Beta turn on Gamma if Alpha offers subsidized labor? Will Mu be able to reconcile with Alpha's central government, or will there be civil war?

The answers to these questions are contingent upon the three countries being able to turn an age-old conflict into a problem-solving opportunity.

Welcome to our world of water diplomacy!

2

CHALLENGING THE CONVENTIONAL WISDOM ABOUT WATER MANAGEMENT

The Evolution of Water Resources Management

In our fictional world of Indopotamia, water was not a scarce resource during the time of our hunter-gatherer ancestors. Early civilization started near water and thrived along rivers. For thousands of years—from Mesopotamian cities, to Mayan civilization, to modern day Boston in the United States—we have figured out ways to move water around to facilitate development and enhance the quality of life. It is common knowledge that fresh water is a finite resource and that the overall supply of fresh water has not changed since the time of the dinosaurs.

Supplying enough water has been the focus of water management for centuries. Boston was founded in 1620 (although there are indications that Native Americans inhabited the area as long ago as 2500 BC). Like many other places, people relied on cisterns, wells, and a spring on Boston Common for their water in those early days. As the city grew, this supply was inadequate and the quality was often poor. In 1848, Boston drew its municipal water supply from Lake Cochituate. Water flowed from there into the Frog Pond on Boston Common. On October 25, 1848, there was a famous water celebration to acknowledge the opening of the Frog Pond. It began with the roar of 100 cannons and that drew over 100,000 people. In 1848, Boston's population was only 127,000. By 1900, it had tripled to over 550,000. To facilitate continued growth over the next 100 years, engineers had to dramatically enhance water availability. Today, the Metropolitan Water Resources Authority provides water to 2.5 million users in 46 cities and towns around Boston.

Despite this long experience in water resources management—from the water works of Mesopotamian cities, to the marvelous water technologies of the Romans, to the optimism of modern engineering, including dam building, waste water treatment and irrigation—the history of supply-focused water management has not always gone smoothly. The development of Boston's water supply is also a story of how natural and societal systems became entangled, and how this interaction created conflict and confusion. The sources Boston tapped in 1848 moved progressively west, from urban areas to the sparsely populated western portion of the state where the water was as yet untainted. Twentieth-century water supply projects, in contrast to earlier water projects, involved the creation of massive storage reservoirs like the Quabbin reservoir, in the middle of Massachusetts, which necessitated the inundation of several communities. To this day, residents in that area resent the fact that their communities were sacrificed to satisfy Boston's growing need for drinking water.

In 1848, it was a remarkable engineering success to bring ten million gallons per day of water from 19 miles west of Boston via the Cochituate aqueduct. At that time, issues of environmental or societal impact were not explicitly considered or discussed. Over the years, multiple water projects were initiated to meet the city's growing demand. Finally, in 1946 construction of the Quabbin Reservoir, about 50 miles west of Boston, was completed. At that time, the 412 billion gallon reservoir was the largest man-made reservoir in the world devoted solely to water supply. In creating the Quabbin Reservoir, four towns—Greenwich, Enfield, Dana, and Prescott—were inundated. Natural and societal systems became intricately coupled when water management focused primarily on supply-side strategies to meet competing and growing needs.

To identify the most appropriate water policies, programs or projects, analysts usually think in terms of the supply-side and the demand-side. The supply-side is structure-oriented, requiring investments in things that need to be built with the help of engineers and technical experts. Supply-side investments typically hinge on cost–benefit analysis (CBA), a key purpose of which is to quantify the advantages and disadvantages of one project or another in terms of dollar costs.

This focus on a predominantly techno-centered search for additional sources of water to meet growing demand—from the remarkable engineering success story of the 1840s that provided Boston with a piped and clean water supply to the complex stories of tradeoff between the destruction and displacement of marginalized communities—is not unique to Boston. Similar supply-focused and engineering-centered water projects have been embraced and celebrated in many cities and communities around the world, from Bombay to Buenos Aires.

Time and situations, however, have changed. Now, there are over six billion humans competing for roughly the same amount of water that was available eons ago. Now we are concerned about an array of demand-side considerations: Is water a property right or a human right? Do fish have more rights to water than corn? Is maximization of economic utility more important than environmental sustainability? How can we reconcile competing cultural and religious values associated with water? How much water do people actually need, and should we adjust our ways of living to reduce overall demand?

There is an increasing realization that the supply-side focus that has dominated the water agenda throughout human history is insufficient to meet the complex challenges of our time. An earlier generation of water development, typified by the construction of large dams, centralized wastewater treatment plants, and irrigation projects, benefited a large share of the Earth's people by helping to reduce starvation and water-related illness and protecting against devastating floods. With continuing economic development, however, this centralized construction-oriented approach has hit its limits. These constraints are exacerbated by the scarcity of new water sources, the costs of developing them, and the pressures of population growth, rapid urbanization, poverty, ecosystem degradation, biodiversity losses, and global climate change.

Integrated Water Resources Management (IWRM)

Until the 1970s, most water managers sought to solve specific localized water problems without worrying about the impacts that water management decisions might have on other components of natural (water quantity, water quality, ecological functions, and services) and societal (economic, cultural, institutional) systems. The United Nations-sponsored 1977 water conference in Mar del Plata, Argentina is viewed by many (Lee 1992; Biswas 2004; Heathcote 2009) as a landmark event in water management. The occasion gave global recognition to the shortcomings of supply-side focused water management.

There was agreement that water managers could not afford to focus on a single sector or a single commodity-oriented approach. Instead, they needed to take a more balanced, people-oriented approach. This shift acknowledged that water bodies are multiple user systems, and that more equitable and sustainable management can only occur when the needs and goals of multiple users are considered in a multidimensional, multisectoral context. Thus, the era of integrated water resources management (IWRM) was born. Over the past four decades, IWRM has been strongly endorsed at other international events in Rio de Janeiro, Dublin, The Hague, Johannesburg, and Kyoto. Most international organizations, including the United Nations and the World Bank, have adopted IWRM as a guiding principle.

It was not until 2000 that IWRM was defined clearly as "a process which promotes the coordinated development and management of water, land and related resources in order to maximize economic and social welfare in an equitable manner without compromising the sustainability of vital ecosystems and the environment" (Global Water Partnership 2000). In addition, five principles to guide water management emerged: (1) water is a finite and vulnerable resource; (2) a participatory approach is necessary; (3) the role of women should be emphasized; (4) the social and economic value of water must be acknowledged; and (5) the three Es of sustainability (economic efficiency, social equity and ecosystem sustainability) must be given priority.

The Global Water Partnership (GWP) that enunciated these principles, presented IWRM as a way to integrate a range of connotations (Saravanan et al 2009), various definitions (Braga 2001; Jonker 2007 Thomas and Durham 2003) and different approaches (Mitchell 1990). In a critical assessment of the GWP definition of IWRM, however, Biswas argued that this approach cannot be implemented because of unresolved operational questions and related difficulties of specifying assessment criteria and metrics. Biswas argues for a focus on operational ("what will be") concerns and suggests disregarding normative ("what ought to be") and strategic ("what can be") concerns (Mitchell 2008).

A key assumption in the GWP definition of IWRM is the idea that actors in any society can and will seek to reach common understandings and coordinate their actions through reasoned argument, consensus, and cooperation (Habermas 1984). Critics argue, however, and illustrate with actual case studies "why and how [such] integration cannot be achieved" (Saravanan et al 2009). Integration is a political process (Allan 2006; Ingram 2011) that requires "realistic analyses" of existing situations (Mollinga et al 2007), as well as the "pooling of explicit and tacit information to create actionable knowledge" (Islam et al 2009; Islam and Susskind 2011).

Many water-related conflicts are, in fact, framed in terms of actors (individuals, communities, businesses, NGOs, states, and countries) competing to protect their own, often conflicting, economic and political interests. Such competition occurs at various scales and often results in gridlock. Water management is further complicated by the existence of natural, societal, political, as well as physical, ecological, and biogeochemical boundaries that create obstacles to integration, while scientists and engineers rely on boundaries to specify initial conditions, conduct controlled experiments, and measure results (Morehouse 2000; Lawford et al 2003). Political actors set geopolitical boundaries for very different reasons. Often, there is a mismatch between political or societal boundaries and natural or geographical boundaries (Varady and Morehouse 2003). There is an ongoing debate about whether IWRM, strongly linked to the engineering community, is a useful frame for water management, given these boundary conflicts.

Reflections on Some Unsettled Problems of Planning in Managing Common Pool Resources

In 1911, Katharine Coman published "Some Unsettled Problems of Irrigation" in the first issue of the *American Economic Review*. She used irrigation systems in the United States west of the one hundredth meridian to describe intricate collective-action problems. This was almost half a century before Hardin identified the so-called "tragedy of the commons" and associated theoretical and implementation problems related to common pool resources (Hardin 1968). The process of converting the desert into farmland demonstrated the challenges of achieving a collective good (in this case, building and running an irrigation system). Nearly all the settlers who relied on private companies did well. Publicly-supported settlers failed, though, because Congress imposed a long-term residency requirement. This prevented "the man with small capital, but possessing those more valuable qualities of brains, pluck, and endurance ... to earn a farm by the labor of his hands." In her reflections on Coman's article, Ostrom summarized it as:

> One gains a general lesson from this analysis that changing the formal governance structure of irrigation is not sufficient to ensure efficient investment in

facilities or that farmers are able to acquire property and make a reasonable living. Building knowledge and trust are, however, essential for solving collective-action problems.

(Ostrom 2011)

Another seminal article by Rittel and Webber, published nearly 40 years ago and cited over 3,360 times suggests that:

The search for scientific bases for confronting problems of social policy is bound to fail, because of the nature of these problems. They are "wicked" problems, whereas science has developed to deal with "tame" problems.

(Rittel and Webber 1973)

Many of our contemporary water problems are "wicked" problems, and we cannot talk about finding optimal or engineered solutions unless a great many non-objective assumptions are imposed. These subjective considerations undermine the credibility of water managers who claim they are relying on purely scientific or technical judgments.

Water Problems are Complex

We make a distinction among three types of water problems: simple, complicated, and complex. Simple problems are characterized as easily knowable, while complicated problems are not simple, but are knowable and predictable. Complex problems, on the other hand, are not easily knowable, and are usually unpredictable. Designing a water-efficient flushing toilet is an example of a simple water problem. Getting water from the Quabbin Reservoir to take a morning shower on the 16th floor of an apartment building in Boston is a complicated problem. Figuring out how water flows, where every pipe goes, how many pumps and other control fixtures are required, and what type of chemicals are needed to keep water potable takes significant engineering ingenuity and creativity. Still, with careful study we can know with (near) certainty what each component of such a water distribution system requires and how to control it. A complicated system is knowable, predictable, and controllable.

Now, think about the inundation of the four towns required to create the Quabbin Reservoir. Consider the task of balancing the competing demands for water for fishing and farming or urban development and wilderness preservation. Suddenly, we go from a complicated to a complex water problem.

No matter how much effort one puts into studying each of the natural and societal elements of a complex water problem, we will never gain the certainty associated with simple and complicated problems. Complex problems are not fully knowable because there are too many variables and because they interact in unpredictable ways. Complex problems are neither predictable nor controllable.

Water management problems are complex because they arise when natural, societal, and political processes and variables interact. They are complex not only because they involve numerous stakeholders (e.g., farmers, industrial users, urban developers, environmental activists, and others) competing for a limited and common resource, but also because they cross multiple boundaries and scales (e.g., physical, disciplinary, jurisdictional, and others). For centuries we have taken nature apart, analyzing its components in ever increasing detail. We realize now that this process of "reductionism" can only take us so far. For example, our most sophisticated scientific understanding of the structure and properties of a water molecule does not help us explain why a collection of these molecules will be liquid just above 0°C and solid just below 0°C.

"Systems engineering" has been central to the practice of IWRM. A system is usually defined as an interconnected set of components organized in a bounded domain to achieve an objective. Systems are often more than the sum of their parts. "Systems engineering" works beautifully when systems are readily bounded and the cause–effect dynamics involved are well understood. Sending Apollo to the moon or optimizing water distribution are brilliant success stories of the systems engineering approach to solving complicated, but well-defined, scientific problems. When system boundaries are ill defined and cause–effect relationships are not well understood, a systems engineering approach may not provide much insight.

We can understand and optimize simple and complicated systems by taking them apart and analyzing their details. However, we cannot understand and manage complex systems by applying the same reductionist strategy. In our view, continued efforts to apply existing analytical tools, like cost–benefit analysis and optimization theory, to complex water problems are not likely to help (Ackoff 1979; Bennis et al 2010; Stiglitz et al 2010).

We argue that water resources might be more effectively managed if we understood more about the interaction and feedback among components of the relevant natural, societal, and political systems. One of the most powerful tools for representing functional relationships among large numbers of interconnected components is network analysis. A network (or graph) is a collection of nodes (vertices) and links (edges) between the nodes. These links can be directed or undirected, and weighted or unweighted. A water network can be described as an interconnected set of nodes representing natural, societal, and political variables. The flow of information among these nodes to update their status makes them dynamic. Our challenge is to identify the mechanisms that define the flow of information among the nodes.

It is in the context of networks that we propose a new approach to water management—we call this the Water Diplomacy Framework (WDF). It is rooted in complexity theory and non-zero-sum negotiation, and seeks to bridge scientific objectivity and contextual understanding. The WDF posits that complex water problems might be more effectively managed by thinking about water as a flexible resource and invoking three key assumptions about water networks: (1) water networks are open and continuously changing as a function of the interactions among natural, societal, and political forces; (2) water network characterization and management must account for uncertainty, nonlinearity, and feedback; and (3) the management of water networks ought to be adaptive and negotiated using a "non–zero–sum" approach.

Water is a Flexible Resource

Traditionally, water has been viewed and managed as a scarce resource. Any water set aside for one group is presumably unavailable to other groups. But this notion demands closer scrutiny. Water is actually becoming an expandable resource. Ongoing and fundamental change in the sharing of knowledge and new technologies now allows the same resource to be used in multiple ways by multiple users. When viewed as a fixed pie, water allocation leads to conflicts. With new knowledge, however, water may become an expandable resource. Three illustrative examples explain why and how: (1) adoption of new technology and emerging practices from one part of the world (e.g., the United States) have already enhanced water productivity in another (e.g., sub-Saharan African) by a factor of four (Kijne et al. 2003); (2) distinctions between blue water (i.e., in lakes, streams, and aquifers) and green water (i.e., moisture in the soil) can substantially change the total amount of available water for competing uses (Falkenmark and Rockstrom 2006); and (3) the use of embedded or "virtual" water (i.e., water used in the production of goods and services) can reverse water shortages in many parts of the world. In other words, a more flexible conceptualization of water has the potential to minimize water conflicts by capitalizing on knowledge sharing. To make such transformation possible, however, we need a knowledge base that allows access to both explicit (scientifically objective) *water information* and tacit (contextually relevant) *water knowledge*. This synthesis is at the heart of the WDF.

Thinking in Terms of Water Networks Rather than Systems

The origin of many of our water management problems stems from our fragmented or bounded view of water as a "natural object" or, at other times, as a "societal issue," or, at still other moments, as a "political construct." The components of each water resource management puzzle can fit together in so many different ways that it is practically impossible to use "reductionist" or traditional "systems engineering" methodologies to resolve water management conflicts.

To contextualize the complexities of water management problems, we offer the example from the Apalachicola-Chattahoochee-Flint (ACF) river basin shared by three states in the southeastern United States (Figure 2.1). The ACF basin drains 19,800 square miles of western Georgia (GA), northern Florida (FL), and eastern Alabama (AL). Nearly 2.6 million people depend on the ACF for their water. From 1960 to the 1970s, the ACF basin went through a cycle of floods and drought. Water supply and irrigation were the primary uses of water in the basin. The management of these resources by the U.S. Army Corps of Engineers (USACE) worked well when water needs were minimal and stakeholders were few (Leitman 2005).

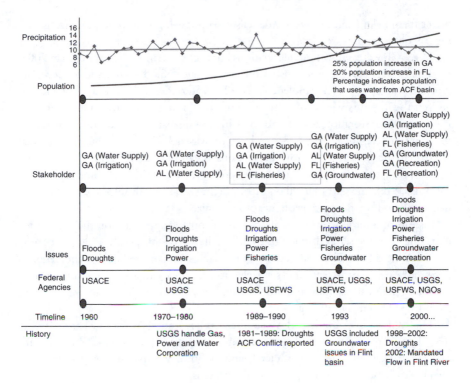

FIGURE 2.1 Evolution of the ACF basin water problem

Conflict grew, however, as the number of competing stakeholders and their needs increased (Figure 2.1). In addition to local socio-economic and environmental changes, the water management context was transformed by the introduction of new laws, such as the National Environmental Policy Act and the Endangered Species Act at the national level. These provided legal ground for Florida, the U.S. Fish and Wildlife Services, and a range of non-governmental organizations to contest decisions made by the USACE. The fragmentation of agency mandates at the national level also created competition and rivalries with state and local governments (Clemons 2004; Leitman 2005; Feldman 2008).

The first incident of interstate water conflict in the ACF basin occurred in 1989. To address increased water demand in Georgia, the USACE proposed to reallocate Lake Lanier's water from hydropower to water supply. This alarmed Alabama because increased withdrawals could stunt economic growth along the Chattahoochee River between Georgia and Alabama. In 1990, Alabama sued the USACE. Florida joined the lawsuit, claiming that reduced water flow would harm the coastal environment, particularly the oyster industry in Apalachicola Bay. Despite 13 years of deliberation, the states were unable to reach agreement on how to allocate the ACF basin's water. Time ran out and the matter went to court (Leitman 2005). In July 2009, a U.S. District Court Judge ruled that the USACE had illegally reallocated water from Lake Lanier to meet Atlanta's urban development requirements, giving Georgia too much water.

The ACF water allocation negotiations highlight (Clemons 2004; Leitman 2005; Feldman 2008) several problems with the way scientific data and socio-economic considerations are often handled. An ACF Water Resources Study intended to address the dispute over possible stream flows (i.e., minimal in-stream flow requirements versus restoration of natural flows in downstream ecosystems) seemed surprisingly limited in its scope. Second, there was no agreement on the data that needed to be collected or how projections regarding future demand should be made. Third, there was a fundamental disagreement about the appropriate allocation formula. While the states agreed on the modeling tools that should be used, they did not agree on how the output ought to be analyzed.

Instead of jointly defining an allocation formula, or agreeing on the criteria for choosing one, the states decided to evaluate alternative proposals using the criteria that they liked best. The lead negotiator for each state changed during the negotiations (as a result of federal and state elections). These deficiencies illustrate how evolving political realities and unresolved scientific questions can make water allocations extremely difficult.

The ACF basin example—as well as the conflict among Alpha, Beta and Gamma described in the opening Indopotamia fable—shows how the links, interactions, and feedback among natural, societal, and political variables make it difficult to characterize and resolve complex water problems.

Acknowledging the Complexity of Water Networks

Many water management problems stem from what we describe as the competition, interconnection, and feedback among natural and societal processes within a political domain (NSPD), as shown in Figure 2.2. Within the natural domain, the interplay among three important variables—water quantity (Q), water quality (P), and ecosystems (E)—can lead to conflict. Within the societal domain, there are equally complex interdependencies and feedback among social values and cultural norms

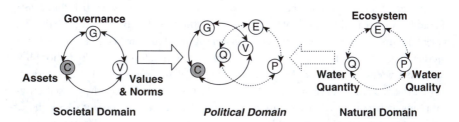

FIGURE 2.2 Interactions among natural and societal processes within a political domain (NSPD)

(V), assets including economic and human resources (C), and governance institutions (G) (Islam, et al 2010; Islam and Susskind 2011).

Our observation that natural and societal domains are linked is not novel. Our argument, however, is that there are strong boundaries between these domains that are likely to get in the way of resolving complex water management problems. Currently, there is no accepted framework to explain how these domains interact or how the interactions among these variables should be managed. A key goal of our work has been to create a better way to characterize (Chapter 3) and manage (Chapter 4) these complex interactions.

A Synthesis of Natural, Societal, and Political Domains

The need to integrate across these three domains has been recognized by many other water scholars and practitioners (Biswas 2004; Lankford and Cour 2005; Pahl-Wostl et al 2007). What we hope to add is the recognition that uncertainty, feedback, and complexity must be addressed explicitly. Each water management problem is highly sensitive to its particular context: because knowledge about the interactions among the three domains is both local and contextual, management interventions that work in one watershed may not be applicable in another.

Thus, an overarching challenge for theory-building is: How do we integrate multiple kinds of knowledge from the natural and societal domains into a diagnostic framework that water professionals can use to deal with the complexity of water management networks in any particular location? More specifically: Why do certain management interventions work in one watershed, but not in another? What can be learned by studying the effects of a particular intervention (e.g., rules of reservoir operation) in different watersheds?

Our guiding hypothesis is: differences in societal processes, natural settings and political contexts are likely to cause the same water management interventions to have different outcomes in different settings. To make appropriate network-specific decisions, we must learn how to characterize the interacting variables and processes (Figure 2.2). We suggest concentrating on the dominant variables in the natural (E, P, and Q) and societal (C, G, and V) domains. The presence or absence of one or more of these variables (C, V, G, Q, P, and E) should help to differentiate water problems and conflicts in different settings and at different scales from each other. For example, the implementation and success of a specific management policy in a particular watershed may depend on the governance arrangements that have emerged in their location. On the other hand, the same governance policy may fail in another watershed where the interactions among C, G, and Q are weak because they depend on ill-conceived incentive instruments (C), leading to ineffective implementation (G) of water allocation plans (Q).

Prioritization of variables and feedback loops in NSPDs is difficult, especially because methods for characterizing these variables are not yet well developed. Water quantity can be measured and characterized with a high level of accuracy using a standard protocol, but it is much more difficult to measure and characterize the effects of governance structures with a similar level of accuracy and certainty. Sometimes inadequate data or simplistic assumptions are used in the absence of a clearer understanding. Often they fail to incorporate adequate diagnostic checks during calibration and validation (Scruggs 2007). One can also find qualitative methods that overstate the generality of particular cases or fail to use well-grounded concepts and theories (Goldthorpe 1997). Clearly, no method is immune to indiscriminate use. Methods of characterizing NSPD relationships need not be discarded because they have been poorly applied in the past. Instead, we must develop improved measurement practices: approaches that combine complementary methods of analysis will probably be superior (Gray et al 2007).

Our focus is in on the need to characterize interactions among the variables and processes in three domains. With the proper analysis in hand, it will be easier to resolve boundary-crossing water disputes, especially when data are difficult to collect and not readily comparable because of disparities in scales and levels. One needs considerable contextual knowledge to grasp the complexities of a particular water network's operations.

Water Diplomacy Framework: Three Propositions

It is in the context of these interactions that we propose a new Water Diplomacy Framework (WDF) rooted in ideas about complexity theory and non-zero-sum approaches to negotiation. Table 2.1 displays the most important elements of the WDF.

There are three key propositions critical to using the WDF.

Water Networks are Open and Continuously Changing

Our first proposition is that water management problems are complex because they arise when natural, societal, and political forces interact. In such networked (i.e., linked) interactions, boundaries are dynamic and porous. This complexity can render the tools of traditional systems engineering and optimization inadequate. In "systems engineering thinking," one seeks to represent interconnected components and their dynamic relationships mathematically. As noted earlier, such an approach works well when systems are bounded, relatively stable, and cause–effect relationships are well understood. When these conditions hold, objective functions (that is, the relationships among variables) can be predicted with confidence. Within the context of coupled water networks, however, when variables are not clearly bounded, and objective functions are more a matter of preference than science (for example, a preference to maximize sustainability as opposed to economic utility), the indiscriminate application of systems engineering or optimization theory will not lead to clear cut solutions.

We need a different way to represent the complex network interactions among variables in the NSPD context. Complexity theory, in our view, is particularly useful for characterizing and managing such relationships in these settings. Perhaps the

TABLE 2.1 Key Elements of the Water Diplomacy Framework (WDF) and How it Differs from Conventional Conflict Resolution Approaches

	The Water Diplomacy Framework (WDF)	Conventional conflict resolution theory (applied to water and other common-pool resources)
Domains and scales	Water crosses multiple domains (natural, societal, political) and boundaries at different scales (space, time, jurisdictional, institutional).	Watershed or river-basin falls within a bounded domain.
Water availability	Virtual or embedded water, blue and green water, technology sharing and negotiated problem-solving that permit re-use can "create flexibility" in water for competing demands.	Water is a scarce resource, and competing demands over fixed availability will lead to conflict.
Water systems	Water networks are made up of societal and natural elements that cross boundaries and change constantly in unpredictable ways within a political context.	Water systems are bounded by their natural components; cause–effect relationships are known and can be readily modeled.
Water management	All stakeholders need to be involved at every decision-making step including problem framing; heavy investments in experimentation and monitoring are key to adaptive management; the process of collaborative problem-solving needs to be professionally facilitated.	Decisions are usually expert-driven; scientific analysis precedes participation by stakeholders; long-range plans guide short-term decisions; the goal is usually optimization, given competing political demands.
Key analytic tools	Stakeholder assessment, joint fact-finding, scenario planning and mediated problem-solving are the key tools.	Systems engineering, optimization, game theory, and negotiation support-systems are most important.
Negotiation theory	The Mutual Gains Approach (MGA) to value creation; multiparty negotiation keyed to coalitional behavior; mediation as informal problem-solving are vital to effective non-zero-sum negotiation.	Hard bargaining informed by prisoner's dilemma-style game theory; principal–agent theory; decision-analysis (Pareto optimality); theory of two-level games.

most profound insight from the study of complex networks is that they are dominated by the unexpected. The smallest causes can have large effects while factors generally considered primary may have no effect at all as perturbations or new information work their way from node to node. Consequently, the message for water managers is: plan as if these networks and interactions among nodes are unpredictable. Effective management of complex water systems requires an approach that takes advantage of the unexpected and assumes an adaptive learning orientation. To that end, we propose viewing water management tasks through the lens of complexity theory with an explicit reliance on adaptive network representation (Barabasi 2003; Bar Yam 2004; Liu et al 2011).

To develop this alternative framework, we begin by assuming that most complex water management problems can be best understood as the product of competition, feedback, and interconnection among natural and societal variables in a political context as shown in Figure 2.2. Given how porous, coupled, and interactive these three domains are, we cannot explain—much less forecast—their behavior without treating them as forces beyond the control of water managers. Framing water management challenges or conflicts in terms of intricately linked natural, societal, and political variables makes clear why they are not readily bounded and change constantly in uncertain ways. Here is an example: an irrigation canal is both a conveyance instrument for water and a device that generates a dependent relationship between upstream and downstream users. Ongoing relationships among the users have as much to do with the management of the canal as they do with the physical or engineering characteristics of the canal. Separating these two attributes of the canal—its "natural" meaning in the environment and its "societal" meaning—and treating them as separate rather than as part of a complex and unpredictable network, will lead to management mistakes.

Reflective water professionals not only need a general understanding of interactions among variables and processes within a water network, but also a method of generating context-specific insights. Local and contextual understanding resides in the experience of a wide array of actors. So, the best method of capturing context-specific information is to involve representatives of all relevant institutional actors and groups. Methods for identifying and ensuring adequate stakeholder and network representation are discussed at length in subsequent chapters.

Water Network Management must Account for Interactions, Non-linearity and Feedback

Recent discussions in the water management community have raised questions about the usefulness of historical statistics for designing water systems for an uncertain future (Milly et al 2008; Stakhiv 2011). A standardized approach in the practice community relies on being able to make precise estimates of the probabilities of events based on historical data. That is, analysts must have confidence that going back in time (or looking at enough previous situations), will yield a clear pattern or tendency that can be relied on to occur, in general, in the future. Once they have a forecast, they try to figure out the most efficient way to achieve their objectives given what the future is likely to hold. However, in the complex world of water management, there is too much uncertainty to forecast with confidence. The emergence of climate change, for example, has already altered the dynamics of the hydrologic cycle, ranging from changes in the frequency of storms to the intensity of rainfall to the timing of snowmelt. It is not possible at the present time to predict the exact impact of climate change on water networks—yet, it is crucial that communities take action now to adapt to the climate change risks they face.

There are tools for managing water resources in the face of uncertainty, but these are quite different from the modeling and forecasting tools that assume stocks and flows of water systems can be modeled with predictive accuracy. Scenario planning, for example, which simulates alternative futures and takes a "portfolio approach" to generating "no regrets" actions offers a way forward (Wright and Cairns 2011). Forecasting models are still important, but they have to be deployed in new ways that focus on a spectrum of potential outcomes under alternative futures rather than on the limited predict–and–choose paradigm that has characterized the past use of models as forecasting tools.

Water Network Management must be Adaptive and use Non-zero-sum Approaches to Negotiation

A dominant assumption in water management has been that the allocation of common-pool resources (another name for public goods like water resources and the ecosystems that support them) is always a win–lose situation. More powerful actors "win" and gain control of resources, less powerful parties "lose" and only have access to water if more powerful nations, groups, water owners, or others permit it. The emergence of non-zero-sum, or mutual gains negotiation theory, over the past few decades, has challenged this win–lose logic by offering a "value-creating" alternative that allows groups with conflicting goals to achieve them simultaneously (Raiffa 1985; Lewicki et al 2010; Mnookin 2010). This mutual gains approach to negotiation rests on the assumption that joint fact-finding, the discovery of interlocking trades, contingent commitments, and an adaptive approach to handling uncertainty can "maximize joint gains." "All-gain" negotiations usually require the assistance of a neutral facilitator or mediator to manage the process of problem solving (Susskind and Cruikshank 1987).

To summarize: water conflicts occur when natural, societal, and political forces interact. Together, these interactions generate what we call water management networks. As population growth, economic development, and climate change impose pressure on finite water resources, interventions at critical nodes and links of these networks will become increasingly important. Science alone is not sufficient to resolve disputes within these networks. Nor is policy-making that does not take science into account likely to yield sustainable solutions. Rather, sustainable solutions are most likely to be found through a negotiated and joint problem-solving approach that blends science, policy, and politics to understand (Chapter 3) and manage (Chapter 4) complex water problems.

Selected Readings with Commentaries

K. Coman "Some Unsettled Problems of Irrigation," (1911)

Introduction

This article appeared in the first issue of the *American Economic Review* 100 years ago in 1911, and focuses on issues of irrigation for the western plains in the United States, describing outcomes achieved by private and public institutional arrangements. Many of the irrigation systems promoted by the federal government failed, while the projects supported through private enterprise did well. By analyzing the relative merits of the two methods of irrigating public land, the author found that the tendency of government officials to focus on legislative details while ignoring more relevant and contextual issues often led to unintended yet disastrous consequences. Another issue she emphasizes is the inadequacies of the water rights systems in the western United States.

We selected this article because the unsettled problems identified by Coman a century ago are relevant and unsettled even today. By holding the ecological systems and geographical region (western states) fixed, she was able to describe how different institutional arrangements and governance structures led to different outcomes for the same management objective. She also describes how our attempts to "tame" nature for societal benefits—in this case, our attempts to make the desert into a garden often lead to very difficult collective action problems because natural, societal, and political systems become intricately coupled.

Our Quest to Tame Nature for Societal Benefit (p. 36)

[Geologist Nathaniel] Shaler called the Cordilleran area, comprising the western third of the United States, the "curse of the Continent." West of the hundredth meridian, precipitation, except for certain favored sections, is insufficient for agriculture or for forest growth, and pasturage can be reckoned on for the spring and early summer only. From the mesas and foothills of the Rockies to the western slope of the Sierras, arid or semi-arid conditions prevail. The average annual rainfall varies from two inches in the deserts of the southwest to 20 inches on the Great Plains, but nowhere except on the north Pacific coast does it furnish a reliance for the farmer. On the western slopes of the mountain ranges, where the moisture-laden winds of the Pacific ascend to colder altitudes, there is considerable rain. The precipitation of autumn and winter is held in vast beds of snow and ice until the fervid suns of May and June release the flow. Then springs and torrents rush down to the lowlands, the rivers overflow their banks, and the valleys are flooded. How to conserve this excess water to serve the needs of summer-grown crops is the problem of arid America.

The first Americans who attempted to farm the desert were the pioneers of the Mormon migration. Brigham Young and the 140 devoted saints who followed his lead across the Wasatch Mountains to the new Zion on the mesa above Great Salt Lake had no knowledge of irrigation; but agriculture was a sine qua non to a settlement so remote from civilization. Within two hours of their arrival they began to plow for a belated planting. "We found the land so dry," wrote Lorenzo Snow, "that to plow it was impossible, and in attempting to do so some of the plow beams were broken. We therefore had to distribute the water over the land before it could be broken."

Our Inability to Distinguish between "Boom and Boomerang" (pp. 37–38)

Water rights were free as air, and every *ranchero* used the streams that the winter rains sent across the lowlands, according to his own convenience. Only in the pueblos, San José and Los Angeles, was there endeavor to treat the scant supply as a common property to be developed by the building of dams and ditches under the direction of the public authorities. The advent of the gold seekers made heavy demands upon the water resources of the Sierras. Running water was a prime necessity in placer mining, the new market afforded by the mining communities induced a far more extensive agriculture, and water rights became, for the first time, a matter of serious concern. The water needed to operate the sluices and long toms had to be conducted to the diggings in flumes, sometimes many miles in length and representing a considerable investment of labor and capital. Mining custom, which the California Legislature formulated into law, established the principle of "first come first served." A notice posted at the point of diversion, stating date of posting and the amount of water to be taken out, constituted a claim to a specified number of miners' inches. The law required that the claim should be registered with the county officer within 60 days of the posting, and the courts later decided that the claimant must prove that the construction of ditches, canals, or flumes had been undertaken immediately and prosecuted with diligence, and that the water drawn off was being put to a beneficial use. Notwithstanding these precautions, the miners' custom, notably when applied to agricultural districts, gave rise to strife and uncertainties that seriously handicapped industrial development. It had become evident that land with no assured water supply was of little value for agriculture. The streams that might be utilized within the resources of individual ranchmen were soon monopolized, and works of greater cost were needed to construct diversion dams, build main canals, and set up pumping machinery for the irrigation of the thousands of acres of fertile land that lay back from the water courses. In the hope of encouraging the development of the Great Valley between the Sierras and the Coast Range, a region all untouched by the Spanish regime, the Legislature (1862) passed an act authorizing the incorporation of canal companies and the construction of canals "for the transportation of passengers and freights, or for the purpose of irrigation and water power, or for the conveyance of water for mining or manufacturing purposes, or for all of such purposes." Under this liberal enactment, several irrigation companies were organized on the basis of claims to the flow of the San Joaquin and its tributary rivers.

The common law doctrine of riparian rights was adopted by the courts of California in the early years of American occupation with no question as to its applicability to the new conditions. Practice vested in the adjacent landowner not only right to a continuous and undiminished flow, but to diversion for his own use and also for sale to agriculturists less fortunately situated. The effect was to give an extraordinary advantage to first comers and to speculators in this most valuable of natural resources. Subsequent cultivators, men who were proposing to convert cattle ranches and wheat fields into orchards and vegetable gardens, were obliged to be content with what was left or to defend their secondary claims by force. Appeal to the courts involved long and costly suits with inconclusive results, since the process of adjudication must be gone over whenever a new claimant appeared or an old claim was revived or extended. Agricultural enterprise in California is still retarded by the uncertainty of water rights and the heavy costs of litigation. Any one of the tens and hundreds of appropriators along a single stream may bring suit to enlarge his share, and the law offers no remedy to the revival of old disputes. Often the small farmer cannot afford the contest, and the victory goes by default to the wealthy ranchman or the well capitalized company. The doctrine of riparian rights even as modified by 50 years of court decisions is hopelessly inadequate, and the water users have had to resort to private agreements with mutual recognition of all the rights appurtenant to the supply in question. The doctrine of appropriation, the only doctrine suited to an arid agriculture, has never been formally recognized in California, much less the superior right of the public to the water supply which is precipitated on the mountain heights, collected in state-owned lakes and rivers, and which is indispensable to the future development of the commonwealth.

By 1875 all the easily irrigated lands of the Cordilleran area were occupied, and it had become evident that the homestead law which had worked so satisfactorily in the Mississippi Valley was quite inapplicable to the conditions presented to the settler in the arid regions. The Desert Land Act was passed (1877) with intent to offer a sufficient reward to induce the man taking up public land to put in irrigating works. A full section of arid land, four times the amount permitted under the preemption or homestead acts, might be acquired by paying 25 cents per acre at the time of entry, redeeming the same by irrigation, within three years, and then paying the final charge of one dollar per acre. Residence on the holding was not required. The Desert Land Act was put through Congress by men who knew little of western conditions, on the assumption that water would always be found upon the land and in such form that it could be readily diverted to the fields; but few regions are so fortunately situated. Irrigation on a scale necessary to utilize distant mountain streams or to pump the subterranean flow required more capital and engineering skill than the ranchmen possessed and usually developed a capacity for watering tens of thousands of acres. The organization of irrigation companies followed. Hydrographic engineering was a new art, construction was not infrequently faulty, water rights uncertain, and the quantity of water available often grossly exaggerated. The promoters of these schemes relied on the fact that the settlers could do nothing toward completing their titles without water, and they devised contracts by which the farmer paid a flat rate per acre, regardless of the amount of water furnished. The temptation to contract to irrigate more land than could be provided for proved irresistible, and many of the settlers were ruined. Such promoters soon discovered, however, that they had killed the goose that laid the golden egg, for without water users there could be no revenue.

Conflicts between Legal and Financial Issues and the Carey Act (1894) (pp. 41–43)

The legal problems of irrigation are thus in a fair way to settlement, but the financial problems remain. The furnishing of water by a private monopoly is no more satisfactory to an agricultural district than to a municipality, and the danger of inadequate supply and exorbitant charges is no less a menace. In southern Spain, where this system obtains and water is sold at auction, the water rates mount in a dry season to an all but prohibitive point. In a wet summer, on the contrary, when the farmers have no need of the artificial supply, they fall so low that the company does not realize enough revenue to offset running expenses. The California law of 1862 empowered water companies "to establish, collect and receive rates, water rents or tolls, which shall be subject to regulation by the board of supervisors of the county or counties in which the work is situated, but which shall not be reduced by the supervisors so low as to yield to the stockholders less than one and one half percent per month on the capital actually invested." This method of adjusting charges has not proven entirely satisfactory to either producer or consumer of the water supply. Eighteen percent, a not unusual rate of profit in the early days of California, is excessive now that there is abundant capital on hand for such investments, while the "capital actually invested" means an overestimate of the present value of the property. No redress was provided in case the supply furnished is insufficient to meet all engagements, and in the not infrequent cases where the canal crosses county lines, the several boards of supervisors may come to different conclusions as to the justice of a given rate. Finally, the courts have ruled that a contract negotiated between company and water user cannot be set aside by the dictum of a public officer, and many farmers have been induced to accept a "contracting-out" clause which renders the arbitrament of the supervisors invalid. However, in the last analysis, the prosperity of the community served is recognized by sane promoters to be the ultimate source of revenue, and charges are regulated by what the farmer can afford to pay rather than by what the water monopoly might possibly extort, while in many districts the waterworks are owned and rates fixed by the farmers themselves.

On the other hand the financial future of the irrigation company is often far from reassuring. Irrigation works are usually built in advance of settlement, and returns sufficient to pay interest on the cost of construction cannot be expected until the number of consumers has reached the full capacity of the flow. Even where all conditions are favorable, water abundant, works adequately built, and soil and climate promising, the promoters of water companies aiming to supply settlers on public lands are often balked of dividends by the "sooners," who seek out each new project in advance of the constructing engineers and locate their claims as soon as the surveyors' stakes are driven. By more or less fraudulent compliance with the homestead act, they manage to get possession of the best land under the prospective canal. They have no intention of developing their holdings and use little or no water for irrigation but hold their patents for a rise in value and thus retard legitimate settlement. An arrangement far more satisfactory both to the farmers and to the purveyor of water obtains where both land and water supply are owned by the same company. Thus the Crocker estate in the San Joaquin valley is being sold in small tracts of five, ten, and 20 acres to actual colonists, and the deed of sale guarantees sufficient water for irrigation at the flat rate of one dollar per acre per year. The same terms are accorded under the Miller & Lux irrigation project and on some of the great California wheat ranches that are now being divided into fruit farms. The system is an admirable one, ensuring to the cultivator the indispensable water supply at a reasonable and unvarying price and to the owner of the works an adequate return on his investment.

Long and intimate acquaintance with the vexations that beset irrigation projects inspired Senator Carey of Wyoming to urge upon Congress the legislation which has finally put the private irrigation of public lands on a rational basis. Under the Carey Act (1894), the federal government offers to make over to any one of the arid states complying with certain provisions as to reclamation and settlement one million acres of the public land or such portion thereof as has been demonstrated by actual survey to be susceptible of irrigation. The land commission of the state participating in this privilege is made responsible for the projects undertaken. The adequacy of the water right, the character of the works, the financial standing of the undertaking company are all passed upon and the prescribed specifications accepted in a written contract before the lands covered by the project may be offered for sale or advertisements issued. The lands are sold by the state officials in tracts of from 20 to 160 acres, at a rate fixed by each state and to bona fide settlers. Persons filing on these lands must furnish proof of at least 30 days' residence and the cultivation of one eighth of the tract before receiving clear title. Furthermore, they must have signed a contract with the water company agreeing to purchase the water right at a specified charge per acre. Ten years is allowed for the water right payment, but this obligation may be anticipated or passed on with the title in case the holding is sold to a later incumbent. The settlers are purchasing not water only but the irrigating system. The price put upon this perpetual possession varies according to the exigencies of construction from $25 to $50 per acre, estimated on the supposition that, all the lands being taken up, the returns will cover the cost of the works and a fair profit on the investment. The capital once recovered, the promoting company proceeds to a new venture, leaving the settlers owners of the works. The water rights are converted into water stock, and a water-users' association is organized in which the farmers hold stock in proportion to their respective acreage. This cooperative company, like the irrigation district, is responsible for the maintenance of canals, the distribution of water, and for any repairs that may prove necessary.

Two Irrigation Experiments with the Same Intended Objective But Very Different Outcomes (pp. 45–46)

We have had sufficient experience of irrigation under the two Congressional enactments to enable us to reach certain conclusions as to the economic problems arising in this new field of agricultural experiment. It has become quite clear that the irrigated farm is not a poor man's proposition. The process of converting the desert into a garden may appear like legerdemain in "before and after" photographs, but in actual experience it is a slow, laborious, and costly affair. The land is the least expensive factor in this situation. The indispensable water right is purchased at from $10 to $90 per acre; the annual cost of keeping up the canals amounts to from one to three dollars per acre, while the grading of the soil and the construction of laterals and ditches calls for a considerable additional outlay. The Reclamation Service recommends that a man undertaking to homestead under one of its irrigation projects bring with him $2,000 to lay out in buildings, stock, and living expenses for the first year or two while the tract is being cleared and graded and made ready for crops. It takes some years to determine what crops are best suited to each variety of soil and climate and for what products a profitable market can be had. On several of the government projects, the Department of Agriculture maintains an experiment station, but even more important to the economic success of the farmers is the service of the project engineer. As soon as the work of construction is accomplished, the hydrographic engineer is succeeded by a man experienced in the problems of soil and climate, the duty of water and the possibilities of cereals, fruits, and vegetables; who has, moreover, sufficient knowledge of human nature to fit him to deal patiently and wisely with the mistakes and discouragements of the novices in irrigation with whom he has to deal. The government has sent some of its best trained men to oversee these reclamation experiments.

The relative merits of the two methods of irrigating the public land now authorized by the federal government may be studied on Snake River, where an area of one million acres is being brought under irrigation, half by private companies operating under the Carey Act, and half by the two government projects of Boise-Payette and Minidoka. The settlers under the Twin Falls Land and Water Company are largely men of some property who have been able to secure their patents within the minimum term, while keeping up water payments of $25 per acre and developing their holdings into flourishing little farms. After securing title to his homestead, the Carey Act farmer may mortgage the land for money with which to make permanent improvements, or, if climate and environment prove unsatisfactory, he can sell to advantage, assigning his water contract to the purchaser, who assumes the unpaid installments. Nearly all the Twin Falls settlers have "proved up," and there has never occurred a cause of failure to meet the annual charges, nor has a single contract been forfeited for arrears.

The Minidoka Project illustrates the virtues and defects of the government method as compared with those of a private company. The long-term residence prescribed by the Reclamation Act is a serious obstacle to enterprising men, far outweighing the

50 cents an acre charge for the state lands, and homesteaders who have means to meet the cash payment gravitate to Twin Falls, notwithstanding the superior quality and lower cost of the government waterworks. The charge for perpetual water rights at Minidoka, determined by the bare cost of constructing and maintaining the irrigating system, is $22 and $30 per acre. The farmers do not come into possession of the works, as under the Carey Act, and this seems, at first blush, an injustice; but community management has its difficulties and dangers. Under the Reclamation Service, the water-users' association must pay for the maintenance of the canals and the distribution of the water, but the dams and the reservoirs remain the property of Uncle Sam, and he is responsible for their integrity. This principle will probably be extended to the electric power developed at the government dam. The current is used for the benefit of landowners under the project (the charge for private lighting is one half a cent per kilowatt hour); but the water-users are responsible only for its economical and equitable distribution. The advantage of having the Reclamation Service assume deterioration charges for these costly plants is evident.

Consequences of a Policy Requirement that had Nothing to do with Irrigation (pp. 47–48)

The wisest and most experienced of the project engineers are agreed that the chief difficulty of the farmer on the government projects is lack of capital. Under the restrictions of the Reclamation Act, the homesteader cannot secure a title to his land until the expiration of the five year residence term. The commutation clause of the Homestead Law by which patent might be secured on payment of the government price of $1.25 per acre was expressly excluded from the Act of 1902. Under these conditions the settler cannot borrow money on his holding during the initial years when his need is greatest. Mr. Thomas H. Means, until recently project engineer at Truckee-Carson, has made a study of the relinquishments among the 544 homesteaders who filed under that project, and his conclusions are based on a personal acquaintance with most of the entrymen. The number of canceled and relinquished entries was 238 or almost half the total filings. Of the failures, 180 or 67 percent were men and women who had filed in the spirit of speculation, with no intention of building a home, but in the hope of selling out their claim to some later comer. From such settlers no genuine effort to make a success of irrigation could be expected. Of the 327 bona fide homesteaders, 88 have relinquished their holdings, and 96 are likely to fail, leaving only 143 or 43 percent in the successful class. In searching for the causes of failure, Mr. Means sets down four to lack of experience, 23 to adverse conditions— sandy or alkali soil, inability to get water in time to save crops, etc., and 34 to lack of capital and 35 to plain discouragement. It would seem that 71 percent of the failures are due to conditions arising out of the long residence requirement. In the unpublished paper from which he permits me to quote, Mr. Means states: "The residence clause is one great stumbling block. A settler is required to move his family on his homestead within six months after his filing. He brings his family out to the sun-blistered desert without shade or grass, and this often does much to discourage his wife and his family. Could this man carry on his improvements and let his family live in town or back east, he could accomplish more and suffer much less hardship. Improvements should be made the test, not residence." Again: "If a settler could obtain some form of title to his place more promptly and so have something on which he could borrow money, he would often make good where he now fails. In reclamation projects, the government would want to retain the first mortgage until the water right was paid for, but the second mortgage owner could at any time remove the first mortgage by paying up the water right charges." The clause in the Reclamation Act requiring "actual and continuous" residence for a term of five years was intended to guard against dummy filings, speculation and the accumulation of large estates—to reserve this last and richest portion of the public lands for the genuine farmer. But the bane of *latifundia* is less to be dreaded on irrigated lands, where intensive farming is alone profitable, than in a grain or cattle region. Congress has made the long-term residence requirement absolute just where it is least needed. If the commutation permissible under the Homestead Act were allowed, and a man might pay down the statutory price of the land, the position of the homesteader would approximate that of the settler under the Carey Act; but even more effective would be Mr. Means' suggestion that the residence requirement be dropped or abated and that salable title be given as soon as a certain amount of improvement has been made. In this way the man with small capital but possessing those more valuable qualities of brains, pluck, and endurance, would be enabled to earn a farm by the labor of his hands, as truly as did his forbears in the humid states east of the Missouri River.

Commentary

This article summarizes the challenges of organizing and implementing effective intervention—in this case, providing an efficient irrigation system to make arid western plains irrigable—for societal benefit. Precipitation variation between 2 and 20 inches for the region west of the one hundredth meridian made it hydrologically difficult to establish productive agriculture in the region. Yet people tried, and the decade from 1880 to 1890 was the boom period of irrigation for the region. The capacities of lakes, streams, and other available water resources were overestimated, "speculation ran riot," and stocks and bonds were sold without testing their legal or financial soundness. The great majority of these projects ultimately failed, and many of the promoters and investors wondered what the difference was between "a boom and a boomerang." Fast forward to our time and think about the "dot-com boom and bust of 2000" or the disastrous financial consequences of the "mortgage-based security" practices of 2007. A key lesson from this article is that situations change, but human responses to those changes are slow to adjust.

Coman (1911) identifies two key problems faced by the farmers in the late nineteenth and early twentieth centuries. First, top-down interventions where government officials lack contextual knowledge and rely on legal instruments to implement policy interventions usually lead to unintended and unanticipated consequences. Second, water rights adopted by western states are inadequate to deal with natural water regimes in the region. For example, when the availability of water is constrained by nature (as in the case of the western plains), appropriative rights systems may provide a more effective legal foundation to encourage irrigation development. By comparing and contrasting private and public implementation of the Carey Act (1894), Coman shows how a policy requiring "actual and continuous residency" essentially led to the failure of many publicly supported irrigation projects, while none of the private projects failed.

K. Ostrom "Reflections on 'Some Unsettled Problems of Irrigation'," (2011)

Introduction

In this article, Ostrom (2011) provides a detailed analysis of Coman's (1911) descriptive and insightful case studies of irrigation experiments and collective action in the late nineteenth and early twentieth centuries in the American West. Ostrom (2011) uses a social–ecological systems (SES) framework to analyze Coman's cases and provides further insight and lessons for addressing collective action problems. A general lesson from her analysis is that changing formal institutional structures may not be sufficient to ensure effective implementation of irrigation projects. Ostrom (2011) emphasizes the importance of knowledge and trust to solve collective-action problems.

We selected this article because our WDF explicitly recognizes the importance of these two attributes—knowledge and trust—to address emerging water problems. In particular, the first two propositions of the WDF describe the type of knowledge necessary to formulate and frame complex water problems, while the third proposition suggests ways to develop trust in creating and using that knowledge for effective management of common pool resources like water.

Challenges and Difficulties Related to Collective Action Problems (pp. 49, 50–51)

Coman described tough collective-action problems half a century before Mancur Olson (1965) and Garrett Hardin (1968) identified the challenging theoretical problem facing many groups. When groups need to cooperate to achieve a collective good (such as building and running an irrigation system), strong temptations exist for participants to "hold out" and not contribute. Holdouts receive the benefits of joint work whether or not they contribute if the others contribute. If most participants adopt the holdout strategy, however, the collective benefit is not produced. The remedy proposed by many theorists to such collective-action problems (also referred to as social dilemmas) is to turn them over to governments to solve (p. 49).

Coman identifies two continuing problems that increased the difficulties farmers faced and the number of failures. The first problem was the tendency of government officials who lacked information about the problems facing participants to focus on legislative details rather than appreciating the need for innovative solutions to difficult problems that are not outlined in formal law. This problem continues into the present time. The second problem was related to overcoming the inadequacies of the water rights systems adopted by many western states. The legal doctrine of riparian water rights—all those living adjacent to a source of water have rights to the "full flow" of water—that originally evolved in the English system of common law had successfully been implemented in the eastern United States, which shared many climatic features with England. Since that was the dominant legal doctrine applied in the nineteenth century, it was adopted by many western state legislatures as the official state law related to water. When water is scarce, however, an appropriative rights system—rights that are assigned to users in terms of their history of use—is recommended by many resource economists as a more appropriate legal foundation for irrigation development. This problem has been tackled through legislation in many western states.

Water producers in others, such as California, must work out arrangements through the court system to overcome the difficulties of relying on riparian water rights (see Blomquist 1992; Blomquist and Ostrom 2008).

The article is filled with descriptions of failures in achieving productive outcomes in an environment characterized by water scarcity and unpredictability. In the midst of failures, however, Coman also identifies several successful efforts. These include the efforts of Mormon migrants to the mesa above the Great Salt Lake to farm the desert and the work of 600 settlers on the Snake River in Idaho to dig 90 miles of bench canals. It is intriguing to read an article in the first issue of *American Economic Review* that describes empirical settings where some participants, but not all, overcame collective-action problems. We now know from the existence of a large number of cases beyond those described by Coman (Berkes 2007; Chhatre and Agrawal 2008; McCay and Acheson 1987; National Research Council 1986) that participants facing collective-action problems do sometimes succeed to overcome the problems contrary to the prediction derived from models of these problems (Clark 2006). We also know that many cases of failure also exist (Myers and Worm 2003). Partly as a result of these failures, many scholars presume that an external government must intervene in order to enforce rules on participants to change their incentives and enable them to achieve productive outcomes.

Coman's article illustrates that the typical solution posed to the problem of collective action—turn the problem over to the government—is not a panacea. Among the failures that she reports are efforts by national and state governments in the United States to construct irrigation systems to provide water for new settlers in sufficient time for them to start growing crops and survive in the harsh environment of the American West. Coman mentions multiple factors that jointly affect the success or failure of diverse efforts to govern and manage complex resource systems. Given her focus on one type of ecological resource—the western plains—she substantially reduces the complexity of analysis by holding the attributes of the ecology broadly similar while describing the outcomes achieved within diverse institutional arrangements. Her article is, however, filled with many descriptive details that are not analytically linked to explain why success is achieved in some settings and failure occurs in others. (pp. 50–51)

Updating the Theory of Collective Action Problems (p. 51)

A central problem facing contemporary scholars is updating the theory of collective action so as to explain the diversity of results noted by Coman and in many other historical records of collective-action problems in the field. What are the factors that enable some groups to achieve difficult collective outcomes while others fail? Over time, we have gained greater insights into the multiple factors affecting the likelihood of individuals (and/or their government) to solve diverse forms of collective action problems—particularly those dilemmas related to natural resources (Poteete, Janssen and Ostrom 2010). To explain behavior and outcomes in social dilemmas we need to adopt two broad strategies: (i) use a behavioral approach for understanding individual action to analyze whether those involved can gain essential trust that the others involved in a repeated situation are trustworthy and willing reciprocators of trust and (ii) develop better frameworks for identifying the working parts of social-ecological systems that citizens and officials face in trying to develop resources productively.

Applying the Social-Ecological Systems (SES) Framework to Irrigation Problems (p. 52)

To gain some understanding of the difficult problems that Coman discusses, we will use the SES framework to organize the complex network of variables that Coman mentions throughout her article. I will first discuss the three action situations that Coman mentions—constructing irrigation systems, keeping them maintained, and sustaining the economic survival of the homesteaders. Then we will discuss the resource systems, resource units, and actors in her examples. In Section III, we will turn to a more in-depth analysis of four of the governance systems she discusses as they interact with the other core variables of the SES framework.

A. Three Action Situations

Throughout her article, Coman refers to the interactions and outcomes that occur in three types of collective-action situations affected by the broader variables shown in Figure 1. The first problematic situation she discusses is the initial construction of an irrigation system adjacent to where farmers are settling. Some of these engineering works were constructed by the national government, but others were constructed by local private entrepreneurs or by local government irrigation districts established by a state government. The speed and reliability of the construction varied substantially. Whether it was a government or a private firm that was responsible for construction, however, was not the major factor affecting the speed and reliability of system construction. Outcomes were affected primarily by incentives of the participants resulting from the specific rules established by the governance system for a particular project and whether these incentives fostered an increase or decrease in the level of trust that actors had in others doing their part to achieve positive outcomes for themselves and others.

The second collective-action problem she discusses is the ongoing maintenance of an irrigation system in this environment. "Nothing goes to wreck more quickly than irrigation works where repairs are not maintained; the ditches fill with sand or silt, the flumes warp in the sun, and the cement dams disintegrate under the alternate action of frost and heat" (Coman 1911 p. 5). In situations where an irrigation system was built on time and generated water for farmers to start production, they were willing to take over the ownership and maintenance of the system (and to collect dues from each other for this service), but on other systems, the project "ended," no one undertook maintenance for it, and the system disintegrated rapidly.

The third problem Coman discusses is sustaining the economic survival of the homesteaders who moved west to increase their economic welfare. One might not think of this as a collective-action problem, but in this setting the survival of farm families was dependent on the availability of water (as a result of the first two action situations mentioned above), on the rules related to length of time and investment that had to be made before the title to land could be turned over to a homesteader, and on the level of reciprocity and trust established in a project to help when a family was facing particular problems.

B. Resource System, Resource Units, and Actors in the Western Plains of the United States

At the time of Coman's research, the western plains were considered to be "the curse of the Continent" (Coman 1911 p. 1) since they were a particularly difficult environment to convert into productive farming country. Thus, throughout her account, Coman focuses on one broad type of resource system, one set of relevant resource units, and one set of actors involved in undertaking irrigation. The attributes of each of these broad variables that she mentions as important factors are listed ... The resource systems and resource units of the plains underlie the biophysical problems she identifies as occurring under each of the diverse governance systems she discusses. They form a relatively constant "environment" in which to study the impact of diverse governance systems as these affect the level of trust achieved among participants and the behavior of farmers and officials ... An important commonality was that most of the homesteaders were newcomers and had little knowledge relevant to farming in the west. Little variation exists in the type of resource system, resource units, and actors involved across the diverse governance systems. Thus, the general attributes of these core variables as she describes them are listed ... and need to be taken into account when analyzing how a governance system affected action situations and outcomes.

Lessons Learned from Coman (1911) and SES Framework (pp. 60–61)

Coman was wise in calling her article "Some *Unsettled* Problems of Irrigation." While many irrigation systems had been successfully organized in the eastern sections of the United States, overcoming the difficulties posed by the resource system and resource units of the American western plains was a far more challenging problem. Coman devotes a substantial portion of her article to describing failures. She does, however, describe some fascinating success stories. Is it possible to derive some general lessons from this analysis using a behavioral approach to explaining human actions and the SES framework? Yes, lessons can be drawn out of this analysis, but they are different from the frequently stated solutions to collective-action problems related to resources—turn the resource over to a government or privatize it. In other words, changing the formal governance system *alone* is not a sufficient solution to difficult collective-action problems.

Building Knowledge and Trust Are Essential for Solving Collective-Action Problems (p. 62)

What we need to learn from the analysis of the cases described by Coman is the importance of knowledge and trust—two attributes related to the individual making decisions that are frequently not included in contemporary analysis. When one uses the classical assumption that individuals have complete information about the situation in which they are making decisions, one is assuming that individuals already have knowledge about how a resource functions, the technical means for managing a resource, and about the trustworthiness of the others involved. That assumption is appropriate in simple settings where participants have acquired knowledge of system structure and dynamics and the others involved over a long period of interaction. But, in the era of homesteading in the American West, both homesteaders and public officials lacked sufficient knowledge about the diversity of specific conditions existing in a particular watershed, or even where the boundaries of the watershed were. The appropriate institutions that one would design to solve complex resource problems involve mechanisms to ensure that participants were either selected for the knowledge they had about this type of system or to provide ways to gain that information rapidly. As Coman points out, farmers that settled in areas where Agricultural Extension Services were available, gained higher returns from their farms and faced less risks of financial ruin.

It is also important for future theoretical work to address whether the structure of action situations created by governance systems enhance or detract from participants' capacities for building trust (Walker and Ostrom 2009). While Arrow (1974) long ago stressed the importance of trust as the most efficient mechanism to govern transactions, this pathbreaking work has not been taken as seriously as needed. In undertaking analysis of potential governance systems to cope more effectively with a problem, we should be asking whether future rules support or undermine the development of trust and reciprocity over time (Bowles 2008; Frey 1994, 1997). Coman was surprised that the Reclamation Service engineer could motivate six hundred men to undertake the hard labor of digging miles and miles of canals. The engineer organized this work, however, so that the farmers could gain trust in each other and see that by working together they gained important outcomes for them all. Thus, an important lesson from Coman's study is the need to self-consciously address whether "solutions" proposed to solve collective-action problems are likely to enhance the knowledge and trust of the actors involved. Simply imposing general solutions, such as government or private property, will rarely work in practice unless those involved gain sufficient knowledge and trust to make the systems work over the long run."

Commentary

This article summarizes the challenges associated with collective action problems by reflecting on and analyzing several irrigation experiments described by Coman (1911). In her reflection, Ostrom (2011) argues that a key stumbling block for the theory of collective action is to explain the diversity of cases—noted by Coman (1911) and many other historical records—within a generalizable framework. In other words, is it possible to identify a set of explanatory variables that enables some groups to achieve desirable collective action outcomes while others in parallel situations fail? To know that individuals are facing a collective action problem is not adequate to predict their likely behavior. Understanding the interactions and feedback individuals face within the context of a particular problem is critical to determining failure or success in achieving collective action. This article, using the SES framework, provides a way to analytically link and explain why success may be achieved in some collective action settings while failure occurs in others. It suggests that over a dozen micro–institutional variables may be at play in determining the likelihood of individuals cooperating. Coman's choice of the American West as an ecological unit made the analysis within the SES framework somewhat simpler because variations in resource systems and resource units (see Table 1 in Ostrom 2011) are not the controlling variables. We note that although many irrigation initiatives have been successfully implemented in the eastern United States, addressing the challenges posed by the American West was far more difficult from both the natural and societal domain perspectives. Consequently, one needs to be cognizant of the applicability and limitations of a given framework while analyzing case studies to derive generalizable principles that can be applied in other regions, domains, and situations.

H.W. Rittel and M.M. Webber "Dilemmas in a General Theory of Planning," (1973)

Introduction

In 1973, Rittel and Weber wrote an article that did not include any citations and yet has been cited over 3,300 times as of now. In this article, the authors argued that there is a class of social planning problems that cannot be successfully tackled using traditional linear, analytical approaches. They called these *wicked problems*, in contrast to *tame problems*. A year later, Ackoff (1974) put forward a similar concept, which he called a "social mess" or "unstructured reality." Hopefully, we are (somewhat) wiser today, and less susceptible to the belief that complex water planning and management problems can be "solved" by linear methods akin to traditional systems engineering. Still, it is instructive to reflect on the distinction between "wicked" and "tame" water problems. A key argument that Rittel and Webber made is that for "wicked" social problems, policy issues cannot be definitively described because in a pluralistic society there are no objective definitions of public good or equity. Consequently, policies that attempt to address social problems in general (and water problems in our case) cannot be objectively categorized as correct or false, and we cannot achieve "optimal solutions" to complex water problems unless severe qualifications are imposed.

Our choice of this article is motivated by our emphasis on understanding the type of water management problems (simple, complicated, and complex) we face. In our assessment, many water management problems are complex and "wicked." Water management problems tend to be complex and "wicked" not only because they cross multiple boundaries, but also because they involve various stakeholders competing for limited and common resources. As a result, there is rarely a readily acceptable or obvious solution involving multiple objectives and competing needs.

Easy Problems to "Wicked" Problems (p. 156)

The professional's job was once seen as solving an assortment of problems that appeared to be definable, understandable and consensual. He was hired to eliminate those conditions that predominant opinion judged undesirable. His record has been quite spectacular, of course; the contemporary city and contemporary urban society stand as clean evidences of professional prowess. The streets have been paved, and roads now connect all places; houses shelter virtually everyone; the dread diseases are virtually gone; clean water is piped into nearly every building; sanitary sewers carry wastes from them; schools and hospitals serve virtually every district; and so on. The accomplishments of the past century in these respects have been truly phenomenal, however short of some persons' aspirations they might have been. But now that these relatively easy problems have been dealt with, we have been turning our attention to others that are much more stubborn. The tests for efficiency, that were once so useful as measures of accomplishment, are being challenged by a renewed preoccupation with consequences for equity. The seeming consensus, that might once have allowed distributional problems to be dealt with, is being eroded by the growing awareness of the nation's pluralism and of the differentiation of values that accompanies differentiation of publics. The professionalized cognitive and occupational styles that were refined in the first half of this century, based in Newtonian mechanistic physics, are not readily adapted to contemporary conceptions of interacting open systems and to contemporary concerns with equity. A growing sensitivity to the waves of repercussions that ripple through such systemic networks and to the value consequences of those repercussions has generated the recent reexamination of received values and the recent search for national goals. There seems to be a growing realization that a weak strut in the professional's support system lies at the juncture where goal-formulation, problem-definition and equity issues meet. We should like to address these matters in turn.

Goal Formulation (pp. 156–158)

The search for explicit goals was initiated in force with the opening of the 1960s. In a 1960 RAND publication, Charles J. Hitch urged that *"We* must learn to look at *our objectives* as critically and as professionally as we look at our models and our other inputs."* The subsequent work in systems analysis reaffirmed that injunction. Men in a wide array of fields were prompted to redefine the systems they dealt with in the syntax of verbs rather than nouns—to ask "What do the systems *do?*" rather than "What are they made of?"—and then to ask the most difficult question of all: "What *should* these systems do?". (pp. 156–157)

A deep-running current of optimism in American thought seems to have been propelling these diverse searches for direction-finding instruments. But at the same time, the Americans' traditional faith in a guaranteed Progress is being eroded by the same waves that are wearing down old beliefs in the social order's inherent goodness and in history's intrinsic benevolence. Candide is dead. His place is being occupied by a new conception of future history that, rejecting historicism, is searching for ways of exploiting the intellectual and inventive capabilities of men.

This belief comes in two quite contradictory forms. On the one hand, there is the belief in the "makeability," or unrestricted malleability, of future history by means of the planning intellect—by reasoning, rational discourse, and civilized negotiation. At the same time, there are vocal proponents of the "feeling approach," of compassionate engagement and dramatic action, even of a revival of mysticism, aiming at overcoming The System which is seen as the evil source of misery and suffering.

The Enlightenment may be coming to full maturity in the late 20th century, or it may be on its deathbed. Many Americans seem to believe both that we can perfect future history—that we can deliberately shape future outcomes to accord with our wishes—and that there will be no future history. Some have arrived at deep pessimism and some at resignation. To them, planning for large social systems has proved to be impossible without loss of liberty and equity. Hence, for them the ultimate goal of planning should be anarchy, because it should aim at the elimination of government over others. Still another group has arrived at the conclusion that liberty and equity are luxuries which cannot be afforded by a modern society, and that they should be substituted by "cybernetically feasible" values.

Professionalism has been understood to be one of the major instruments for perfectability, an agent sustaining the traditional American optimism. Based in modern science, each of the professions has been conceived as the medium through which the knowledge of science is applied. In effect, each profession has been seen as a subset of engineering. Planning and the emerging policy sciences are among the more optimistic of those professions. Their representatives refuse to believe that planning for betterment is impossible, however grave their misgivings about the appropriateness of past and present modes of planning. They have not abandoned the hope that the instruments of perfectability can be perfected. It is that view that we want to examine, in an effort to ask whether the social professions are equipped to do what they are expected to do. (pp. 157–158)

Problem Definition (p. 159)

We have come to think about the planning task in very different ways in recent years. We have been learning to ask whether what we are doing is the *right* thing to do. That is to say, we have been learning to ask questions about the *outputs* of actions and to pose problem statements in valuative frameworks. We have been learning to see social processes as the links tying open systems into large and interconnected networks of systems, such that outputs from one become inputs to others. In that structural framework it has become less apparent where problem centers lie, and less apparent *where* and *how* we should intervene even if we do happen to know what aims we seek. We are now sensitized to the waves of repercussions generated by a problem-solving action directed to any one node in the network, and we are no longer surprised to find it inducing problems of greater severity at some other node. And so we have been forced to expand the boundaries of the systems we deal with, trying to internalize those externalities.

This was the professional style of the systems analysts, who were commonly seen as forebearers of the universal problem-solvers. With arrogant confidence, the early systems analysts pronounced themselves ready to take on anyone's perceived problem, diagnostically to discover its hidden character, and then, having exposed its true nature, skillfully to excise its root causes. Two decades of experience have worn the self-assurances thin. These analysts are coming to realize how valid their model really is, for they themselves have been caught by the very same diagnostic difficulties that troubled their clients.

By now we are all beginning to realize that one of the most intractable problems is that of defining problems (of knowing what distinguishes an observed condition from a desired condition) and of locating problems (finding where in the complex causal networks the trouble really lies). In turn, and equally intractable, is the problem of identifying the actions that might effectively narrow the gap between what-is and what-ought-to-be. As we seek to improve the effectiveness of actions in pursuit of valued outcomes, as system boundaries get stretched, and as we become more sophisticated about the complex workings of open societal systems, it becomes ever more difficult to make the planning idea operational.

Many now have an image of *how* an *idealized* planning system would function. It is being seen as an on-going, cybernetic process of governance, incorporating systematic procedures for continuously searching out goals; identifying problems; forecasting uncontrollable contextual changes; inventing alternative strategies, tactics, and time sequenced actions; stimulating alternative and plausible action sets and their consequences; evaluating alternatively forecasted outcomes; statistically monitoring those conditions of the publics and of systems that are judged to be germane; feeding back information to the simulation and decision channels so that errors can be corrected—all in a simultaneously functioning governing process. That set of steps is familiar to all of us, for it comprises what is by now the modern-classical model of planning. And yet we all know that such a planning system is unattainable, even as we seek more closely to approximate it. It is even questionable whether such a planning system is desirable.

"Tame" and "Wicked" Problems (p. 160)

The kinds of problems that planners deal with—societal problems—are inherently different from the problems that scientists and perhaps some classes of engineers deal with. Planning problems are inherently wicked.

As distinguished from problems in the natural sciences, which are definable and separable and may have solutions that are findable, the problems of governmental planning—and especially those of social or policy planning—are ill-defined; and they rely upon elusive political judgment for resolution. (Not "solution." Social problems are never solved. At best they are only re-solved—over and over again.) Permit us to draw a cartoon that will help clarify the distinction we intend.

The problems that scientists and engineers have usually focused upon are mostly "tame" or "benign" ones. As an example, consider a problem of mathematics, such as solving an equation; or the task of an organic chemist in analyzing the structure of some unknown compound; or that of the chess player attempting to accomplish checkmate in five moves. For each the mission is clear. It is clear, in turn, whether or not the problems have been solved. Wicked problems, in contrast, have neither of these clarifying traits; and they include nearly all public policy issues—whether the question concerns the location of a freeway, the adjustment of a tax rate, the modification of school curricula, or the confrontation of crime.

There are at least ten distinguishing properties of planning-type problems, i.e., wicked ones, that planners had better be alert to and which we shall comment upon in turn. As you will see, we are calling them "wicked" not because these properties are themselves ethically deplorable. We use the term "wicked" in a meaning akin to that of "malignant" (in contrast to "benign") or "vicious" (like a circle) or "tricky" (like a leprechaun) or "aggressive" (like a lion, in contrast to the docility of a lamb). We do not mean to personify these properties of social systems by implying malicious intent. But then, you may agree that it becomes morally objectionable for the planner to treat a wicked problem as though it were a tame one, or to tame a wicked problem prematurely, or to refuse to recognize the inherent wickedness of social problems.

Planning to Address "Wicked" Problems (pp. 168–169)

In a setting in which a plurality of publics is politically pursuing a diversity of goals, how is the larger society to deal with its wicked problems in a planful way? How are goals to be set, when the valuative bases are so diverse? Surely a unitary conception of a unitary "public welfare" is an anachronistic one.

We do not even have a theory that tells us how to find out what might be considered a societally best state. We have no theory that tells us what distribution of the social product is best—whether those outputs are expressed in the coinage of money income, information income, cultural opportunities, or whatever. We have come to realize that the concept of *the* social product is not very meaningful; possibly there is no aggregate measure for the welfare of a highly diversified society, if this measure is claimed to be objective and non-partisan. Social science has simply been unable to uncover a social-welfare function that would suggest which decisions would contribute to a societally best state. Instead, we have had to rely upon the axioms of individualism that underlie economic and political theory, deducing, in effect, that the *larger-public* welfare derives from summation of individualistic choices. And yet, we know that this is not necessarily so, as our current experience with air pollution has dramatized.

We also know that many societal processes have the character of zero-sum games. As the population becomes increasingly pluralistic, inter-group differences are likely to be reflected as inter-group rivalries of the zero-sum sorts. If they do, the prospects for inventing positive non-zero-sum development strategies would become increasingly difficult.

Perhaps we can illustrate. A few years ago there was a nearly universal consensus in America that full-employment, high productivity, and widespread distribution of consumer durables fitted into a development strategy in which all would be winners. That consensus is now being eroded. Now, when substitutes for wages are being disbursed to the poor, the college student, and the retired, as well as to the more traditional recipient of nonwage incomes, our conceptions of "employment" and of a full-employment economy are having to be revised. Now, when it is recognized that raw materials that enter the economy end up as residuals polluting the air mantle and the rivers, many are becoming wary of rising manufacturing production. And, when some of the new middle-class religions are exorcising worldly goods in favor of less tangible communal "goods," the consumption-oriented society is being challenged—oddly enough, to be sure, by those who were reared in its affluence.

What was once a clear-cut win-win strategy, that had the status of a near-truism, has now become a source of contentious differences among subpublics.

Commentary

This article summarizes the challenges and difficulties associated with "wicked" problems, particularly in the context of social policy-making. The authors make a distinction between "tame" and "wicked" problems, and suggest why it is important to recognize this distinction in formulating effective policy options. Within the context of water management, a tame problem has a relatively well-defined and stable problem statement as well as a solution that can be objectively evaluated.

Wicked problems are different. Wicked problems are ill-defined, ambiguous, and associated with strong moral, political, and professional values. Complex water management problems are wicked problems where certainty of solutions and degree of consensus vary widely. In fact, there is often little consensus about what the problem is, let alone how to resolve it. Furthermore, wicked problems are constantly changing because of the complex interactions among the natural, societal and political forces involved.

References

Ackoff, R.L. 1974. *Redesigning the Future*. New York: John Wiley & Sons.

Ackoff, R.L. 1979. The future of operational research is past, *Journal of the Operational Research Society, 30*(2): 93–104.

Allan, T. 2006. IWRM: The new sanctioned discourse, in P.P. Mollinga, A. Dicit, and K. Athukorala (eds) *Intergrated Water Resources Management: Global Theory, Emerging Practices and Local Needs* (pp. 38–63). New Delhi: Sage Publications.

Arrow, K.J. 1974. *The Limits of Organization*. New York: W.W. Norton.

Barabasi, A. L. 2003. *Linked: How Everything is Connected to Everything Else and What it Means for Business, Science, and Everyday Life*. New York: Plume Publisher.

Bar-Yam, Y. 2004. Multiscale variety in complex systems, *Complexity, 9*: 37–45.

Bennis, W.M., Medin, D.L., and Bartels, D M. 2010. The cost and benefits of calculation and moral rules, *Perspectives on Psychological Science, 5*(2): 187–202.

Berkes, F. 2007. Community based conservation in a globalized world, *Proceedings of the National Academy of Sciences, 104*(39): 15188–15193.

Biswas, A.K. 2004. Integrated water resources management: A reassessment, *Water International, 29*(2): 248–256.

Blomquist, W. 1992. *Dividing the Waters: Governing Groundwater in Southern California*. San Francisco: Institute for Contemporary Studies Press.

Blomquist, W. and Ostrom, E. 2008. Deliberation, learning, and institutional change: The evolution of institutions in judicial settings, *Constitutional Political Economy, 19*(3): 180–202.

Bowles, S. 2008. Policies designed for self interested citizens may undermine the moral sentiments: Evidence from economic experiments, *Science, 320*(5883): 1605–1609.

Braga, B.P.F. 2001. Integrated urban water resources management: A challenge into the 21st century, *Water Resources Development, 17*: 581–599.

Chhatre, A. and Agrawal, A. 2008. Forest commons and local enforcement, *Proceedings of the National Academy of Sciences, 105*(36): 13286–13291.

Clark, C.W. 2006. *The Worldwide Crisis in Fisheries: Economic Models and Human Behavior*. New York: Cambridge University Press.

Clemons, J. 2004. Interstate water disputes: A road map for states, *Southeastern Environmental Law Journal, 12*: 2.

Coman, K. 1911. Some Unsettled Problems of Irrigation, *American Economic Review 1*(1): 1–19. [Reprinted in *American Economic Review, 101* (February 2011: 36–48)].

Falkenmark, M. and Rockstrom, J. 2006. The new blue and green water paradigm: Breaking new ground for water resources planning and management, *Journal of Water Resources Planning and Management, (May/June 2006)*, 129–134.

Feldman, D. L. 2008. Barriers to adaptive management: Lessons from the Apalachicola-Chattahoochee-Flint Compact, *Society & Natural Resources, 21*(6), 512–525.

Frey, B.S. 1994. How intrinsic motivation is crowded out and in, *Rationality and Society* 6(3): 334–352.

Frey, B.S. 1997. A constitution for knaves crowds out civic virtues, *Economic Journal* 107(443): 1043–1053.

Global Water Partnership (GWP). 2000, "Integrated Water Resources Management," Global Water Partnership Technical Advisory Committee, Background Paper no. 4

Goldthorpe, J.H. 1997. Current issues in comparative macrosociology: A debate on methodological issues, *Comparative Social Research, 16*, 1–26.

Gray, P., Williamson, J., Karp, D., and Dalphin. J. 2007. *The Research Imagination: An Introduction to Qualitative and Quantitative Methods*. New York: Cambridge University Press.

Habermas, J. 1984. *The Theory of Communicative Action: Reason and the Rationalization of Society – volume 1*, Boston: Beacon Press.

Hardin, G. 1968. The tragedy of the commons, *Science, 162*(3859): 1243–1248.

Heathcote, I.S. 2009. *Integrated Watershed Management: Principles and Practice*. New Jersey: John Wiley & Sons.

Hitch, C.J. 1960. *On the Choice of Objectives in Systems Studies*, Santa Monica, CA: The RAND Corporation.

Ingram, H. 2011. Beyond Universal Remedies for Good Water Governance: A Political and Contextual Approach, in A. Garrido and H. Ingram (eds) *Water for Food in a Changing World*. New York: Routledge.

Islam, S. et al. 2009. AquaPedia: Building Capacity to Resolve Water Conflicts. 5th World Water Forum, March 16–22, Istanbul, Turkey.

Islam, S., Gao, Y., and Akanda, A. 2010. Water 2100: A synthesis of natural and societal domains to create actionable knowledge through AquaPedia and water diplomacy, *Hydrocomplexity: New tools for solving wicked water problems, International Association of Hydrological Science Publication 338*.

Islam, S. and Susskind, L. 2011. Water Diplomacy: Managing the science, policy, and politics of water networks through negotiation; Presented at the European Geophysical Union, Vienna, Austria.

Jonker, L., 2007. Integrated water resources management: The theory-praxis-nexus-A South African perspective. *Physics and Chemistry of the Earth, 32*: 1257–1263.

Kijne, J.W., Barker, R., and Molden, D. 2003. Water productivity in agriculture: Limits and opportunities for improvement, *Comprehensive assessment of water management in agriculture series, 1*, CABI Pub.

Lankford, B.A. & Cour J. 2005. From integrated to adaptive: A new framework for water resources management of river basins, in *The Proceedings of the East Africa River Basin Management Conference, Morogoro, Tanzania, 7–9 March*.

Lawford, R. et al. 2003. *Water: Science, Policy, and Management*. Water Resources Monograph, American Geophysical Union, USA.

Lee, T. 1992. Water management since the adoption of the Mar del Plata Action Plan: Lessons for the 1990s, *Natural Resources Forum, 16*(3): 202–211

Leitman, S. 2005. Apalachicola-Chattahoochee-Flint Basin: Tri-state Negotiations of a Water Allocation Formula, in J.T. Scholz and B. Stiftel (eds.) *Adaptive Governance and Water Conflict: New Institutions for Collaborative Planning* (pp. 74–88) Washington D.C.: Resources for the Future.

Lewicki, R.J., Barry, B., and Saunders, D.M. 2010. *Essentials of Negotiation, 5th Edition*, New York: McGraw Hill Higher Education.

Liu, J. et al. 2007: Complexity of coupled human and natural systems, *Science, 317*: 1513–1518.

Liu, Y., Slotine J., and Barabasi, A.L. 2011. Controllability of complex networks, *Nature,* *473*: 167–173.

McCay, B.J and Acheson, J.M. eds. 1987. *The Question of the Commons: The Culture and Ecology of Communal Resources,* Tucson, AZ: University of Arizona Press.

Milly, P.C.D., Betancourt, J., Falkenmark, M., Hirsch, R.M., Kundzewicz, Z.W., Lettenmaier, D.P., et al. 2008. Climate change: Stationarity is dead: Whither water management? *Science, 319*(5863), 573–574.

Mitchell, B. (ed.), 1990. *Integrated Water Management: International Experiences and Perspectives.* London: Belhaven Press.

Mitchell, B. 2008. Resource and environmental management: Connecting the academy with practice, *Canadian Geographer, 52*: 131–145.

Mnookin, R. 2010. *Bargaining with the Devil: When to Negotiate, When to Fight.* New York: Simon & Schuster.

Mollinga, P., Meinzen-Dick, R.S., and Merrey, J.D. 2007. Politics, plurality and problemsheds: A strategic approach for reform of agricultural water resources management. *Development Policy Review, 25*: 699–719.

Morehouse, B.J. 2000. "Boundaries in climate science: water resource discourse," Presented at the Symposium on Climate, Water, and Transboundary Challenges in the Americas, University of California, Santa Barbara, Santa Barbara, CA, 16–19 July.

Myers, R. and Worm, B. 2003. Rapid Worldwide Depletion of Predatory Fish Communities, *Nature, 423*(5): 280–283.

National Research Council. 1986. *Panel on Common Property Resource Management.* Proceedings of the Conference on Common Property Resource Management, April 21–26, 1985, Washington DC: National Academy Press.

Olson, M. 1965 *The Logic of Collective Action: Public Goods and the Theory of Groups* (Revised edition). Cambridge, MA: Harvard University Press.

Ostrom, K. 2011. Reflections on "Some Unsettled Problems of Irrigation," *American Economic Review, 101*: 49–63.

Pahl-Wostl, C., Craps, M., Dewulf, A., Mostert, E., Tabara, D., and Taillieu, T. 2007. Social learning and water resources management. *Ecology and Society, 12*(2): 5.

Poteete, A.R., Janssen, M.A., and Ostrom E. 2010. *Working Together: Collective Action, the Commons and Multiple Methods in Practice.* Princeton, NJ: Princeton University Press.

Raiffa, H. 1985. *The Art and Science of Negotiation.* Cambridge, MA: Belknap Press.

Rittel, H.W.J., & Webber, M.M. 1973. Dilemmas in a general theory of planning. *Policy Sciences, 4*(2): 155–169.

Saravanan, V.S., McDonald, G.T., and Mollinga, P.P. 2009. Critical review of Integrated Water Resources Management: Moving beyond polarized discourse. *Natural Resources Forum, 33*: 76–86.

Scruggs, Lyle. 2007. What's Multiple Regression Got to Do with It? *Comparative Social Research 24*: 309–323.

Stakhiv, E.Z. 2011. Pragmatic approaches for water management under climate change uncertainty. *Journal of the American Water Resources Association, 47*(6): 1183–1196.

Stiglitz, J.E., Sen, A., and Fitoussi, J.-P . 2010. *Mismeasuring Our Lives: Why GDP Doesn't Add Up.* New York: New Press.

Susskind, L. and Cruikshank, J. 1987. *Breaking the Impasse: Consensual Approaches to Resolve Public Disputes.* New York, NY: Basic Books Inc.

Thomas, J.S. and Durham, B. 2003. Integrated water resources management: Looking at the whole picture, *Desalination, 156*: 21–28.

Varady, R.G. and Morehouse B.J. 2003. Moving borders from the periphery to the center: River basins, political boundaries, and water management policy, in R. Lawford,

D. Fort, H. Hartman, and S. Eden, *Water: Science, Policy, and Management,* Water Resources Monograph, 16.

Walker, J. and Ostrom E. 2009. Trust and Reciprocity as Foundations for Corporations, in K.S. Cook, M. Levi, and R. Hardin (eds) *Whom Can We Trust?: How Groups, Networks, and Institutions Make Trust Possible* (pp. 91–124). New York: Russell Sage Foundation.

Wright, G. and Cairns, G. 2011. *Scenario Thinking: Practical Approaches to the Future.* New York: Palgrave/McMillan.

3

UNDERSTANDING AND CHARACTERIZING COMPLEX WATER MANAGEMENT PROBLEMS

"Humans appear to have about the same number of genes, with similar sequence, and we both like cheese. So why aren't mice more like us?" (Gunter and Dhand 2005). The answer probably lies in the way genes are networked and regulated in mice and in humans.

Since Descartes, all the giants of science have tried to produce the simplest possible explanation of observed phenomena. At its core, reductionism is an approach to understanding a system by examining the sub-systems that make it up. Relationships among sub-systems are not given much attention. For example, consider a water molecule. It is composed of hydrogen and oxygen atoms. Once the positions of these atoms are specified; so, too, are the relationships among the atoms. Once the relative locations of genes are specified in a human or a mouse, the relationships among these genes are also specified. A purely reductionist approach does not focus on relationships, it merely specifies that sub-systems are present. We realize now that reductionism of this sort can provide only limited insight.

We also recognize that a system is more than sum of its parts, and "systems engineering" only works when systems are bounded and cause–effect dynamics are well-understood. Many observed phenomena—particularly those at the interface between natural and societal domains—pose overwhelming challenges to reductionist as well as systems engineering approaches. For example, the intricacies of climate change, and the dynamics of poverty alleviation, are too complex for us to model and explain given what we cannot know. As Barbasi puts it:

> We have learned that nature is not a well-designed puzzle with only one way to put it back together. In complex systems the components can fit in so many different ways that it would take billions of years for us to try them all.

(Barabasi 2003)

When relatively simple components collectively give rise to very complex behavior—for example, when mice and humans show differences despite having a similar number of genes; or when the micro-scale behavior of individuals generates an unpredictable macro-scale economy; or when three wetlands in the same region evolve differently because of small contextual differences in natural and societal processes—we need a different approach to understanding and managing such complexity. This is particularly evident from emerging studies of coupled human-natural systems (Liu et al 2007; Narayanan and Venot 2009).

At the most fundamental level, complex systems challenge the idea that by understanding a system at its component level, we will be able to understand the system as a whole. To put it in simpler terms, and borrowing the words of Miller and Page: "One and one may well make two, but to really understand two we must know both about the nature of *one* and the meaning of *and*" (Miller and Page 2007). Unless we understand the nature of all the different components (i.e., the nature of *one*) as well as their evolving and continuously changing relationships (i.e., the meaning of *and*), we will not be able to understand or manage such systems.

Water problems are complex because they arise from the interaction of numerous natural, societal, and political forces. The complexity of water resource management is the result of the presence of a great many stakeholders with competing needs who interact on multiple levels and scales simultaneously. Competition among these stakeholders creates conflicts of various kinds. This competition also shapes access to and allocation of most contested waters.

The origin of many of our water management difficulties stems from our fragmented or bounded view of water as a "natural object," or at other times, as a "societal issue," or at still other moments as a "political construct." The components of a water resource management puzzle can fit together in so many different ways that it is practically impossible to use "reductionist" or traditional "systems engineering" methodologies to resolve complex water management problems.

We need a different approach that takes account of these complexities. In our view, instead of thinking about managing systems that are bounded and made up of components that interact in predictable ways, it is more helpful to think in terms of complex water networks.

The Properties of Complex Systems and Water Networks

Complex Collective Behavior and Emergence

Water systems comprise large interconnected networks that operate in multiple domains (e.g., natural, societal, political) simultaneously. They also operate at multiple scales (e.g., spatial, temporal, jurisdictional, institutional) and at different levels (e.g., local, regional, global). Our understanding of water management problems for a given domain, scale, and level is not easily transferrable to other scales, domains, or levels. It is the collective action of large numbers of interacting components that gives rise to the complex behavior of water systems.

Indeed, different elements of complex systems can interact in apparently random ways, and these interactions change the behavior of the system itself. For example, the success of a specific watershed management plan may depend on the support of key stakeholders. The same policy or approach may fail in another watershed where the interactions among economic assets (C), governance structures (G) and water quantity (Q) are weak, even though stakeholders in the second location may have participated in the decision-making. The incentives (C) were successful in the first basin, but did not drive the allocation of water (Q) in the second basin in the same way.

In complexity theory, such a property of a complex system is known as emergence. Emergence helps us understand why a given intervention (e.g., building a dam) to achieve a particular water management objective (e.g., hydropower generation) may lead to unexpected outcomes even though that strategy has worked in other locations to achieve the desired purpose. Our awareness of emergence in water management comes from detailed case studies of local actions that show how aggregate-level responses produce seemingly random results. Emergent phenomena challenge our cognitive bias: we expect the same actions to have the same results. Emergence raises serious questions about how we ought to manage complex water networks. We cannot assume that an aggregate level response in one location will repeat itself in other locations until we understand the nuances of local and contextual variation, as well as the impact that local interactions and variations are likely to have.

Complex Systems do not Behave in Predictable Ways

Perhaps the most important insight from the study of complex systems over the last several decades is that they are characterized by the unexpected. There is no meaningful notion of "expectation" in such systems. A very small perturbation in a water system can create a very large effect due to nonlinearity and feedback loops of various kinds. In other words, we cannot accurately predict even the evolution of one of the simplest and most widely studied deterministic systems (e.g., a one variable nonlinear logistic map with one parameter combining the effects of birth and death rate) because of its sensitivity to initial conditions. In chaos theory, this is popularly known as the butterfly effect: a small change at one point in a nonlinear system can result in large differences at a later stage. In theory, the movement of a butterfly's wings in China could, in fact, actually affect weather patterns in New York City, thousands of miles away. As far as complex water networks are concerned, it is likely that small perturbations can produce unpredictable and sometimes drastic results by triggering increasingly significant events.

Boundary Crossing is Pervasive and Water Systems are Open

Another source of complexity in water systems arises from the boundary-crossing nature of water. For example, when a water droplet evaporates from the ocean, it becomes public property. It travels through the atmosphere and remains public property, and a common resource, until it falls as rain on the Quabbin reservoir. Then, it moves through a series of pipes and pumps to arrive at a residence in Boston where it becomes private property. It leaves that house as used water. When it goes into Boston Harbor, it becomes public property again. During this journey, the water droplet crosses multiple scales, domains, and boundaries. Such boundary crossing is pervasive in all water networks. The complexity involved is a function of the interplay between processes, people, and institutions at multiple levels and scales.

These interactions produce and use matter, energy, and information drawn from both the internal and external environment. This means that the water networks are typically "open" to external influences. A closed system would not exchange matter, energy, or information with anything external—almost like an airtight bottle. Truly closed systems are rare. Yet, most of our fundamental theories of physics are based on the notion of a closed system. To make the "hammer of physics" applicable to the "nails" of the real world, we often use the idea of "externalities": the unintended spillover effects. In complex systems, externalities co-evolve with the system itself; consequently, we cannot easily separate "internal" from "external," so we must treat complex water networks as "open."

Co-evolution is Inherent in Water Networks

Complex water networks exist within their own environment, and they are also part of that environment. As their environment changes, they evolve to ensure their survival. But, because they are part of their environment, when they evolve, they change their environment as well. This process of evolution is dynamic and continuous. For such complex evolving situations, learning from each change is essential to understanding and influencing the evolution of the network. For example, knowing exactly how Alpha, Beta, and Gamma managed their water resources in Indopotamia when there were hunters and gatherers, or even later when they learned how to cultivate rice, would not be of much help in developing an implementable water management strategy for the twenty-first century. The situation has changed, and interactions among the key variables have changed. Over the last 100 years, Alpha has monopolized access to the river, insisting that only after its needs are met will Beta and Gamma be allowed more water. In desperation, Beta and Gamma decided to join forces and build a dam. Alpha is quite unhappy about that idea. In recent decades, glacial melting and associated sea-level rise have caused salt-water intrusion, affecting rice cultivation. In some ways we are talking about the same environment throughout that entire time period, but the interactions among the natural, societal, and political variables in Indopotamia today are radically different from the patterns that existed hundreds of years ago.

Flash forward to the real world. The three wetlands studied by Narayanan and Venot (2009) are all experiencing high population pressure, competition between aquaculture and agriculture, conflicting political and local interests, and a disconnect between global conservation initiatives and local concerns. Yet the manifestation of these variables, processes, and drivers are quite different for each of the three wetlands even though they share similar geographical and climatic attributes. Such intricate couplings among the natural, societal, and political domains requires a different way of thinking about and representing what is going on when networks are open, context dependent, and continuously changing.

Water Networks

Even when coupled natural–societal–political systems appear to act in apparently unpredictable ways, there are patterns at work. We define these complex and unpredictable interconnections in terms of water networks. A network is a collection of nodes in which some pairs are connected by links. These links and the interactions among them across a range of scales make the idea of a network particularly useful for the study of complex water systems.

One of the most powerful ways to represent functional relationships among large numbers of interconnected components that cross multiple boundaries and scales is network analysis. A network (or graph) is a collection of nodes (vertices) and links (edges) between the nodes. The links can be directed or undirected and weighted or unweighted. A water network can be described as an interconnected set of nodes representing variables from the natural, societal, and political domains. The flow of information among these variables through these links is what enables the nodes to update their status and makes them dynamic. The challenge is to identify the mechanisms that define and improve the flow of information among the nodes. It is in the context of networks that we propose a new water management approach called the Water Diplomacy Framework (WDF). It is rooted in complexity theory and ideas about non-zero-sum negotiation. These tools can be used to synthesize scientific knowledge with contextual understanding to manage complex water problems within a political setting.

Water networks are characterized by a multiplicity of relationships among the variables and processes that are connected at different scales and levels. Appropriate network representation allows us to track the flow of information across variables, processes, and scales. When we talk about "connectedness" in a complex water network, we need to think about two related things. The first is *structural* connectedness: who (what) is linked to whom (what). The other is connectedness in terms of *action*: because each action (for example, building a dam to generate hydropower) affects the current state of the system and its future evolution. And, just as the underlying structure of a water network can be complex, so too can the coupled behavior of different actions at different scales and levels. Consequently, to understand and manage a water network we must be reflective and strategic. By thinking carefully about the structural characteristics of water networks and their interactions, we should be able to produce the results we want. To understand and characterize these structural features, we need to define what we mean by domains, scales, and levels.

Domains, Scales, and Levels

As noted above, water management is difficult because water networks cross domains (natural, societal, and political), scales (space, time, jurisdictional, institutional, knowledge, etc.) and levels (area: patches to global; knowledge: a specific case to generalized principles). Following Gibson et al (2000) and Cash et al (2006), we define "scale" as the spatial, temporal, quantitative, or analytical dimensions used to measure and study any phenomenon. We define "levels" as the units of analysis located at different positions on each scale (e.g., levels refer to seconds, days, seasons, decades, etc. for temporal scales).

Many water management problems stem from what we describe as the competition, interconnection, and feedback among NSPD as shown in Figure 2.2 and discussed in Chapter 2. The causes and consequences of each water management intervention need to be measured at different levels and along multiple scales. While the natural sciences have long understood the importance of scale, approaches to thinking about and measuring what is happening at different scales and levels in the social sciences have been less explicit and more variable (Gibson et al 2000). The growing need for interdisciplinary work across the natural, societal, and political domains, however, demands that we adopt some common language about scales and levels when we talk about complex water management networks and the problems of redesigning or changing them.

Scales and Levels

Scale and level considerations are increasingly important in managing complex water networks, yet they are mostly taken for granted by decision-makers. The scale and level at which a water management problem is defined, as well as the scale and level at which it can be addressed, are potentially contentious issues. The framing of a problem as local, regional, national, or global is not without consequences. A water policy formulated at a regional scale will affect and be affected by what happens at other scales and levels. Moreover, what is appropriate at a regional level may well be considered inappropriate at a local or global level. Often, actors in the same water management dispute refer to different and conflicting scales and levels when they frame their questions and concerns.

Linking scale and level issues from the natural domain (for example, flood forecasting at a local scale or seasonal water availability for irrigation at the regional scale) to the societal domain (for example, responses to floods or planning for food shortages due to a predicted drought) is important to the improvement of current water management practices. Currently, theory and practices in the natural domain tend to focus on adaptive behavior and aspects of spatial and temporal data related to water quantity, water quality, and ecosystem function and services. While in the societal domain, water management focuses more on levels and scales of behavior in institutional behavior and temporal concerns. Political reality demands cross-scale and cross-level interactions for effective water management. In general, the lack of fit between the societal domain (e.g., scale of water governance) and the natural domain (e.g., water allocations at each scale) creates serious scaling problems for water management (Cash et al 2006; Young 2006).

Issues of Scale and Related Challenges for Water Networks

It is evident from our review of the literature that terms such as scale and level are used interchangeably, and key concepts related to scale are defined quite differently across disciplines. For the sake of clarity, we will adopt the definitions used by Gibson et al (2000). They refer to scale as the spatial, temporal, quantitative, or analytical dimensions used to measure and study objects, variables, and processes. Level, on the other hand, refers to locations along a scale. Many scales are characterized by hierarchy; that is, they are either exclusive or nested. One form of hierarchy—characteristic of many complex systems—is a nested or constitutive hierarchy. In this configuration, one level can combine and influence another and, in so doing, can create new functions, services, and emergent properties. In complex, constitutive hierarchies, characteristics of larger units are not simple combinations of attributes of smaller units, but can take on new and often unexpected behavior affected by context and local variations. A detailed description and definitions of related terms are provided in selected readings and commentaries at the end of this chapter.

Traditionally, spatial and temporal scales are well understood in water engineering and hydrology. As shown in Figure 3.1, hydrologic processes can span many spatial levels (less than a kilometer to global) as well as temporal scales (from seconds to billions of years). The spatial and temporal characteristics of the basic water balance (solid, liquid, and vapor) in all compartments of the global system: atmosphere, oceans, and lands can span a range of process scales (National Research Council 1991).

Hydrologic processes are often observed and modeled at short timescales, while estimates are needed for very long timescales (e.g., the lifetime of a dam). Figure 3.2 (adapted from Bloschl and Sivapalan 1995) provides examples ranging from flood warning to design of large dams where different estimates from different types of hydrological models are required. The associated timescales range from minutes to hundreds of years. Similarly, models and theories developed in small space-scale laboratory experiments for infiltration and runoff are expected to work at the large scale of catchment areas. Conversely, sometimes large-scale model outputs are used to downscale at the watershed scale. This invariably involves transferring information (upscaling: from small to large levels; or downscaling: from large to small levels in time or space). This is called *scaling* in the water science literature, and the problems associated with it are called *scale issues*. For example, developing a very detailed understanding of water processes at a given level (e.g., infiltration of water at the hill slope level or development of a thunderstorm) may not provide relevant information for management of water resources at a different level (e.g., design of a large dam). In this example, note that we are only talking about one variable

GLOBAL
10000 Km
1000 Km
100 Km
10 Km
1 Km
LOCAL
SPACE
GLOBAL WEATHER SYSTEMS
GLOBAL CO_2 VARIATIONS
DEVELOPMENT OF MAJOR RIVER BASINS
SOIL FORMATION
RUNOFF CYCLE
MESOSCALE WEATHER SYSTEMS (FLOODS)
DRAINAGE NETWORK FORMATION
MESOSCALE SOIL MOISTURE VARIATION
SOIL EROSION
SHALLOW GROUND WATER CIRCULATION
NUTRIENT CYCLES
THUNDER STORM
TIME
LOG SEC
2
4
6
8
10
12
14
16
SEC
MIN
DAY
YEAR
CENTURY
ONE MILLION YEARS
ONE BILLION YEARS

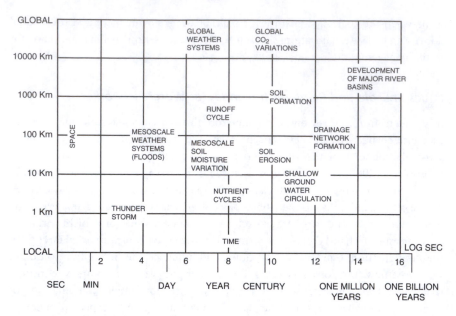

FIGURE 3.1 Illustrative range of hydrologic process scales
Source: (adapted from Figure 2.9 of NRC 1991)

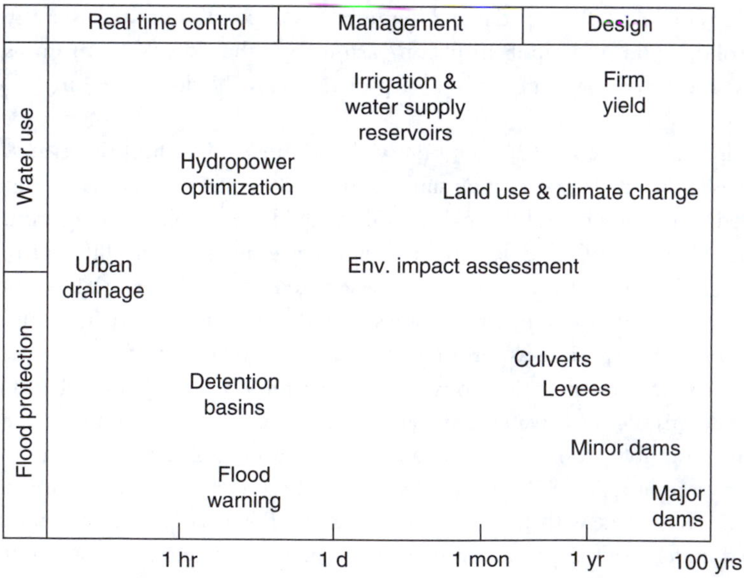

Real time control
Management
Design
Water use
Flood protection
Irrigation & water supply reservoirs
Firm yield
Hydropower optimization
Land use & climate change
Urban drainage
Env. impact assessment
Culverts
Levees
Detention basins
Minor dams
Flood warning
Major dams
1 hr
1 d
1 mon
1 yr
100 yrs

FIGURE 3.2 Real-time control to design of water resources at different temporal levels
Source: (adapted from Figure 1 of Bloschl and Sivapalan 1995)

(water quantity) from one domain (natural) crossing levels in two scales (space and time). This is a complicated problem but it can be solved with adequate data, model development, calibration, and validation.

The spatial scale is perhaps the best studied scale for a discipline like geography. Closely related to spatial scale is jurisdictional scale, which is central in many studies of governance. There are other scales (e.g., institutional, management, networks, knowledge) worth considering in water network management (Cash et al 2006). Figure 3.3 adapted from Cash et al (2006) shows several scales and associated levels. Interactions will occur within or across scales and levels, leading to substantial complexity in managing a water network. "Cross-level" interactions among levels within a scale (flood forecasting and water management) and "cross-scale" interactions across different scales (spatial and jurisdictional scales) or "multi-level" and "multi-scale" interactions (see Figure 3.4) may lead to differences in the way water networks operate, even when they have similar structural properties. Changes in water networks may arise from these interactions or be caused by still other variables.

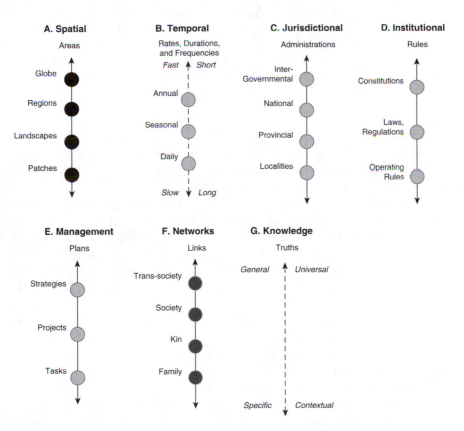

FIGURE 3.3 Different scales and levels that are critical to understanding and addressing complex water problems

Source: (adapted from Cash et al 2006)

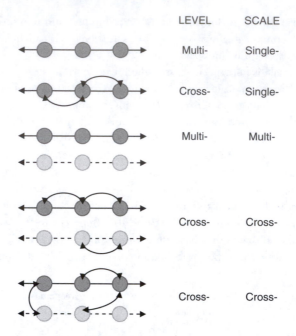

FIGURE 3.4 Cross-level, cross-scale, multilevel and multiscale interactions for water problems

Source: (adapted from Cash et al 2006)

For example, decentralization of government operations can produce periods of strong interaction among high-level national institutions and those at the local government level at one point in time, and then settle into a much more modest and steady level of interaction later on (Young 2006).

Following the work of Gibson et al (2000), we categorize important issues related to scale into four areas relevant to natural, societal, and political domains: (1) how the extent and resolution of different scales affect the identification of patterns; (2) how different levels on a scale affect the explanation of an observed phenomena in the natural domain (weather versus climate on a temporal scale but different level) or the societal domain (micro-scale behavior of an individual and macro-scale response of the economy); (3) what theoretical propositions derived about phenomena at one level on a spatial, temporal, or quantitative scale may tell us about its generalization potential at another level (e.g., a detailed understanding of movement of a water droplet on a creek may not be easily scalable to predict flow of water in the Ganges); and (4) how and which processes can be optimized at particular scale to design effective water management strategies.

We define the "scale challenge" as a situation in which a combination of cross-scale and cross-level interactions create confusion and undermine the effectiveness of water network management. This challenge includes: (1) the failure to recognize important scale and level considerations in a given water network; (2) the persistence of mismatches between levels and scales in observation, modeling, intervention, and implementation; and (3) a failure to recognize differences in the way that scales are perceived and valued by different actors, even at the same level. Cash et al (2006) refer to these challenges as "ignorance," "mismatch," and "plurality." We will discuss their relevance for managing water networks in the next chapter. What follows is a discussion of another level of mismatch between watersheds (a commonly accepted unit for water management), "problem-sheds" and "policy-sheds," and the challenges these mismatches pose for water management.

Watersheds are Disconnected from Problem-sheds and Policy-sheds

A watershed is a hydrologic unit: *an area of land draining into a common body of water such as a lake, river, or ocean.* It is sometimes called a "river basin" or a "catchment." We will use the term "watershed" or "river basin" to refer to a hydrologic unit throughout this book. There are 2,110 watersheds in the continental United States that cross county, state, and national boundaries. For example, the Mississippi River watershed is the largest in North America. With its many tributaries, the Mississippi's watershed drains all or parts of 31 states and eventually reaches the Gulf of Mexico. Contrast this definition of watershed as a hydrologic unit with the definition proposed by geographer John Powell as: "that area of land, a bounded hydrologic system, within which all living things are inextricably linked by their common water course and where, as humans settled, simple logic demanded that they become part of a community" (Powell 1875). In Powell's definition, the natural domain (a bounded hydrologic system) is coupled with the societal domain (living things are linked and become part of a community). Over the last 10,000 years, Native Americans have lived along the Mississippi. Most were hunter–gatherers or herders and some formed prolific agricultural societies similar to the inhabitants of our fictional Indopotamia river basin. The arrival of Europeans in the 1500s forever changed the structure of the water network in the Mississippi watershed. With increased scientific understanding of how the global water cycle is affected by warming and cooling of the Pacific Ocean during El-Niño or La-Niña over the last several decades, we now know that the Mississippi watershed is not a bounded hydrologic system—even from a natural domain perspective. It is affected by global teleconnections including El-Niño and La-Niña. Yet, the use of the watershed as a unit of analysis is common in both the natural and societal domains (e.g., "watershed user association" as a governance unit).

Cohen and Davidson (2011) argue that there are at least five recognized challenges associated with a watershed approach to water resource management: the difficulties of boundary choice; accountability; public participation; and asymmetries with "problem-sheds" and "policy-sheds." First, even when the boundary choice from a hydrologic perspective is clear, the choice about which boundary to use for governance is often unclear. The choice of which watershed boundary to use is a political decision as much as a scientific one (Blomquist and Schlager 2005). Indeed, the nested nature of watersheds (ordering of streams, sub-watersheds and tributaries) lends itself to a number of different boundary options. Decisions about which hydrologic boundary to use for the purposes of governance or management should probably be based on a combination of natural, societal, and political considerations. Second, watersheds are usually not aligned with electoral or jurisdictional boundaries. Consequently, the usual means of ensuring accountability do not necessarily apply. Watershed-scale organizations may not show answerability or responsiveness to those living and working within the watershed (Blomquist and Schlager 2005). Third, the watershed approach represents both a scaling up from the local level and a scaling down from the national or state level, but the notion of working at the watershed level has largely been framed in terms of decentralization (Kemper et al 2007). It is not clear why this is the case. There does not appear to be anything inherently participatory or empowering about re-scaling (Cohen and Davidson 2011). Fourth, asymmetries between problem-shed and watershed may create challenges for water network managers. A problem-shed is defined as a "geographic area that is large enough to encompass management issues, but small enough to make implementation feasible" (Griffin 1999). Watersheds frequently influence—and are influenced by—factors outside their boundaries. Consequently, it is important to identify an appropriate level (e.g., choice of problem-shed may be more appropriate than watershed) when we think about assigning water network management responsibilities. Fifth, another set of challenges arises from asymmetries between watershed boundaries and traditional administrative scales (i.e., policy-sheds). These can be compounded by policy gaps and overlaps among different policy-sheds (Cohen and Davidson 2011).

Difficulties created by these challenges and asymmetries complicate data collection, monitoring, and assessment of policy interventions. A case in point is the ACF river basin shared by three US states as discussed in Chapter 2. In addition to evolving socioeconomic and environmental concerns at the local level, the context for the ACF was transformed by the introduction of new legislation at the national level, including the National Environmental Policy Act and the Endangered Species Act. These provided legal ground for Florida, and the U.S\. Fish and Wildlife Service, and a range of non-governmental organizations to contest water allocation proposals made by the U.S. Army Corps of Engineers (Clemons 2004; Leitman 2005).

The challenges discussed above related to the watershed approach to decision-making create significant obstacles to effective water management. Efforts to address these challenges could involve adjusting boundaries as each water management problem arises in an attempt to obtain an accountable and participatory result. This is not to suggest that the watershed is not a useful boundary. In some situations, watersheds may work well. However, questions should be raised about how and when to use the watershed boundary and how watersheds are linked to problem-sheds and policy-sheds. For example, what decisions (interventions) should be made at the watershed level and which (interventions) should be made at other levels? In this context, we will explore the use of a negotiated approach to managing complex water networks in the next chapter.

Selected Readings with Commentaries

J. Liu et al "Complexity of Coupled Human and Natural Systems," (2007)

Introduction

This review article provides a synthesis of six case studies to show that interactions between human and natural systems vary across space, time, and institutional levels. These case studies are chosen from a variety of ecological, socioeconomic, political, and cultural settings. They highlight the key attributes of coupled natural and human systems, which exhibit: (1) feedback among different components; (2) nonlinear interactions and thresholds (transition points between alternate states); (3) surprises and unintended consequences due to policy interventions; (4) legacy effects from prior interventions varying in duration from decades to centuries; (5) varying degrees of resilience depending on the context and type of human interventions; and (6) heterogeneity across a range of scales and levels. These case studies provide critical insights into the complexity of coupled natural and societal domains and highlight the importance of coordinated long-term efforts across multiple domains, scales, and levels.

Complex Interactions and Feedback Loops in Coupled Natural and Societal Systems (p. 1513)

Coupled human and natural systems are integrated systems in which people interact with natural components. Although many studies have examined human–nature interactions, the complexity of coupled systems has not been well understood. The lack of progress is largely due to the traditional separation of ecological and social sciences. Although some scholars have studied coupled systems as complex adaptive systems, most of the previous work has been theoretical rather than empirical. An increasing number of interdisciplinary programs have been integrating ecological and social sciences to study coupled human and natural systems (e.g., social-ecological systems and human-environment systems). Here, we synthesize six case studies to demonstrate the approaches used and results found.

These studies are on five continents: the Kenyan Highlands in Africa (Kenya); the Wolong Nature Reserve for giant pandas in China (Wolong); Central Puget Sound of Washington (Puget Sound) and Northern Highland Lake District of Wisconsin (Wisconsin) in the United States; an area near Altamira, State of Pará, Brazil (Altamira); and Kristianstads Vattenrike of Sweden (Vattenriket). They include urban (Puget Sound), semi-urban (Vattenriket), and rural areas (Altamira, Kenya, Wisconsin, and Wolong), and they are in developed countries (Puget Sound, Wisconsin, and Vattenriket) and developing countries (Altamira, Kenya, and Wolong). These studies are in different ecological, socioeconomic, political, demographic, and cultural settings, and they encompass a variety of ecosystem services and environmental problems.

These studies share four major features. First, they explicitly address complex interactions and feedback between human and natural systems. Unlike traditional ecological research that often excluded human impacts or social research that generally ignored ecological effects, these studies consider both ecological and human components as well as their connections. Thus, they measure not only ecological variables (e.g., landscape patterns, wildlife habitat, and biodiversity) and human variables (e.g., socioeconomic processes, social networks, agents, and structures of multilevel governance) ... but also variables that link natural and human components (e.g., fuel wood collection and use of ecosystem services). Second, each study team is interdisciplinary, engaging both ecological and social scientists around common questions. Third, these studies integrate various tools and techniques from ecological and social sciences as well as other disciplines such as remote sensing and geographic information sciences for data collection, management, analysis, modeling, and integration. Fourth, they are simultaneously context specific and longitudinal over periods of time long enough to elucidate temporal dynamics. As such, these studies have offered unique interdisciplinary insights into complexities that cannot be gained from ecological or social research alone.

Nonlinearity and Thresholds (p. 1514)

Numerous relationships in coupled systems are nonlinear. In Wisconsin, for instance, fallen trees that provide critical fish habitat in lakes and streams drastically decrease when housing density exceeds about seven houses per kilometer of shoreline. Bird richness in the Puget Sound landscape with single-family housing and fragments of native forest increases nonlinearly with forest cover and peaks when 50 to 60 percent of the land is forested.

Thresholds [transition points between alternate states] are common forms of nonlinearity. In Vattenriket, an intentional participatory process mobilized stakeholders laying the groundwork for a shift from conventional management to adaptive co-management ... Cultural values and environmental concerns prompted local stakeholders to build new knowledge, develop new visions and goals, and create new social networks. The result of these community activities was a new and more suitable governance system of adaptive co-management of the landscape.

System behaviors shift from one state to another over time (temporal thresholds) and across space (spatial thresholds).

Surprises (pp. 1514–1515)

When complexity is not understood, people may be surprised at the outcomes of human nature couplings. For example, smelt (Osmerus mordax) was initially introduced to Wisconsin as a prey species for game fish such as walleyes (Stizostedion vitreum), but smelt ate juvenile walleyes leading to loss of walleye populations. In Puget Sound, growth management policy has caused urban density to intensify inside the urban growth boundary while unintentionally facilitating sprawl outside the urban growth boundary. Conservation policies can also generate unintended perverse results. In Wolong, for instance, high-quality panda habitat degraded faster after the area was established as a reserve than before the reserve's creation. To prevent further degradation, a natural forest conservation program was introduced in 2001 for local residents to monitor illegal harvesting. Unexpectedly, a large number of new households formed in 2001 because many households decided to split into smaller ones to more effectively capture subsidies (20 to 25% of the average household income) given to households as part of the program. The household proliferation and reduction in household size (number of people in a household) increased demand for fuelwood and land for house construction.

Some ecosystems can only be sustained through human management practices, whereas many conservation efforts preclude such human interference. For example, the wetland site under the Ramsar Convention (an international treaty for the conservation and sustainable use of wetlands) in Vattenriket was set aside for conservation purposes, but the wetland became overgrown when grazing was halted. This unintended consequence led to an understanding of grazing as essential to maintaining this wetland system.

Some Generalizable Findings from Diverse Case Studies (p. 1516)

Results such as those reviewed here benefit from and help advance the integration of ecological and social sciences. The approaches used and the results from these studies can be applied to many other coupled systems at local, national, and global levels. For instance, the finding that the number of households increased faster than the human population size in Wolong over the past three decades has led to the discovery that this trend is global and is particularly profound in the 76 countries with biodiversity hotspots. The Lake Futures Project in Wisconsin was a prototype used to develop approaches for the Millennium Ecosystem Assessment scenarios.

Comparison of these studies provides important insights into diverse complex characteristics that cannot be observed in a single study. The types of surprises found in the case studies differ, although all of them originated from the interactions between human and natural systems. All six studies have demonstrated legacy effects, but legacy durations varied from decades to centuries. Because of the independent nature of these studies, information from one study is not necessarily available in or transferable to other studies. To increase the extent of generalizing from case studies, future research on coupled systems must include not only separate site specific studies but also coordinated, long-term comparative projects across multiple sites to capture a full spectrum of variations. Furthermore, all the studies in this review focus on interactions within the system, rather than interactions among different coupled systems. As globalization intensifies, there are more interactions among even geographically distant systems and across scales. Thus, it is critical to move beyond the existing approaches for studying coupled systems, to develop more comprehensive portfolios, and to build an international network for interdisciplinary research spanning local, regional, national, and global levels.

Commentary

This article summarizes some of the difficulties associated with characterizing and managing coupled natural and human systems. It attempts to synthesize our understanding of such couplings by engaging ecological and social scientists around common questions. The case studies integrate a range of existing tools and techniques from several disciplines to provide context-specific and longitudinal narratives over long enough time scales to elucidate the importance of temporal dynamics. In coupled systems, humans and nature interact reciprocally, and create complex feedback. Understanding these feedback loops and their impacts on the co-evolution of the coupled systems is critical to implementing effective management strategies. When complexity is not anticipated, one may be surprised at the outcomes and unintended consequences created by policy interventions. Lessons learned from these case studies have particular relevance to water management. For example, the wetland site in Vattenriket (Sweden) was set aside for conservation; but it became overgrown when grazing was halted. Such unintended consequences, and related surprises, are common for coupled natural and societal systems. A key lesson we may derive from this study is the importance of understanding and characterizing water management problems in terms of coupled natural and human systems. A careful analysis of reciprocal effects and feedback loops, nonlinearity and thresholds, surprises, legacy effects and time lags, resilience, and heterogeneity allow us to diagnose how complexity might best be handled.

P. Allen "What is Complexity Science? Knowledge of the Limits of Knowledge," (2001)

Introduction

This paper summarizes key ideas from complexity science. It argues that complex systems must deal with continuously changing outcomes that result from interactions among the various elements of such systems. Such outcomes are usually not predictable unless it is possible to make assumptions that reduce the degrees of freedom of the elements in the systems. Instead of focusing on solving problems, the author argues for developing a capacity to understand the limits of knowledge and under what circumstances knowledge is transferable across domains, scales, and levels.

Assumptions and Implications Related to Reducing "Complexity" to "Simplicity" (pp. 24–27)

In a recent article (Allen, 2000), the underlying assumptions involved in the modeling of situations were systematically presented. In essence, we attempt to trade the "complexity" of the real world for the "simplicity" of some reduced representation. The reduction occurs through assumptions concerning: (1) The relevant "system" boundary (exclude the less relevant); (2) The reduction of full heterogeneity to a typology of elements (agents that might be molecules, individuals, groups, etc.); (3) Individuals of average type; (4) Processes that run at their average rate.

If all four assumptions can be made, our situation can be described by a set of deterministic differential equations (system dynamics) that allow clear predictions to be made and "optimizations" to be carried out. If the first three assumptions can be made, then we have stochastic differential equations that can self-organize, as the system may jump between different basins of attraction that reflect distinct patterns of dynamical behavior. With only assumptions 1 and 2, our picture becomes one of adaptive evolutionary change, in which the pattern of possible attractor basins alters and a system can spontaneously evolve new types of agent, new behaviors, and new problems. In this case, naturally we cannot predict the creative response of the system to any particular action we may take, and the evaluation of the future "performance" of a new design or action is extremely uncertain (Allen, 1988, 1990, 1994a, 1994b). In summary, a previous article (Allen, 2000) showed how the science of complex systems could lead to a table concerning our limits to knowledge. If we now take the different kinds of knowledge in which we may be interested concerning a situation, we can number them according to: (1) What type of situation or object we are studying (classification: "prediction" by similarity); (2) What it is "made of" and how it "works."; (3) Its "history" and why it is as it is.; (4) How it may behave (prediction).; (5) How and in what way its behavior might be changed (intervention and prediction).

We can then establish [Table 3.1], which therefore in some ways provides us with a very compressed view of the science of complexity. The idea behind the "modeling" approach is not that it should create true representations of "reality." Instead, it is seen as one method that leads to the provision of causal "conjectures" that can be compared with and tested against reality. When they appear to fit that reality, we may feel temporarily satisfied; when they disagree with reality, we can set about examining why the discrepancy occurred. This model is our "interpretive framework" for sense making and knowledge building. It will almost certainly change over time as a result of our experiences. It is developed in order to answer questions that are of interest to the developer, or the potential user, and both the model and the questions will change over time. The questions that are addressed influence the variables that are chosen for study, the mechanisms that are supposed to link them, the boundary of the system considered, and the type of scenarios and events that are explored. The model is not reality, but merely a creation of the modeler that is intended to help reflect on the questions that are of interest. The process involved is not telling us whether our current model is true or false, but rather whether it appears to work or not. If it does, then it will be useful in answering the questions asked of it. If it doesn't, then it is telling us that we need to rethink our interpretive framework, and that some new conjecture is needed.

TABLE 3.1 Systematic Knowledge Concerning the Limits to Systematic Knowledge

Assumptions	2	3	4	5
Type of model	Evolutionary	Self-organizing	Nonlinear dynamics (including chaos)	Equilibrium
Type of system	Can change structurally	Can change its configuration and connectivity	Fixed	Fixed
Composition	Can change qualitatively	Can lead to new, emergent properties	Yes	Yes
History	Important in all levels of description	Is important at the system level	Irrelevant	Irrelevant
Prediction	Very limited. Inherent uncertainty	Probabilistic	Yes	Yes
Intervention and prediction	Very limited. Inherent uncertainty	Probabilistic	Yes	Yes

Source: (from Allen 2000)

The usefulness may well come down to a question of the spatiotemporal scales of interest to the modeler or user. For example, if we compare an evolutionary situation to one that is so fluid and nebulous that there are no discernible forms, and no stability for even short times, we see that what makes an evolutionary model possible is the existence of quasi-stable forms for some time at least. If we are only interested in events over very short times compared to those usually involved in structural change, it may be perfectly legitimate and useful to consider the structural forms as fixed. This doesn't mean that they are, it just means that we can proceed to do some calculations about what can happen over the short term, without having to struggle with how forms may evolve and change. Of course, we need to remain conscious that over a longer period forms and mechanisms will change and that our actions may well be accelerating this process, but nevertheless it can still mean that some self organizing dynamic is useful.

Equally, if we can reasonably assume not only that system structure is stable but in addition that fluctuations around the average are small, we may find that prediction using a set of dynamic equations provides useful knowledge. If fluctuations are weak, it means that large fluctuations capable of kicking the system into a new regime/attractor basin are very rare and infrequent. This gives us some knowledge of the probability of this occurring over a given period. So, our model can allow us to make predictions about the behavior of a system as well as the associated probabilities and risk of an unusual fluctuation occurring and changing the regime. An example of this might be the idea of a 10-year event and a 100-year event in weather forecasting, where we use historical statistics to suggest how frequent critical fluctuations are. Of course, this assumes an overall "stationarity" of the system, supposing that processes such as climate change are not happening. Clearly, when 100-year events start to occur more often, we are tempted to suppose that the system is not stationary, and that climate change is occurring. However, this would be "after the facts," not before.

These are examples of the usefulness of different models, and the knowledge with which they provide us, all of which are imperfect and not strictly true in an absolute sense, but some of which are useful. Systematic knowledge therefore should not be seen in absolute terms, but as being possible for some time and some situations, provided that we apply our "complexity-reduction" assumptions honestly. Instead of simply saying that "all is flux, all is mystery," we may admit that this is so over the very long term (who wants to guess what the universe is for?). However, for some times of interest and for some situations, we can obtain useful knowledge about their probable behavior, and this can be updated by continually applying the "learning" process of trying to "model" the situation.

Complexity: Interaction and Feedback of Multiple Objectivities and Subjectivities (pp. 27–29)

The essence of complex systems is that they represent the "joining" of multiple subjectivities—multiple dimensions interacting with overlapping but not identical multiple dimensions. In a traditional "system dynamics" view, a flow diagram represents a series of reservoirs that are connected by pipes and a unidimensional flow takes place between them. Typically, some simple laws express the rate of flow between the reservoirs, possibly as a function of the height of water. Instead of this, we find that the real world consists of connected entities that have their own perceptions, inner worlds, and possibilities of action. We may contrast a flow diagram of money or water with an "influence" diagram that notes that one component "affects" another.

The water flowing through (a) in [Figure 3.5] is totally different and subject to accountability rules compared to system (b), which shows how a company, its products, and its potential customers interact. Let us consider briefly the three "simple" arrows of system (b):

(1) Potential customers influence the company to design a product that they think will be successful. But, this requires the company to "seek" information about potential customers, and therefore to think of ways to do this. It requires the generation of "knowledge" or "belief" about what kind of customers exist, and what they may be looking for, and this essentially must be based on a series of "conjectures" that the company must make about the nature of the different subjectivities in the environment. In essence, the company must "gamble" that its conjectures about possible customers and their desires are sufficiently correct to make enough of them buy the product.

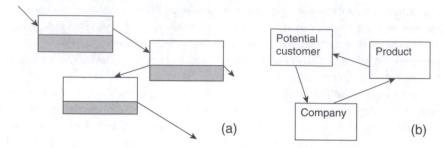

FIGURE 3.5 Comparison between a simple and a complex system
Source: (from Allen 2000)

(2) The company produces a particular product. This results from the beliefs at which the company has arrived concerning the kind of customers that are "out there" and the qualities they are looking for at the price they are willing to pay. The arrow therefore encompasses the way in which the marketing people in the company have been able to affect the designer and the new product development process to try to produce the "right thing at the right price." Secondly, it implies that the designer knows how to put components together in such a way that the product or service "delivers" what was hoped.

(3) The product then attracts, or fails to attract, potential customers. However, this requires the potential customers to "look at" the product and see what it might do for them, and at what cost. Customers must be able to translate the attributes of the product or service into the fulfillment of some need or desire that they have. The "interaction" between the product and the customer must be "engineered" by the company so that there will be some encounter with potential customers. The locations and timings must therefore be suited to the movements and attention of the potential target customers.

(4) The arrow from the customer to the company finally consists of a flow of money that occurs when a potential customer becomes an actual customer. This part is physical and real and can be stored easily on a database. However, it is the "result" of a whole lot of less palpable processes, of conjecture, of characterization, and of information analysis, most of which do not correspond to flows of anything accountable.

What is important is that inside each "box" there are multiple possible behaviors. Ultimately, this comes down to the existence of internal micro-diversity that gives each box a range of possible responses. These are tried out and the results used either to reinforce or to challenge their use. Each box is therefore trying to make sense of its environment, which includes the other boxes and their behaviors.

The real issue is that the boxes marked "company" and "potential customer" actually enclose different worlds. The dimensions, goals, aims, and experiences in these boxes are quite different. Most importantly, each box contains a whole range of possible behaviors and beliefs, and the agents "inside" may have mechanisms that enable them to find out which ones work in the environment. What this means is dealt with in the next section. When the boxes interact, therefore, as reflected in some simple "arrows" of influence, what we really have is the communication of two different worlds that inhabit different sets of dimensions. However, to "successfully" connect two different "boxes" so as to achieve some joint task requiring cooperation demands some human intervention to "translate" the meaning that exists in one space into the language and meaning of the other.

The simple "arrow" of connection, therefore, will not be a "simple connection," but instead may require a person who is capable of speaking across the boxes and who possibly has experience of both worlds. It is also the core reason explaining the need for interdisciplinary studies. Each discipline takes a partial and particular view of a situation, and in so doing promotes analysis and expertise at the expense of the ability of integration, synthesis, and a holistic view. Management is a domain in which a multidisciplinary, integrative approach is required if we are to get real results in dealing with a real-world problem. The science of complex systems is extremely important for management and for policy, since only with an integrated, multidimensional approach will advice be related successfully to the real-world situation. This may indeed spell the limits to knowledge and turn us from the attractive but misleading mirage of prediction.

Complex Systems Science: Limits of Knowledge and Process of Adaptive Learning (pp. 39–41)

The science of complex systems is about systems whose internal structure is not reducible to a mechanical system. In particular, it is about connected complex systems, for which the assumptions of average types and average interactions are not appropriate and are not made. Such systems co-evolve with their environment, being "open" to flows of energy, matter, and information across whatever boundaries we have chosen to define. These flows do not obey simple, fixed laws, but instead result from the internal "sense making" going on inside them, as experience, conjectures, and experiments are used to modify the interpretive frameworks within.

Because of this, the behavior of the systems with which each system is coevolving are necessarily uncertain and creative, and is not best represented by some predictable, fixed trajectory. This takes some steps toward the "postmodern" point of view. But as Cilliers (1998) indicates, the original definition of postmodernism (Lyotard, 1984) does not take us to the situation of total subjectivity where no assumptions can be made, but rather to the domain of evolutionary complex systems discussed in this article.

Instead of a fixed landscape of attractors, and a system operating in one of them, we have a changing system, moving in a changing landscape of potential attractors. Provided that there is an underlying potential of diversity, then creativity and noise (supposing that they are different) provide a constant exploration of "other" possibilities. Our simple model only supposed 20 possible underlying behaviors, but obviously in any realistic human situation the number would be very large. In dealing with a changing environment, therefore, we find a "law of excess diversity" in which system survival in the long term requires more underlying diversity than would be considered requisite at any time. Some of these possible behaviors mark the system and alter the dimensions of its attributes, leading to new attractors and new behaviors, toward which the system may begin to move but at which it may never arrive, as new changes may occur "on the way." The real revolution is not therefore about a neoclassical, equilibrium view as opposed to nonlinear dynamics having cyclic and chaotic attractors, but instead is about the representation of the world as a non-stationary situation of permanent adaptation and change. The picture we have arrived at here is one that Stacey et al. (2000) refer to as a "transformational teleology," in which potential futures (patterns of attractors and of pathways) are being transformed in the present. The landscape of attractors we may calculate now is not in fact where we shall go, because it is itself being transformed by our present experiences.

The macro-structures that emerge spontaneously in complex systems constrain the choices of individuals and fashion their experience. Behaviors are being affected by "knowledge" and this is driven by the learning experience of individuals. Each actor is coevolving with the structures resulting from the behavior and knowledge/ ignorance of all the others, and surprise and uncertainty are part of the result. The "selection" process results from the success or failure of different behaviors and strategies in the competitive and cooperative dynamical game that is running.

However, there is no single optimal strategy. What emerge are structural attractors, ecologies of behaviors, beliefs, and strategies, clustered in a mutually consistent way, and characterized by a mixture of competition and symbiosis. This nested hierarchy of structure is the result of evolution and is not necessarily "optimal" in any way, because there is a multiplicity of subjectivities and intentions, fed by a web of imperfect information and diverse interpretive frameworks. In human systems, at the microscopic level, behavior reflects the different beliefs of individuals based on past experience, and it is the interaction of these behaviors that actually creates the future. In so doing, it will often fail to fulfill the expectations of many of the actors, leading them either to modify their (mis)understanding of the world, or alternatively simply leave them perplexed. Evolution in human systems is therefore a continual, imperfect learning process, spurred by the difference between expectation and experience, but rarely providing enough information for a complete understanding.

Although this sounds tragic, it is in fact our salvation. It is this very "ignorance" or multiple misunderstandings that generates micro-diversity, and leads therefore to exploration and (imperfect) learning. In turn, the changes in behavior that are the external sign of that "learning" induce fresh uncertainties in the behavior of the system, and therefore new ignorance (Allen, 1993). Knowledge, once acted on, begins to lose its value. This offers a much more realistic picture of the complex game that is being played in the world, and one that our models can begin to quantify and explore.

In a world of change, which is the reality of existence, what we need is knowledge about the process of learning. From evolutionary complex systems thinking, we find models that can help reveal the mechanisms of adaptation and learning, and that can also help imagine and explore possible avenues of adaptation and response. These models have a different aim from those used operationally in many domains. Instead of being detailed descriptions of existing systems, they are more concerned with exploring possible futures and the qualitative nature of these. They are also more concerned with the mechanisms that provide such systems with the capacity to explore, to evaluate, and to transform themselves over time. They address the "what might be," rather than the "what is" or "what will be." It is the entry into the social sciences of the philosophical revolution that Prigogine wrote about in physics some 25 years ago. It is the transition in our thinking from "Being to Becoming." It is about moving from the study of existing physical objects using repeatable objective experiments, to methods with which to imagine possible futures and with which to understand how possible futures can be imagined. It is about system transformation through multiple subjective experiences, and their accompanying diversity of interpretive, meaning-giving frameworks.

> Reality changes, and with it experiences change also. In addition, however, the interpretive frameworks or models people use change, and what people learn from their changed experiences is transformed. Recognizing these new "limits to knowledge," therefore, should not depress us. Instead, we should understand that this is what makes life interesting, and what life is, has always been, and will always be about.

Commentary

Allen shows how the science of complex systems can provide a way to understand the limits on building systematic knowledge. He argues that the science of complexity is about the limits of systematic knowledge in situations where problem descriptions are "intricate or hard to unravel." In our water context, we call this class of problems "messy" or "wicked." Allen suggests two underlying reasons for this messiness: either the problem description contains a large number of interacting elements, or nonlinear interactions among system elements lead to multiple futures and creative/surprising responses. This reasoning is in sharp contrast with the classical Newtonian view in which the laws of interaction allow us to predict and intervene to achieve desired outcomes. This simple and comforting view of the "predictable future" is being challenged by our increased understanding of nonlinear, coupled systems where predictability is limited and emergence characterizes the systems. To understand and manage such systems, as Allen elegantly summarizes, "the answer will be neither that of 'full prediction' nor that of 'no prediction,' not 'total control' nor 'totally out of control.' As most people have secretly suspected, the answer lies somewhere in between. And now science is coming to this conclusion as well" (Allen 2000). Our approach to understanding and managing complex water management problems is also motivated by this search to find the answers drawing on ideas from complexity theory and non-zero-sum negotiations.

E. Mitleton-Kelly "Ten Principles of Complexity and Enabling Infrastructures," (2003)

Introduction

This book chapter summarizes ideas about complexity in human organizations. Complexity theory in this context offers a framework for understanding the behavior of social (human) systems. It discusses four principles of complexity— emergence, connectivity, interdependence, and feedback—and emphasizes their inter-relationship. It suggests that it is not adequate to isolate one principle or characteristic, such as emergence, and concentrate on it to the exclusion of the others. Mitleton-Kelly argues for a deeper understanding of complex systems that might help to anticipate and explain socio-ecological interactions. She also examines the extent to which complexity principles developed for natural systems are applicable to societal systems. The chapter highlights that "theories of complexity provide a conceptual framework, a way of thinking, and a way of seeing the world."

Principles of Complexity Theory for Societal Systems (pp. 2–4)

By comparison with the natural sciences there was relatively little work on developing a *theory* of complex *social* systems despite the influx of books on complexity and its application to management in the past 6–7 years (an extensive review of such publications is given by Maguire & McKelvey 1999). The notable exceptions are the work of Luhman on autopoiesis, Arthur in economics and the work on strategy by Lane & Maxfield (1997), Parker & Stacey (1994) and Stacey (1995, 1996, 2000, 2001). A theory in this context is interpreted as an *explanatory framework that helps us understand the behaviour of a complex social (human) system.* (The focus of the author's work and hence the focus of this chapter is on human organisations. Other researchers have concentrated on non-human social systems, such as bees, ants, wasps, etc.) Such a theory may provide a different way of thinking about organisations, and could change strategic thinking and our approach to the creation of new organizational forms—that is, the structure, culture, and technology infrastructure of an organisation. It may also facilitate, in a more modest way, the emergence of different *ways of organizing* within a limited context such as a single department within a firm. The case study at the end of this chapter describes how a different way of organising emerged in the Information Technology Department in the London office of an international bank.

The chapter will discuss each principle in turn, providing some of the scientific background and describing in what way each principle may be *relevant* and *appropriate* to a human system. Regarding the five areas of research listed on the left hand side of [Figure 3.6] dissipative structures are discussed at length as part of the "far-from equilibrium" and "historicity" principles; complex adaptive systems research underlies most of the other principles and the work of Kauffman is referred to extensively; autopoiesis is not discussed in this chapter but it has played an important role in the thinking underlying the current work (for the implications and applications of autopoiesis see Mingers 1995); chaos theory is given a separate section, but the discussion is not extensive; and Arthur's work on increasing returns is discussed under the 'path-dependence' principle.

The four principles grouped together in [Figure 3.6] of emergence, connectivity, interdependence, and feedback are familiar from systems theory. Complexity builds on and enriches systems theory by articulating additional characteristics of complex systems and by emphasising their inter-relationship and interdependence. It is not enough to isolate one principle or characteristic such as self-organisation or emergence and concentrate on it in exclusion of the others. The approach taken by this chapter argues for a deeper understanding of complex systems by looking at several characteristics and by building a rich inter-related picture of a complex social system. It is this deeper insight that will allow strategists to develop better strategies and organisational designers to facilitate the creation of organisational forms that will be sustainable in a constantly changing environment.

The discussion is based on *generic principles,* in the sense that these principles or characteristics are common to all natural complex systems. One way of looking at complex human systems is to examine the generic characteristics of natural complex systems and to consider whether they are *relevant or appropriate to social systems.* But there is one limitation in that approach, which is to understand that such an

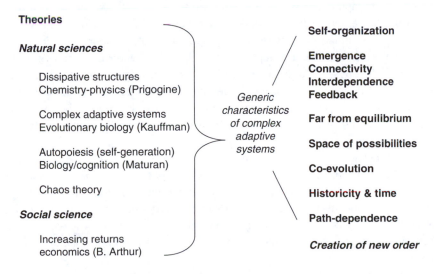

FIGURE 3.6 Five areas of theoretical research and generic characteristics of complex adaptive systems

Source: (from Mitleton-Kelly 2003)

examination is merely a starting point and not a mapping, and that social systems need to be studied in their own right.

This limitation is emphasized for two reasons: (a) although it is desirable that explanation in one domain is consistent with explanation in another and that these explanations honour the *Principle of Consistency* (Hodgson 2001, p90), characteristics and behaviour cannot be mapped directly from one domain to another, without a rigorous process of testing for appropriateness and relevance. Not only may the unit of analysis be quite different, but scientific and social domains may also have certain fundamental differences that may invalidate direct mapping. For example humans have the capacity to reflect and to make deliberate choices and decisions among alternative paths of actions. This capacity may well distinguish human behaviour from that of biological, physical or chemical entities; (b) a number of researchers consider the principles of complexity only as metaphors or analogies when applied to human systems. But metaphors and analogies are both limiting and limited and do not help us understand the fundamental nature of a system under study. This does not mean that neither metaphor nor analogy may be used. We use them as "transitional objects" all the time in the sense that they help the transition in our thinking when faced with new or difficult ideas or concepts. The point being emphasised, is that using metaphor and analogy is not the *only* avenue available to us in understanding complexity in an organisational or broader social context. Since organisations are, by their very nature, complex evolving systems, they need to be considered as complex systems in their own right.

Another way of looking at complexity is that suggested by Nicolis and Prigogine (1989 p8) "It is more natural, or at least less ambiguous, to speak of *complex behavior* rather than complex systems. The study of such behavior will reveal certain common characteristics among different classes of systems and will allow us to arrive at a proper understanding of complexity." This approach both honours the Principle of Consistency and avoids the metaphor debate. It may however upset some sociologists who do not find "arguments from science" convincing. But this is to miss Nicolis's and Prigogine's point, when they put the emphasis on the behaviour or characteristics of *all* complex systems. Nicolis and Prigogine are not behaviourists; they study the behaviour of complex systems in order to understand their deeper, essential nature.

This provides us with the underlying reason for studying complexity. *It explains and thus helps us to understand the nature of the world—and the organisations—we live in.* The term *"complexity"* will be used to refer to the *theories of complexity* (in the literature the plural "theories" is reduced to the singular for ease of reference and this practice will be used here) and *"complex behaviour"* to the behaviour that arises from the interplay of the characteristics or principles of complex systems.

> Complexity is not a methodology or a set of tools (although it does provide both). It certainly is not a "management fad". The **theories of complexity**

provide a conceptual framework, **a way of thinking**, and **a way of seeing the world**.

Principles of Complexity: Implications for Management (pp. 26–27)

This chapter introduces some of the principles of complexity based on the generic characteristics of all complex systems. It uses the logic of complexity to argue for a different approach to managing organisations through the identification, development, and implementation of an *enabling infrastructure*, which includes the cultural, social, and technical conditions that facilitate the day-to-day running of an organisation or the creation of a new organisational form.

Enabling conditions are suggested using the principles of complexity. Complex systems are not "designed" in great detail. They are made up of interacting agents, whose interactions create emergent properties, qualities, and patterns of behaviour. It is the actions of individual agents and the immense variety of those actions that constantly influence and create emergent macro patterns or structures. In turn the macro structure of a complex ecosystem influences individual entities, and the evolutionary process moves constantly between micro behaviours and emergent structures, each influencing and recreating each other.

The complexity approach to managing is one of fostering, of creating enabling conditions, of recognising that excessive control and intervention can be counterproductive. When enabling conditions permit an organisation to explore its space of possibilities, the organisation can take risks and try new ideas. Risk taking is meant to help find new solutions, alternative ways to do business, to keep evolving through established connectivities while establishing new ways of connecting (Mitleton-Kelly 2000).

This approach implies that all involved take responsibility for the decisions and actions they carry out on behalf of the organisation. They should not take unnecessary risks, nor are they blamed if the exploration of possibilities does not work. It is in the nature of exploration that some solutions will work and some will not.

Thus, another aspect of an enabling infrastructure is the provision of space, both in the metaphorical and actual senses. A good leader provides psychological space for others to learn, but also physical space and resources for that learning to take place. Individual and group learning is a prerequisite for adaptation, and the conditions for learning and for the sharing of knowledge need to be provided.

Complexity's great strength is that it crosses the boundaries of disciplines in both the natural and social sciences. It may one day provide us with a unified approach capable of linking those disciplines, because understanding the behaviour of complex systems in other subjects helps one gain deeper insights into phenomena in one's own field. Much work now being done on complexity in a variety of fields, from anthropology and psychology to economics and organisational science, will in due course change the way we see organisations, will help us understand their nature as complex systems, and ultimately will change the way that we manage organisations.

Commentary

This chapter uses the logic of complexity from natural systems to argue for a different approach to managing organizations through the identification, development, and implementation of an *enabling infrastructure*. Such enabling conditions include cultural, social, and technical considerations that facilitate the day-to-day operation of an organization or the creation of new organizational forms. To survive and thrive, an organizational entity needs to explore its space of possibilities and to generate variety. Complexity also suggests that the search for a single "optimum" strategy may neither be possible nor desirable. Any strategy can only be optimum under certain conditions, and when those conditions change, the strategy may no longer be ideal. To survive, an organization needs to be flexible and adaptive. Flexible adaptation also requires new connections and new ways of *seeing* things.

A key finding from complexity theory is that the future is unknowable. Stacey (2007) summarizes the consequences of this finding for planning and management: (1) analysis loses its primacy; (2) contingency (cause and effect) loses its meaning; (3) long-term planning becomes nearly impossible; and (4) statistical relationships become dubious. A main lesson one can derive from this unknowable future and related consequences is the need for continuous learning and adaptive management as situations change. Several management experts—despite working in very different fields—have reached comparable conclusions. Unfortunately, there is very little formally validated evidence to demonstrate that the complexity theory-based prescriptions for management produce the desired results (e.g., long-term survival of an organization). Although conceptually appealing, what follows from this is that complexity theory (primarily developed for natural systems) is an emerging area within which some intriguing results have been found that raise interesting conjectures about societal systems. What does *not* follow is that such findings apply to managing coupled natural and societal systems.

N.C. Narayanan and J.P. Venot "Drivers of Change in Fragile Environments: Challenges to Governance in Indian Wetlands," (2009)

Introduction

This paper uses "polycentric governance" as an analytical frame to examine the management of three wetlands in India. It highlights how coupled processes of natural and societal changes can shape governance patterns. It concludes by emphasizing that power and politics in the governance of natural resources need to be taken into account for effective decision-making.

Three Indian Wetlands: Cross-cutting Drivers to Environmental Degradation and Social Changes (p. 321)

Environmental and social changes are intertwined processes that both result from, and shape, governance patterns. Understanding these dynamics and their drivers is important for appreciating the scope for democratic and polycentric governance regimes and their outcomes. Three Indian wetlands of international importance under the Ramsar Convention, the Chilika, Kolleru and Vembanad lakes, have been selected for this study. Results are based on a comprehensive literature review, multiple field visits since 2003, and interviews with farmers, fishermen, and public servants.

The three study sites represent different socio-environments with their own trajectories, yet the Ramsar Convention lists "good practices" for their equitable and sustainable uses ... The analytical framework ... is used to understand how the wetlands are governed on the ground, and what the social and environmental dynamics at stake are. The three lakes each have unique stories of environmental degradation and social changes, with commonalities in their trajectories including: the enclosure of the commons and subsequent capitalization of resources and social marginalization, conflicting interests and intense local politics, a disconnect between global conservation discourses and local concerns, weak institutional arrangements, and global economic forces. The three wetlands are not affected by these dynamics in the same way ... but it is important to look upon each of those processes when designing governance regimes and assessing their scope for sustainable environmental management.

Lessons from Three Indian Wetlands: Issues of Scales and Complexity (pp. 331–332)

The case studies of three Indian wetlands of international importance under the Ramsar Convention, Chilika, Kolleru and Vembanad, were used to illustrate the challenges to implementing polycentric governance regimes for the management of natural resources in the Indian context. This investigation was motivated by the twin processes of environmental and social changes that reveal the politics of access to, use and control of natural resources in the three wetlands.

Polycentric governance proves useful as an analytical tool to study the stories of the wetlands. The three lakes have their unique stories with commonalities in their trajectories including: the enclosure of the commons and subsequent capitalization of resources and social marginalization, conflicting interests and intense local politics, a disconnect between global conservation discourses and local concerns, weak institutional arrangements, and global economic forces.

Our main concern however was to explore the feasibility of an oft-advocated polycentric governance regime that would recognize the multiple uses and values of wetlands and the multiple demands of the population for an integrated management of natural resources. Polycentric governance is defined as "a regime that possesses a number of specific institutional attributes capable of providing and producing essential collective goods and services to the citizens in that regime [... and] that provides choice alternatives among a network of different governance structures" (Andersson and Ostrom, 2008). As acknowledged by the tenants of polycentric governance, the three case studies clarified that such a mode of governance essentially remains a theoretical construct (Andersson and Ostrom, 2008). Recognizing multiple resources claims and reconciling stakeholders' values and objectives are not easy tasks. Differences in values, objectives and interests—and related conflicts among actors—persist even after agreement on specific actions is reached (Wollenberg et al., 2001). Incentives at some scales may also be incompatible with goods and services produced at a different scale (Andersson and Ostrom, 2008). A polycentric approach to governance reminds us that giving more attention to these *differences* could help when designing mechanisms for the sustainable management of natural resources.

But when it comes to managing the resources, calls for the creation of meso-level coordinating bodies become pervasive. In the context we studied, such institutional arrangements (as the CDA) were found to be mainly technocratic, dominated by local elites and insensitive to the needs of the communities directly dependent on natural resources for their livelihood. The concerns that they address are generally global (environmental conservation, economic development) and beyond the rhetoric, local participation remains limited. However, two case studies (Chilika and Vembanad) have also illustrated that the marginalized can become active participants through political mobilization. Although the degree of success is varied, the scope of such collective action illustrated the possibility of influencing intermediate platforms of governance that now function mainly as bureaucratic bodies. This calls for more work to understand "the distribution of costs and benefits over time regarding multiple linkages. Also fundamentally important is the question of how to deal with differences in power within networks and among groups at different levels of organization" (Berkes, 2007: 15193).

Commentary

This paper focuses on the challenges of handling competing interests and multiple needs in three wetland case studies in India. The authors argue that although multiple and often irreconcilable interests are a reality in common pool resource management, the politics of access and the competing stakes of different actors are not well understood. They refer to "polycentric governance"—drawing on the notion of pluralism—to highlight the presence of multiple competing needs in natural resources management. A key feature in this approach is the recognition that multi-level analyses are needed to demonstrate how actors at different levels interact and influence each others' decision making.

C.C. Gibson, E. Ostrom, and T.K. Ahn "The Concept of Scale and the Human Dimensions of Global Change: A Survey," (2000)

Introduction

This paper explicitly recognizes the pervasive nature of scales and their importance in understanding and managing common pool resources. The authors define scales and levels explicitly. Such a vocabulary is important to enhancing communications and interactions among researchers and practitioners from different domains.

Scales and Levels (p. 218–220)

It is clear that terms such as level and scale are frequently used interchangeably and that many of the key concepts related to scale are used differently across disciplines and scholars. Thus, we present in [Table 3.2] definitions of key terms that we have come to use after reading the literature cited in our bibliography and struggling with the confusion created by different uses of the same word. (p. 218)

We use the term scale to refer to the spatial, temporal, quantitative, or analytical dimensions used by scientists to measure and study objects and processes [see Table 3.2]. Levels, on the other hand, refer to locations along a scale. Most frequently, a level refers to a region along a measurement dimension. Micro, meso, and macro levels refer broadly to regions on spatial scales referring to small-, medium-, and large-sized phenomena. Levels related to time, for example, could involve short, medium, and long durations. Scaling problems can be related to issues of scale and or level. All scales also have extent and resolution, although these may not be explicitly noted in a particular study. Extent refers to the magnitude of a dimension used in measuring a phenomenon. In regard to time, extent may involve a day, a week, a year, a decade, a century, a millennium, or many millennia. In regard to space, extent may range from a meter to millions of square meters or more. In regard to quantity, the number of individuals considered by the observer to be involved in a social relationship may vary from two to billions, as may the quantity of goods and the other entities of interest to social scientists. (p. 219)

Many scales are closely related to the concept of hierarchy. A hierarchy is a conceptually or causally linked system for grouping phenomena along an analytical scale. For political scientists, the concept of hierarchy is frequently limited to a system of personnel ranking that defines the authority of individuals dependent upon their formal position within a hierarchy. Generals command captains who command lieutenants and so on, down to the privates who can be commanded by anyone of higher rank. This is an example of an exclusive hierarchy, whereby the objects at the higher level do not contain the objects at a lower level, i.e. they are not nested.

There are many other examples of exclusive hierarchy where the concept of command and control is absent. One example is that of the organisms ranked in the food chain whereby the top carnivores eat carnivores who eat grazers who eat plants (Allen and Hoekstra, 1992, p. 33).

TABLE 3.2 Definitions of Key Terms Related to the Concept of Scales

Term	Definition
Scale	The spatial, temporal, quantitative, or analytical dimensions used to measure and study any phenomenon.
Extent	The size of the spatial, temporal, quantitative, or analytical dimensions of a scale.
Resolution (grain)	The precision used in measurement.
Hierarchy	A conceptually or causally linked system of grouping objects or processes along an analytical scale.
Inclusive hierarchy	Groups of objects or processes that are ranked as lower in a hierarchy are contained in or subdivisions of groups that are ranked as higher in the system (e.g. modern taxonomic classifications—kingdom, phylum, subphylum, class, family, genus, species).
Exclusive hierarchy	Groups of objects or processes that are ranked as lower in a hierarchy are not contained in or subdivisions of groups that are ranked as higher in the system (e.g. military ranking systems—general, captain, lieutenant, sergeant, corporal, private).
Constitutive hierarchy	Groups of objects or processes are combined into new units that are then combined into still new units with their own functions and emergent properties.
Levels	The units of analysis that are located at the same position on a scale. Many conceptual scales contain levels that are ordered hierarchically, but not all levels are linked to one another in a hierarchical system.
Absolute scale	The distance, time, or quantity measured on an objectively calibrated measurement device.
Relative scale	A transformation of an absolute scale to one that describes the functional relationship of one object or process to another (e.g., the relative distance between two locations based on the time required by an organism to move between them).

Source: Turner et al., (1989a), p. 246; Mayr (1982), p. 65; Allen and Hoekstra (1992); Gibson et al., (2000).

In contrast, there are two types of nested hierarchies: inclusive and constitutive. Inclusive hierarchies involve orderings whereby phenomena grouped together at any one level are contained in the category used to describe higher levels, but having no particular organization at each level. Major analytical classification systems are usually inclusive hierarchies. One of the best-known examples is the Linnaen hierarchy of taxonomic categories. Most inclusive hierarchies are classificatory rather than explanatory devices: the units at a lower level (e.g. the species of a genus) do not interact configurally to produce emergent properties of a new higher-level unit.

The second type of nested hierarchy—most characteristic of complex systems—is a constitutive hierarchy. In this type of hierarchy, the lower level can combine into new units that have new organizations, functions, and emergent properties (Mayr, 1982, p. 65). All living organisms and most complex, nonliving, systems are linked in constitutive hierarchies, e.g. molecules are contained in cells that are contained in tissues that are contained in organisms that are contained in populations. These levels are on a conceptual scale based on functional relationships rather than on a spatial or temporal scale.

The concept of emergence is important when trying to understand constitutive hierarchies. In complex, constitutive hierarchies, characteristics of larger units are not simple combinations of attributes of smaller units, but can show new, collective behaviors. According to Baas and Emmeche, (1997, p. 3), some important examples of emergent properties include the general situation of a client and a server (with the interactive help from the server, the client may perform tasks that none of them could do separately); and consciousness (not a property of individual neurons, it is a natural emergent property of the interactions of the neurons in the nervous system) (Baas and Emmeche, 1997, p. 3; see also Baas, 1996).

Many phenomena associated with global change are linked together in constitutive hierarchies. Individual humans are contained in families that are contained in neighborhoods, which are contained in villages or cities, which are contained in regions, which are contained in nations, which are contained in international organizations. In such systems, there is no single "correct" level to study. Phenomena occurring at any one level are affected by mechanisms occurring at the same level, and by levels below and above. (p. 220)

Four Issues Related to Scales (p. 221)

The most important issues related to scale can be grouped into four theoretical areas, each of which is fundamental to the task of explanation in all sciences: (1) how scale, extent, and resolution affect the identification of patterns; (2) how diverse levels on a scale affect the explanation of social phenomena; (3) how theoretical propositions derived about phenomena at one level on a spatial, temporal, or quantitative scale may be generalized to another level (smaller or larger, higher or lower); and (4) how processes can be optimized at particular points or regions on a scale.

Issues of Scales in Natural and Societal Domains (p.236)

With increasing amounts of data that demonstrate a clear human "thumbprint" on small and large ecological phenomena, more and louder calls are being made for the inclusion of the social sciences in the global change research agenda. Although many diagrams of the causes of certain ecological outcomes possess only one large box labeled "human action," the marriage between the physical sciences and the social sciences is far from trivial.

In this paper, we survey one of the most important conceptual challenges to that union—the concept of scale. We argue that common definitions do not exist for scale—even within disciplines— and especially in the social sciences. Because the social sciences focus their attention on social phenomena that encompass many scales and many levels, many social scientists' understanding of the importance of scale tends to be underdeveloped. By contrast, some of the fundamental issues related to scale in the physical sciences were resolved with the development of a unified theory of mechanics, explaining the acceleration of small bodies in free fall as well as the orbit of large planetary bodies.

On the other hand, many social scientists have contributed to our understanding of scale in social phenomena. Geographers, urban analysts, sociologists, economists, and political scientists have taken scale seriously as they explore their substantive areas. And interest continues to grow among social scientists and philosophers of science (e.g. Popper, 1968; Giddens, 1984; Bueno de Mesquita, 1985) to develop of a unifying theory "capable of explaining political behavior at various scales of social activity" (Clark, 1996, p. 284).

Commentary

Gibson et al (2000) emphasize that terms such as scale and level are used interchangeably, and that key concepts related to scale are defined quite differently across disciplines. While natural scientists use scales in a hierarchical sense (primarily in terms of time: seconds, minutes, days, etc., and space: mm, cm, km, etc.), social scientists use them with somewhat less precision. Thus, it is important to develop a common vocabulary. We have adopted their definition to facilitate interactions and communications among reflective practitioners from natural, societal, and political domains.

G. Bloschl and M. Sivapalan "Scale Issues in Hydrological Modeling: A Review," (1995)

Introduction

This review paper provides a summary of scale issues and scaling with a focus on hydrology and water resources management. It distinguishes among three types of scales as they apply to hydrology: process scale, observation scale, and modeling scale.

An Overview of Scale Issues in Hydrology (pp. 251–252)

A framework is provided for scaling and scale issues in hydrology. The first section gives some basic definitions. This is important as researchers do not seem to have agreed on the meaning of concepts such as scale or upscaling. "Process scale", "observation scale" and "modelling (working) scale" require different definitions. The second section discusses heterogeneity and variability in catchments and touches on the implications of randomness and organization for scaling. The third section addresses the linkages across scales from a modelling point of view. It is argued that upscaling typically consists of two steps: distributing and aggregating. Conversely, downscaling involves disaggregation and singling out. Different approaches are discussed for linking state variables, parameters, inputs and conceptualizations across scales. This section also deals with distributed parameter models, which are one way of linking conceptualizations across scales. The fourth section addresses the linkages across scales from a more holistic perspective dealing with dimensional analysis and similarity concepts. The main difference to the modelling point of view is that dimensional analysis and similarity concepts deal with complex processes in a much simpler fashion. Examples of dimensional analysis, similarity analysis and functional normalization in catchment hydrology are given. This section also briefly discusses fractals, which are a popular tool for quantifying variability across scales. The fifth section focuses on one particular aspect of this holistic view, discussing stream network analysis. The paper concludes with identifying key issues and gives some directions for future research.

This review is concerned with scale issues in hydrological modelling, with an emphasis on catchment hydrology. Hydrological models may be either predictive (to obtain a specific answer to a specific problem) or investigative (to further our understanding of hydrological processes) (O'Connell, 199 1; Grayson et al., 1992). Typically, investigative models need more data, are more sophisticated in structure and estimates are less robust, but allow more insight into the system behaviour. The development of both types of model has traditionally followed a set pattern (Mackay and Riley, 1991; O'Connell, 1991) involving the following steps: (a) collecting and analysing data; (b) developing a conceptual model (in the researcher's mind) which describes the important hydrological characteristics of a catchment; (c) translating the conceptual model into a mathematical model; (d) calibrating the mathematical model to fit a part of the historical data by adjusting various coefficients; (e) and validating the model against the remaining historical data set. If the validation is not satisfying, one or more of the previous steps needs to be repeated (Gutknecht, 1991a). If, however, the results are sufficiently close to the observations, the model is considered to be ready for use in a predictive mode. This is a safe strategy when the conditions for the predictions are similar to those of the calibration/validation data set (Bergstrom, 1991). Unfortunately, the conditions are often very different, which creates a range of problems. These are the thrust of this paper.

Conditions are often different in their space or time *scale*. The term *scale* refers to a characteristic time (or length) of a process, observation or model. Specifically, processes are often observed and modelled at short time scales, but estimates are needed for very long time-scales (e.g. the life time of a dam). Models and theories developed in small space-scale laboratory experiments are expected to work at the large scale of catchments. Conversely, sometimes large-scale models and data are used for small-scale predictions. This invariably involves some sort of extrapolation, or equivalently, transfer of information across scales. This transfer of information is called *scaling* and the problems associated with it are *scale issues.*

In the past few years scale issues in hydrology have increased in importance. This is partly due to increased environmental awareness. However, there is still a myriad of unresolved questions and problems. Indeed, "… the issue of the linkage and integration of formulations at different scales has not been addressed adequately. Doing so remains one of the outstanding challenges in the field of surficial processes" (National Research Council, 1991: 143).

Scale issues are not unique to hydrology. They are important in a range of disciplines such as: meteorology and climatology; (Haltiner and Williams, 1980; Avissar, 1995; Raupach and Finnigan, 1995); geomorphology; (de Boer, 1992); oceanography (Stommel, 1963); coastal hydraulics (deVriend, 1991); soil science (Hillel and Elrick, 1990); biology (Haury et al., 1977); and the social sciences (Dovers, 1995). Only a few papers have attempted a review of scale issues in hydrology. The most relevant papers are Dooge (1982; 1986), Klemes (1983), Wood et al. (1990), Beven (1991) and Mackay and Riley (1991). Rodriguez-Iturbe and Gupta (1983) and Gupta et al. (1986a) provide a collection of papers related to scale issues and Dozier (1992) deals with aspects related to data.

Commentary

This review paper highlights the importance of scale issues and scaling for a range of natural science disciplines. It does not, however, discuss many other types of scales (e.g., institutional, management, networks, knowledge) worth considering as part of water management.

D.W. Cash, et al "Scale and Cross-Scale Dynamics: Governance and Information in a Multilevel World," (2006)

Introduction

In this paper, interactions across scales and levels are discussed in detail with regard to governance and information. The authors identified three challenges: (1) the failure to recognize important scale and level interactions; (2) the persistence of mismatches between levels and scales; and (3) a failure to recognize heterogeneity in the way that scales are perceived and valued by different actors, even at the same level.

Scales and Levels (p. 2)

Following Gibson et al. (2000) we define "scale" as the spatial, temporal, quantitative, or analytical dimensions used to measure and study any phenomenon, and "levels" as the units of analysis that are located at different positions on a scale.

The best-studied scale is geographical space or the spatial scale (See Figure 3.3 in this chapter). Environmental, geophysical, and ecological phenomena occur over a continuous range of levels, although particular levels may be more important for particular processes. For example, complex cellular processes govern the decomposition of plant matter lying across a cleared patch of forest, releasing carbon dioxide into the atmosphere. Once released to the atmosphere, molecules of carbon dioxide rapidly merge into a somewhat uniform global mix of gases regulating the Earth's greenhouse effect. Global climate change may result from an amplified greenhouse effect. Thus, global systematic changes and phenomena are linked to and regulated by a complex mix of local processes and vice versa.

Just as spatial scale can be thought of as divided into different "levels," temporal scale can be thought of as divided into different "time frames" related to rates, durations, or frequencies. Thus, biogeophysical phenomena happen at a range of different time frames. Examples are the coexistence of fast cellular metabolism, slow genetic changes, population dynamics that happen over generations, or events with an extremely rapid onset, such as volcanic eruptions or hurricanes. Similarly, phenomena of extremely long duration in global climate dynamics, such as sea temperatures, manifest themselves in changes in relatively short lived hurricane regimes. Social phenomena also happen over a range of time frames: the 24-hr news cycle, electoral events that happen on the order of multiple years, the lifetime of bureaucratic agencies, or the long time frame of large cultural shifts in religion or in dominant economic paradigms and ideologies.

Closely related to spatial scale are jurisdictional scales defined as clearly bounded and organized political units, e.g., towns, counties, states or provinces, and nations, with linkages between them created by constitutional and statutory means. Institutional arrangements, for example, not only have specific jurisdictional charac-teristics but also fall into a hierarchy of rules, ranging from basic operating rules and norms through to systems of rules for making rules or constitutions (Ostrom et al 1999).

Although most attention given to scale in studies of human-environment interactions has focused on spatial, temporal, and jurisdictional issues, there are other scales that may be worth considering in particular cases.

Many environmental management plans and "actions," for example, can be grouped into hierarchical sets ranging from tasks through projects and strategies. Although these relationships are not conventionally framed as a scale issue, we would argue that some of the challenges relating to mismatches may not always have so much to do with space as with the "scale" of management response and change. Some social networks may be "scale free" (e.g., Pastor-Satorras and Vespignani 2001), but others clearly have internal structures that may not be closely correlated with spatial scales. Hence, networks in markets and industries, through clans and religions, or even through professions and voluntary associations may be unrelated to political or geographic space.

Finally, there are benefits to portraying aspects of knowledge as a scale. First, there is often a gap between the highly generalized and generalizable understanding produced by formal science and the experientially and practice-based understanding embedded in both "modern" local and in "traditional" ecological knowledge. This gap can be framed as a lack of cross-level interaction in the knowledge system. Second, although knowledge of processes is useful at larger spatial and temporal scales, often it can only be applied by accepting a lower resolution and application of general processes to the local specific cases.

Cross Scale and Cross Level Interactions (pp. 2–3)

Interactions may occur within or across scales, leading to substantial complexity in dynamics. Although a more precise terminology for scale is not always essential, it does matter to the discussions in this paper and volume. "Cross-level" interactions refer to interactions among levels within a scale, whereas "cross-scale" means interactions across different scales, for example, between spatial domains and jurisdictions (see Figure 3.4 in this Chapter). "Multilevel" is used to indicate the presence of more than one level, and "multiscale" the presence of more than one scale, but without implying that there are important cross-level or cross-scale interactions.

Cross-scale and cross-level interactions may change in strength and direction over time. We refer to this type of changing interaction as the dynamics of the cross-scale or cross-level linkages. Changes may arise from the consequences of those interactions or be caused by other variables. For example, decentralization reforms can produce periods of strong interaction among high-level national institutions and those at the local government level during struggles involving power, responsibilities, and accountability relationships but then settle into a much more modest and steady level of interaction (Lebel et al 2005, Young 2006).

Challenges

In this paper, a "scale challenge" is defined as a situation in which the current combination of cross-scale and cross-level interactions threatens to undermine the resilience of a human-environment system. Three common challenges faced by society are: (1) the failure to recognize important scale and level interactions altogether, (2) the persistence of mismatches between levels and scales in human environment systems, and (3) the failure to recognize heterogeneity in the way that scales are perceived and valued by different actors, even at the same level. We call these the scale challenges of "ignorance," "mismatch," and "plurality."

Cross Scale and Cross Level Dynamics: Implications for Governance (pp. 2–3)

Our understanding of patterns of scale and cross-scale dynamics in linked human-environment systems has advanced substantially in the past decade. There is now an impressive diversity of tools, approaches, and measures for studying scale and scale-related phenomena. The papers in this special issue illustrate that cross-scale and cross level interactions are pervasive, sometimes extremely important, and susceptible to identification and analysis. Against this progress, however, the papers in this special issue acknowledge that there is still relatively little understanding of the dominant mechanisms of cross-scale interaction, especially when analyses go beyond the more conventionally studied spatial, temporal, and jurisdictional scales.

From a management perspective, evidence is accumulating that supports the hypothesis that those systems that more consciously address scale issues and the dynamic linkages across levels are more successful at (1) assessing problems and (2) finding solutions that are more politically and ecologically sustainable. Whether the model is one of institutional interplay, co-management, boundary bridging organizations, or an integration of all three, a core proposition is that in a world increasingly recognized as being multilevel, solutions must be as well. The opposite poles of top-down approaches, which are too blunt and insensitive to local constraints and opportunities, and bottom-up approaches, which are too insensitive to the contribution of local actions to larger problems and the resulting potential for tragedies of the commons, are clearly inadequate in providing both socially robust information (Gibbons 1999) and viable management solutions. A middle path that addresses the complexities of multiple scales and multiple levels is much more difficult, but also what is required.

Commentary

The authors argue that cross–scale and cross–level interactions create significant challenges for understanding and managing common pool resources. They discuss three challenges: ignorance, mismatch, and plurality. Interactions may occur within or across scales as well as within or across levels leading to substantial complexity Cross–scale and cross–level interactions may change in strength and direction over time. Adaptive learning for complex water problems can be conceptualized as a long–term process of co–producing explicit and tacit water knowledge by actors working in different domains, scales, and levels. Within this context, a clear recognition and understanding of cross–scale and cross–level dynamics are essential for managing complex water networks.

A. Cohen and S. Davidson "An Examination of the Watershed Approach: Challenges, Antecedents, and the Transition from Technical Tool to Governance Unit," (2011)

Introduction

This paper discusses issues related to the watershed as a governance unit. It makes a distinction between "watershed," "problem-shed" and "policy-shed" and outlines several challenges associated with a watershed approach to water management.

Watershed Approach to Water Governance (pp. 1–2)

This paper addresses the "watershed approach" to water governance—that is, policy frameworks using watersheds as governance units. Watersheds, defined as areas of land draining into a common body of water (USEPA, 2008a), are a popular scale for water governance initiatives (Baril et al., 2006; Koehler and Koontz, 2008). Although proponents have touted the advantages of using watershed boundaries over their jurisdictional predecessors (e.g. Mitchell, 1990a; Montgomery et al., 1995;

McGinnis, 1999), a number of more recent papers have questioned the benefits of this approach to water governance and have identified significant challenges with its implementation (e.g. Griffin, 1999; Fischhendler and Feitelson, 2005; ... Ferreyra and Kreutzwiser, 2007; Warner et al., 2008; Norman and Bakker, 2009). The emerging debate has typically focused on specific elements of the watershed approach, particularly participation and accountability. This paper broadens the debate by examining themes within the challenges associated with the use of water-shed boundaries and seeking to understand these in light of the emergence and evolution of the concept of watersheds. In so doing, the paper argues that the recognised challenges associated with a watershed approach are symptoms of the conflation of watersheds with other governance tools such as integration and public participation, as well as a conflation of watersheds with Integrated Water Resources Management (IWRM).

The first part of the paper outlines five recognised challenges associated with the watershed approach: the challenges of boundary choice, accountability, public participation, and asymmetries with "problem-sheds" and "policy-sheds". The second part of the paper traces the development and evolution of the watershed concept: its grounding in hydrology and scientism, and the expansion to its use as a policy framework. This evolution highlights the disjuncture between the development of the watershed as a technical tool and its uptake as a governance unit—a disjuncture which has led to a conflation of watersheds with other govern-ance tools as well as with IWRM. In particular, the conceptual jump from technical tool to governance unit was made without an attendant focus on the broader components of water governance; the paper suggests that the effects of this jump are the challenges that have arisen with the increasingly popular watershed approach. Finally, the third section of the paper speaks to the implications of this argument by calling for an analysis of watersheds in and of themselves. Examining watersheds as separate from IWRM and as separate from the suite of governance tools they have come to represent would allow for inquiry into questions that speak to some of the challenges. When, for example, are watersheds useful or appropriate scales to use, and when might other scales (e.g. municipalities or regions) be a better fit? What kinds of decisions are best made at the watershed scale and what kinds of decisions are best made elsewhere? Inquiry into these kinds of questions would be helpful to water managers and environmental scholars seeking to better understand the implications of this popular governance scale.

Watersheds as Tools and Framework (pp. 5–7)

The challenges noted above present significant obstacles for water governance. Efforts to tackle these challenges would involve altering boundaries for each prob-lem in an attempt to obtain an accountable, participatory system that integrates the factors within and outside of a given watershed's boundaries and coordinating these with existing governmental and non-governmental institutional boundaries. Governance at any scale—including the watershed—involves trade-offs between these factors. To assume that watersheds are somehow exempt from these trade-offs is perhaps unrealistic; as Lane et al. (2009) note, "rescaling governance and management is no panacea for the 'wicked' problems of institutional complexity and fragmentation." Or, as Brun (2009) notes, "management on a watershed basis is not a miracle solution." Moreover, the challenges described above are some of the very problems that watershed-based governance models were designed to solve, but instead have perpetuated. This is not to say there is no use for the watershed boundary; there are many situations where watersheds can be extremely useful tools. However, the challenges do prompt interesting questions about how and when to use a watershed boundary. For example, what decisions are best made at the watershed scale and what kinds of decisions might best be made elsewhere? What are the relationships between watersheds and the tools and frameworks with which they have become conflated? This paper does not attempt to provide a full resolution to these questions, but puts forth some potential paths of analysis that may prove fruitful.

Development and Evolution of the Watershed Concept

Addressing the above questions requires inquiry beyond current water governance debates and into the development and evolution of watershed boundaries. This line of investigation is a nonlinear one, because the concept of watershed boundaries veers in, out, and across multiple water dialogues. While the uptake of the watershed as a governance unit is a relatively new phenomenon, recognition of the utility of the hydrologic boundary is not. Some evidence exists of watershed mapping extending as far back as the third century BCE China (Molle, 2009) and drainage areas were mapped in Spain and France in the mid-1800s (Blomquist and Schlager, 2005; Molle, 2009). By the 20th century, managing water within hydrologic boundaries had become increasingly common. Up to this point, the use of hydrologic boundaries was primarily driven by expertise in hydrology and engineering, with an emphasis on efforts towards flood control, irrigation and drainage, and power (Cervoni et al., 2008). In the 1950s, the incorporation of human use and the distribution of costs and benefits into this hydrologic model (Molle, 2009) led to a reframing of the dominant water management paradigm, which was coined as Integrated Water Resources Management (IWRM) in the 1950s (White, 1957). The reinvention and re-emergence of IWRM in the early 1990s included a broadened scope to include both natural and human components (Jønch-Clausen and Fugl, 2001), largely due to the increasing recognition of the need to integrate economic, social, and natural resources under a single framework (GWP, 2000).

Some scholars have argued that the focus on IWRM in the 1990s did not introduce a new concept, but was rather "the rediscovery of a basically more than 60 year old concept" (Biswas, 2004). Nevertheless, the 1990s saw the "new" IWRM become increasingly mainstream through its adoption into international water dialogues (Rahaman and Varis, 2005; Warner et al., 2008) and government planning (McGinnis, 1999; Leach and Pelkey, 2001). IWRM reached such widespread acceptance that it was suggested it had become the "orthodoxy of water resources management" (Jeffrey and Gearey, 2006), part of the "holy trinity of water governance" (Warner et al., 2008) (which also includes river basin planning and multi-stakeholder platforms), and that it "enjoyed a 'near hegemony' as the language of international water policy" (Conca, 2006). Through all of these transitions, proponents of IWRM maintained that watershed boundaries were the scale at which IWRM should ideally be implemented (Jønch-Clausen and Fugl, 2001; Jeffrey and Gearey, 2006; Cervoni et al., 2008). The "old" watershed idea was thus reinvigorated through its suffusion into IWRM (Molle, 2009) as IWRM became increasingly popular.

Between Science and Policy: From Watersheds as Tools to Watersheds as Frameworks

At this point, the watershed narrative takes a twist: watersheds expanded from a mapping and planning tool to a governance framework. The adoption of international IWRM water dialogues by regional, national, and sub-national government agencies and water policy planners appears to have been fixated on watershed boundaries. Rather than as an arm of IWRM or a technical tool (as framed by IWRM's antecedents), watersheds were recast as frameworks; the watershed approach became an umbrella under which other features of IWRM, such as participation and integration, fell. The United States Environmental Protection Agency (USEPA), for example, defines a watershed approach as one that is hydrologically defined, includes all stakeholders, and "strategically addresses priority water resource goals" (USEPA, 2008b). The USEPA framing exemplifies the chasm that formed between the way in which watersheds were framed by IWRM and its antecedents and the way(s) they were reframed by implementing organisations as watersheds moved from their conceptualisation as a technical or planning tool to being conceptualised as a policy framework in the form of the new watershed approach.

A thorough examination and understanding of this conceptual slippage is outside the scope of this paper, but one element of watersheds' uptake bears particular mention here. The concept of watersheds as tools emerged in the context of 19th century scientism (Molle, 2009; Saravanan et al., 2009), during the rise of hydrological sciences (Linton, 2008) and the triumphs of hydraulic regimes and high modernism. These technical origins and focus may have obscured, or at least drawn attention away from, the procedural or governance components of this "new" watershed approach. It is thus perhaps more than coincidental that the foundations of the watershed approach are technical and its core challenges are not. In other words, the challenges identified here can be seen as symptoms of watersheds' jump from a predominantly technical tool to a governance framework within the water landscape. As the concept of watershed boundaries was adopted into water governance efforts, this technical tool—which was not designed to address the broader components of water governance—became a governance unit, but without an attendant focus on the governance or procedural elements of the new watershed approach. The effects of this slip from technical tool to governance framework can be seen in the commonalities between the challenges identified in the first part of the paper.

The challenges of boundary choice, accountability, participation, and the asymmetries between watersheds, problem-sheds and policy-sheds are challenges of governance; they are not scientific or technical challenges relating to issues such as the need for more data or better instrumentation. Though not unwelcome, enhanced monitoring, mapping, or data do not address the roots of the challenges associated with watershed-based governance approaches, all of which relate to social, political, and economic decision-making, but have come to be associated with a hydrologic boundary. Air-sheds' influence on watersheds, for instance, can be framed as a challenge with the watershed approach, but might be usefully re-framed as a governance challenge resulting from affording primacy to one governance unit over another.

Rethinking Watershed Boundaries as a Governance Unit (pp. 10–11)

The development and uptake of the concept of watershed boundaries was in part an effort to address environmental governance more effectively, yet challenges associated with the unit's uptake continue to beset its implementation. Through an examination of these challenges in light of the emergence and development of the concept of watersheds, this paper has argued that the jump from watersheds as technical tools to watersheds as policy frameworks has been problematic. In particular, we suggest that the challenges identified in the first part of the paper may be symptoms of the lack of attention to governance issues in the transition from tool to framework. As such, the challenges might not necessarily lie with watersheds *per se*, but with the governance tools and paradigms with which watersheds have been conflated under the rubric of the "watershed approach".

By untangling watersheds from other concepts with which they have been conflated, watersheds can be re-framed as tools, or choices, that can be marshalled in support of particular policy goals, rather than as mandatory, unquestionable starting points for effective water governance. Teasing apart watersheds in this way facilitates a foray into an analysis of watersheds in and of themselves. Such an analysis is outside the scope of this paper, but paths for future analysis and discussion are suggested. Three lines of inquiry were suggested as starting points to such an analysis: when are watersheds appropriate or useful? What kinds of decisions are best made at the watershed scale? What is the relationship between IWRM and watersheds? Using these questions as starting points, we suggested that watersheds might be appropriate in cases where challenges are hydrologically defined and where strong governance mechanisms already exist. Further, we suggested that watersheds might be less useful than other governance scales for setting water quality standards or in cases where existing governance mechanisms are weak, public interest is low, or analysis shows it is best left out of integration efforts.

Overall, we encourage the thoughtful consideration of the uptake of watersheds as governance units—flaws and all—before embarking on re-scaling efforts.

Commentary

This paper argues that there are at least five challenges associated with the watershed approach to water resource management: the difficulties of boundary choice, accountability, public participation, and asymmetries with "problem-sheds" and "policy-sheds." These challenges create significant barriers to effective water management. Efforts to address these challenges could involve altering boundaries for each water management problem in an attempt to obtain an accountable and participatory outcome. It is important to recognize the boundary-crossing nature of water issues, and then decide how and when to use the watershed boundary and how watersheds are linked to problem-sheds and policy-sheds. Within this context, the use of water networks as interacting sets of nodes and links offers a possible alternative to characterizing complex evolving water systems.

References

Allen, P.M. 1988. Evolution: Why the whole is greater than the sum of its parts, in W. Wolff, C.-J. Soeder, and F. R. Drepper (eds) *Ecodynamics*. Berlin: Springer.

Allen, P.M. 1990. Why the future is not what it was, *Futures, July/August*: 555–569.

Allen, P.M. 1993. Evolution: Persistent ignorance from continual learning, in R.H. Day and P. Chen (eds) *Nonlinear Dynamics and Evolutionary Economics* (pp 101–112). Oxford: Oxford University Press.

Allen, P.M. 1994a. Coherence, chaos and evolution in the social context, *Futures, 26*(6): 583–597.

Allen, P.M. 1994b. Evolutionary complex systems: Models of technology change, in L. Leydesdorff and P. van den Besselaar (eds.) *Chaos and Economic Theory*. London: Pinter.

Allen, P. M. 2000. Knowledge, learning, and ignorance, *Emergence, 2*(4): 78–103.

Allen, T.F.H. and Hoekstra, T.W. 1992. *Toward a Unified Ecology*. New York: Columbia University Press.

Andersson, K.P. and Ostrom, E. 2008. Analyzing decentralized resource regimes from a polycentric perspective, *Policy Sciences, 41*: 71–93.

Avissar, R. 1995. Scaling of land-atmosphere interactions: an atmospheric modelling perspective, *Hydrological Processes, 9*(5/6), 679–695.

Baas, N.A. 1996. A framework for higher order cognition and consciousness. in S. Hameroff, A. Kaszniak, and A. Scott (eds.) *Towards a Science of Consciousness* (pp. 633–648). Cambridge, MA: The MIT Press.

Baas, N.A. and Emmeche, C. 1997. *On Emergence and Explanation*. Working paper 97–02–008. Sante Fe Institute, Santa Fe.

Barabasi, A.L. 2003. *Linked: How everything is connected to everything else and what it means for buisness, science, and everyday life*. New York: Plume Publisher.

Baril, P., Maranda, Y., and Baudrand, J. 2006. Integrated watershed management in Quebec (Canada): A participatory approach centred on local solidarity. *Water Science and Technology, 53*(10): 301–307.

Bergstrom, S. 1991. Principles and confidence in hydrological modeling, *Nordic Hydrology, 22*: 123–136.

Berkes, F. 2007. Communities-based conservation in a globalized world, *Proceedings of the National Academy of Sciences, 104*(39): 15188–15193.

Beven, K.J. 1991. Scale considerations, in D.S. Bowles and P.E. O'Connell (eds.) *Recent Advances in the Modeling of Hydrologic Systems* (pp. 357–371). Dordrecht: Kluwer.

Biswas, A.K. 2004. Integrated water resources management: A reassessment. *International Water Resources Association 29*(2): 248–256.

Blomquist, W. and Schlager, E. 2005. Political pitfalls of integrated watershed management. *Society and Natural Resources, 18*(2): 101–117.

Bloschl, G. and Sivapalan, M. 1995. Scale issues in hydrological modeling: A review, *Hydrological Processes, 9*: 251–290.

Brun, A. 2009. L'approche par bassin versant: Le cas du Québec. *Policy Options, 39*(7): 36–42.

Bueno de Mesquita, B. 1985. Toward a scientific understanding of international conflict: A personal view. *International Studies Quarterly, 29*: 121–136.

Cash, D.W., Adger, W., Berkes, F., Garden, P., Lebel, L., Olsson, P., Pritchard, L., and Young, O. 2006. Scale and cross-scale dynamics: governance and information in a multi-level world, *Ecology and Society, 11*(2): 8.

Cervoni, L., Biro, A., and Beazley, K. 2008. Implementing integrated water resources management: The importance of cross-scale considerations and local conditions in Ontario and Nova Scotia. *Canadian Water Resources Journal, 33*(4): 333–350.

Cilliers, P. 1998. *Complexity and Post-Modernism*. London and New York: Routledge.

Clark, W.R. 1996. Explaining cooperation among, within, and beneath states. *Mershon International Studies Review, 40*: 284–288.

Clemons, J. 2004. Interstate Water Disputes: A Road Map for States, *Southeastern Environmental Law Journal, 12*(2):115–142.

Cohen, A. and Davidson, S. 2011. The watershed approach: Challenges, antecedents, and the transition from technical tool to governance unit, *Water Alternatives, 4*(1): 1–14.

Conca, K. 2006. *Governing Water: Contentious Transnational Politics and Global Institution Building*. Cambridge, MA: The MIT Press.

de Boer, D.H. 1992. Hierarchies and spatial scale in process geomorphology: a review, *Geomorphology, 4*: 303–318.

deVriend, H.J. 1991. Mathematical modelling and large-scale coastal behaviour, *Journal of Hydraulic Research, 29*: 727–740.

Dooge, J.C.I. 1982. Parameterization of hydrologic processes, in P.S. Eagleson (ed.) *Land Surface Processes in Atmospheric General Circulation Models* (pp. 243–288). London: Cambridge University Press.

Dooge, J.C.I. 1986. Looking for hydrologic laws, *Water Resources Research, 22*: 46s–58s.

Dovers, S.R. 1995. A framework for scaling and framing policy problems in sustainability, *Ecological Economics, 12*: 93–106.

Dozier, J. 1992. Opportunities to improve hydrologic data, *Reviews of Geophysics, 30*: 315–331.

Ferreyra, C. and Kreutzwiser, R. 2007. *Integrating Land and Water Stewardship and Drinking Water Source Protection: Challenges and Opportunities*. Newmarket, ON: Conservation Ontario.

Fischhendler, I. and Feitelson, E. 2005. The formation and viability of a non-basin water management: The US – Canada case. *Geoforum, 36*(6): 792–804.

Gibbons, M. 1999. Science's new social contract with society. *Nature, 402* (Supplement): C81–C84.

Gibson, C.C., Ostrom, E. and Ahn, T.K. 2000. The concept of scale and the human dimensions of global change: a survey, *Ecological Economics 32*: 217–239.

Giddens, A. 1984. *The Constitution of Society: Outline of the Theory of Structuration*. Berkeley: University of California Press.

Grayson, R.B., Moore, I.D., and McMahon, T.A. 1992. Physically based hydrologic modelling: 2. Is the concept realistic? *Water Resources Research, 26*: 2659–2666.

Griffin, C.B. 1999. Watershed councils: An emerging form of public participation in natural resource management. *Journal of the American Water Resources Association, 35*(3): 505–518.

Gunter, C. and Dhand, R. 2005. The Chimpanzee Genome, *Nature, 437*: 47.

Gupta, V.K., Rodriguez-Iturbe, I., and Wood, E.F. (eds.). 1986a. *Scale Problems in Hydrology.* Dordrecht: D. Reidel.

Gutknecht, D. 1991a. On the development of "applicable" models for flood forecasting, in F. H. M. Van de Ven, D. Gutknecht, D. P. Loucks, and K. A. Salewicz (eds.) *Hydrology for the Water Management of Large River Basins* (pp. 337–345). Oxford: International Association of Hydrological Sciences

GWP (Global Water Partnership, Technical Advisory Committee). 2000. *Integrated water resources management.* Stockholm, Sweden: Global Water Partnership.

Haltiner, G.J. and Williams, R.T. 1980. *Numerical Prediction and Dynamic Meteorology.* New York: Wiley.

Haury, L.R., McGowan, J.A., and Wiebe, P.H. 1977. Patterns and processes in the time-space scales of plankton distributions, in J. H. Steele (ed.) *Spactial Pattern in Plankton Communities* (pp. 277–327). New York: Plenum Press.

Hillel, D., and Elrick, D E. 1990. Scaling in Soil Physics: Principles and Applications. *Soil Science Society of America, 25.*

Hodgson, G.M. 2001. Is Social Evolution Lamarckian or Darwinian? in J. Laurent and J. Nightingale (eds.) *Darwinism and Evolutionary Economics* (pp. 87–118). Cheltenham: Edward Elgar.

Jeffrey, P. and Gearey, M. 2006. Integrated water resources management: Lost on the road from ambition to realisation? *Water Science and Technology, 53*(1): 1–8.

Jønch-Clausen, T. and Fugl, J. 2001. Firming up the conceptual basis of integrated water management. *International Journal of Water Resources Development, 17*(4): 501–510.

Kemper, K., Blomquist, W.A., and Dinar, A. 2007. *Integrated River Basin Management through Decentralization.* New York: Springer.

Klemes, V. 1983. Conceptualisation and scale in hydrology, *Journal of Hydrology, 65*: 1–23.

Koehler, B. and Koontz, T.M. 2008. Citizen participation in collaborative watershed partnerships, *Environmental Management, 41*(2): 143–154.

Lane, D.A. and Maxfield, R. 1997. Foresight, complexity and strategy, in B.W. Arthur, S. Durlauf, and D.A. Lane (eds.) *The Economy As an Evolving Complex System II: Proceedings,* (Santa Fe Institute Studies in the Sciences of Complexity, Vol. 27). Reading, Mass: Addison-Wesley/Perseus Books.

Lane, M., Robinson, C., and Taylor, B. (eds). 2009. *Contested Country: Local and Regional Resources Management in Australia.* Australia: Commonwealth Scientific and Industrial Research Organisation Publishing.

Leach, W.D. and Pelkey, N. 2001. Making watershed partnerships work: A review of the empirical literature. *Journal of Water Resources Planning and Management, 127*(6): 378–385.

Lebel, L., Garden, P., and Imamura, M. 2005. The politics of scale, position and place in the management of water resources in the Mekong region. *Ecology and Society, 10*(2): 18.

Leitman, S. 2005. Apalachicola-Chattahoochee-Flint Basin: Tri-state negotiations of a water allocation formula, in J.T. Scholz and B. Stiftel (eds.) *Adaptive Governance and Water Conflict: New Institutions for Collaborative Planning* (pp. 74–88), Washington D.C.: Resources for the Future.

Linton, J. 2008. Is the hydrological cycle sustainable? A historical-geographical critique of a modern concept. *Annals of the Association of American Geographers, 98*(3): 630–649.

Liu, J., Dietz, T., Carpenter, S.R., Alberti, M., Folke, C., Moran, E., Pell, A.N., Deadman, P., Kratz, T., Lubchenco, J., Ostrom, E., Ouyang, Z., Provencher, W., Redman, C.L., Schneider, S.H., and Taylor, W.W. 2007. Complexity of coupled human and natural systems, *Science, 317*: 1513–1516.

Lyotard, J.-F. 1984. *The Post-Modern Condition: A Report on Knowledge.* Manchester, UK: Manchester University Press.

Mackay, R. and Riley, M. S. 1991. The problem of scale in the modelling of groundwater flow and transport processes, in *Chemodynamics of Groundwaters* (pp. 17–51). Proc. Workshop November 1991, Mont Sainte-Odile. France. EAWAG, EERO, PIR "Environment" of CNRS, IMF Université Louis Pasteur Strasbourg.

Maguire, S. and McKelvey, B. (eds.). 1999. Special Issue on Complexity and Management: Where Are We? *Emergence 1*(2).

Mayr, E. 1982. *The Growth of Biological Thought: Diversity, Evolution, and Inheritance.* Cambridge, MA: Belknap Press.

McGinnis, M.V. 1999. Making the watershed connection, *Policy Studies Journal 27*(3): 497–501.

Miller, J.H., and Page, S.E. 2007. *Complex Adaptive Systems: An Introduction to Computational Models of Social Life.* Princeton, NJ: Princeton University Press.

Mingers, J. 1995. *Self-Producing Systems: Implications and Applications of Autopoiesis.* New York: Plenum Press.

Mitchell, B. 1990a. Integrated water management, in B. Mitchell (ed.) *Integrated Water Management: International Experiences and Perspectives* (pp. 1–21). London: Belhaven Press.

Mitleton-Kelly, E. 2000. Complexity: Partial support for BPR? in P. Henderson (ed.) *Systems Engineering for Business Process Change* (pp. 24–37), London: Springer-Verlag.

Mitleton-Kelly, E. 2003. Ten principles of complexity and enabling infrastructures, in E. Mitleton-Kelly (ed.) *Complex Systems and Evolutionary Perspectives on Organizations: The Applications of Complexity Theory of Organizations* (pp. 23–50). Oxford: Elsevier.

Molle, F. 2009. River-basin planning and management: The social life of a concept. *Geoforum, 40*(3): 484–494.

Montgomery, D.R., Grant, G.E., and Sullivan, K. 1995. Watershed analysis as a framework for implementing ecosystem management, *Journal of the American Water Resources Association, 31*(3): 369–386.

Narayanan, N.C. and Venot, J.P. 2009. Drivers of change in fragile environments: Challenges to governance in Indian wetlands, *Natural Resources Forum, 33*: 320–333.

National Research Council. 1991. *Opportunities in the Hydrologic Sciences.* Washington, DC: National Academy Press.

Nicolis, G. and Prigogine, I. 1989. *Exploring Complexity: An Introduction.* New York: W.H. Freeman.

Norman, E. and Bakker, K. 2009. Transgressing scales: Water governance across the Canada–U.S. Borderland, *Annals of the Association of American Geographers, 99*(1): 99–117.

O'Connell, P.E. 1991. A historical perspective, in D.S. Bowles and P.E. O'Connell (eds.) *Recent Advances in the Modeling of Hydrologic Systems* (pp. 3–30). Dordrecht: Kluwer.

Ostrom, E., Burger, J., Field, C.B., Norgaard, R.B., and Policansky, D. 1999. Revisiting the commons: Local lessons, global challenges. *Science, 284*: 278–282.

Parker, D. and Stacey, R.D. 1994. *Chaos, Management and Economics: the Implications of Non-Linear Thinking.* Hobart Paper 125, Institute of Economic Affairs.

Pastor-Satorras, R. and A. Vespignani. 2001. Epidemic spreading in scale-free networks. *Physical Review Letters, 86*: 3200–3203.

Popper, K. 1968. *The Logic of Scientific Discovery*. New York: Harper & Row.

Powell, J.W. 1875. *The Exploration of the Colorado River and Its Canyons*. New York: Penguin Classics.

Rahaman, M.M. and Varis, O. 2005. Integrated water resources management: Evolution, prospects and future challenges. *Sustainability: Science, Practice and Policy, 1*(1): 15–21.

Raupach, M.R. and Finnigan, J.J. 1995. Scale issues in boundary layer – meteorology: surface energy balances in heterogenous terrain, *Hydrological Processes, 9*: 589–612.

Rodriguez-Iturbe, I. and Gupta, V. K. (eds.). 1983. Scale problems in hydrology, *Journal of Hydrology, 65*.

Saravanan, V.S., McDonald G.T., and Mollinga, P.P. 2009. Critical review of integrated water resources management: Moving beyond polarised discourse. *Natural Resources Forum, 33*(1): 76–86.

Stacey, R.D. 1995. The science of complexity: An alternative perspective for strategic change processes, *Strategic Management Journal, 16*(6): 477–495.

Stacey, R.D. 1996. *Complexity and Creativity in Organizations*. San Francisco: Berrett-Koehler

Stacey, R., Griffen, D., and Shaw, P. 2000. *Complexity and Management*. London: Routledge.

Stacey, R.D. 2001. *Complex Responsive Processes in Organisations*. London: Routledge.

Stacey, R.D. 2007. *Strategic Management and Organizational Dynamics: The Challenge of Complexity*, 5th Edition. New York: Prentice Hall.

Stommel, H. 1963. Varieties of oceanographic experience, *Science, 139*: 572–576.

Turner, M.G., Dale, V.H., and Gardner, R.H. 1989a. Predicting across Scales: Theory Development and Testing. *Landscape Ecology, 3*(3:4):245–252.

USEPA (United States Environmental Protection Agency). 2008a. *What is a watershed?* www.epa.gov/owow/watershed/what.html (accessed 28 August 2010).

USEPA. 2008b. What is a watershed approach? www.epa.gov/owow/watershed/framework/ch2.html (accessed 28 August 2010).

Warner, J., Wester, P., and Bolding, A. 2008. Going with the flow: River basins as the natural units for water management? *Water Policy, 10*(2): 121–138.

White, G.C. 1957. A perspective of river basin development, *Law and Contemporary Problems, 22*(2): 157–187.

Wollenberg, E., Anderson, J., and Edmunds, D. 2001. Pluralism and the less powerful: Accommodating multiple interests in local forest management, *International Journal of Agricultural Resources, Governance and Ecology, 1*(3/4): 199–222.

Wood, E.F., Sivapalan, M., and Beven, K. 1990. Similarity and scale in catchment storm response, *Reviews of Geophysics, 28*: 1–18.

Young, O. 2006. Vertical interplay among scale-dependent resource regimes, *Ecology and Society, 11*(1): 27.

4

ADDRESSING COMPLEX WATER MANAGEMENT PROBLEMS

From Certainty and Consensus to Uncertainty and Disagreement

Until the middle of the twentieth century, most water resource planning decisions were guided by the notion of maximizing the efficiency of water use. This was done by providing marketable goods and services without much regard for associated impacts. Water resource planning and management is a dramatically different enterprise now (e.g., Boland and Baumann 2009; Kiang et al 2011). Boland and Baumann (2009) point to three changes that have most dramatically influenced water resource decision-making over the last 50 years: (1) evolution of demand; (2) evolution of governance; and (3) evolution of analysis.

A growing emphasis on societal values since the 1960s has led to increased demand for analytical techniques that will allow us to predict the outcomes of interventions in coupled natural and human systems. In many instances, advances in modeling have outpaced the creation of the monitoring systems needed to calibrate and validate these models. In other instances, analytical methods, laws, and policies appropriate for yesterday's problems are being applied to today's problems with limited or no success. Rogers et al (2009) point out that many water professionals view governance as a set of exogenous laws, regulations, and institutions—yet, practical experience suggests something quite different. We now recognize that the driving force in formulating and framing water problems is the politics of competing interests and the values of different groups. Consequently, it is important to move beyond the familiar framing of water problems as technical tasks that can be passed along to experts who will find nonpolitical solutions (Stone 2002). Instead, many water conflicts are best described in terms of differences in values and how to translate those values to policies and actions within the political domain (Stone 2002; Layzer 2006).

Efforts to theorize about water systems have been vast, but the tools and techniques available to pursue and implement these theories in practice have often led to science that is "smart but not wise." This is because we do not have what we need to integrate "scientific learning" with the "contextual reality" of water conflicts that are filled with uncertainty, politics, ambiguity, nonlinearity and feedback. Solutions to most real-world water problems demand integration of such contextual realities. Consequently, water managers who attempt to solve these problems cannot easily translate solutions derived from scientific findings into the messy context of the political world, where natural and societal dynamics must be considered.

Policy science emerged over the past 50 years with the hope that increasing analytical capabilities in the social sciences would lead to smart prescriptive advice. Continued efforts to refine analytical tools like cost–benefit analysis and optimization theory, however, have not yet made it possible to bridge the divide between theory and practice (e.g., Ackoff 1979, Bennis et al 2010, Sheer 2010, Stiglitz et al 2010). The most significant insight that policy science has generated is an understanding that the problems that must be addressed in a public policy context are not like the problems for which the tools of systems engineering or optimization were designed. This realization is often conveyed in terms of the "messiness" or "wicked" nature of the problems that arise when science, policy, and politics are intertwined with uncertainty, ambiguity, nonlinearity, and feedback (e.g., Rittel and Webber 1973; Bar-Yam 2004; Milly et al 2008; Orlove and Caton 2010).

Studies of coupled human–natural systems (Liu et al 2007) suggest that we need to distinguish among three types of water management problems: simple, complicated, and complex. Simple problems are amenable to optimization because they involve easily identifiable and neatly-bounded elements of water management that respond in predictable ways and pose challenges about which there is almost complete agreement with regard to means and ends. Complicated water management problems are those that involve water networks that operate in somewhat unpredictable ways, and about which the means and ends of action are highly contested (and likely will remain that way regardless of how much scientific work is done). Complex water management problems lie in the zone between certainty and uncertainty and between agreement and disagreement.

Referring to the examples of simple, complicated, and complex problems from the second chapter, the design of a water-efficient flushing toilet is an example of a simple water problem. Getting water from the Quabbin reservoir to a shower on the 16th floor of an apartment building in Boston is a complicated problem because with careful study we can know with near certainty what each component of this distribution network does and how to control it. A complicated system is knowable, predictable, and controllable. Consequently, the management of simple and complicated water problems can be guided by rational and objective analyses. But complex water management problems are those in which natural and societal forces get intricately linked, for example, the inundation of four towns that was required to create the Quabbin reservoir. In this book, we are mostly concerned about this third type of boundary-crossing—complex water management problems that fall in the Zone of Complexity. Managing in this Zone requires the use of network theory and the new theory of mutual gains negotiation.

Identifying Issues that Fall into the Zone of Complexity

Water management problems such as allocating supplies among contending interest groups or deciding whether to deploy a new technology that might "expand" water supplies, but at the same time could jeopardize water quality, fit into the complex category. Approaches to addressing many complex water management problems lie somewhere between certainty and uncertainty. When one examines such a continuum and tries to synthesize order (think of a perfectly predictable periodic signal) with disorder (think of a purely random noise), it is possible to find a region called the Zone of Complexity. The dynamics in this Zone are neither entirely certain nor uncertain. Agreement about means and ends is far from clear, but within reach if the right procedures are followed. Complexity theory—drawing on assumptions about nonlinearity, emergence, interaction, feedback, mutation, and adaptation—can be used to identify the parameters of management problems that fall into the Zone of Complexity. Problem-solving in this Zone is distinct from what works in simple and complicated settings (Figure 4.1).

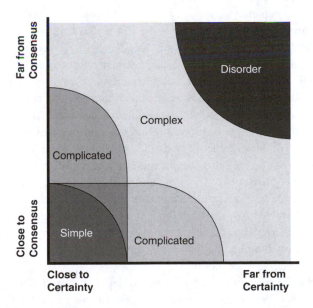

FIGURE 4.1 Degree of certainty and consensus: A lens to organize simple, complicated, complex and disordered water networks
Source: (adapted from Stacey 2007)

Some systems, though they are constantly changing, are stable and predictable, with a high degree of certainty about them. Think of a clock pendulum or the solar system. Other systems lack such stability and predictability: think of the trajectory of a hurricane from its birth in the ocean to landfall. Unstable systems move further and further away from their starting conditions. Some systems are linear and deterministic, and their evolution is perfectly predictable; other systems are random and their evolution must be characterized statistically. Stable and unstable behaviors, or deterministic and random descriptions, are quite well understood in the natural sciences. Such representations work well for bounded systems belonging to simple and complicated classes of water management problems.

Allen (2000) described four underlying assumptions related to modeling simple and complicated water problems: (1) the relevant "system" boundary is fixed; (2) the heterogeneity of different system components can be reduced; (3) system dynamics can be described in terms of their average behavior; and (4) relevant processes can be assumed to evolve at their average rate. If all four assumptions are valid for a given water management system, then system dynamics can be described by a set of deterministic differential equations that will allow predictions to be made with certainty and "optimization" to be carried out. For complex water systems, on the other hand, these assumptions do not apply. Dynamics are controlled by nonlinear interactions and feedback among different components. More importantly, emergence and co-evolution are inherent in complex water management systems (Chapter 3). Consequently, differential equations and traditional systems engineering approaches are not helpful when we try to model system dynamics in the Zone of Complexity. We need a different approach. We will use complexity theory, including the theory of emergence and network theory, to describe the behavior of water management systems in this Zone.

Under certain conditions, water management networks may perform in regular, predictable ways, while under other conditions they may exhibit behavior in which regularity and predictability are entirely lost. Differences in initial conditions may lead to divergent outcomes even when the same management solutions are applied. The most vivid example of this sensitive dependence on initial conditions—in complexity theory and chaos literature (e.g., Kauffmann 1993; Mitchell 2009)—is the idea that the flapping of a butterfly's wing in Beijing can create a hurricane in Boston. There are problems, of course, with this oft-cited example and its application to most complex systems. For example: Which butterfly was responsible? At what time and in what way(s) must the butterfly flap its wings? How many butterflies must be involved? The answer is: it depends. This is not a particularly illuminating or helpful response from the standpoint of a water manager who needs to take action.

Most of the time nothing a butterfly does in Beijing will be linked to a hurricane in Boston. This is because the particular conditions necessary for the butterfly's actions to affect all the forces involved in hurricane formation in just the right way are not likely to occur. For example, hurricanes can hit Boston, but they do not occur very often, and only during hurricane season. Moreover, the particular conditions that must exist to make the hurricane sensitive to small disturbances may well make the hurricane sensitive to other possible disturbances as well. This means that a single butterfly in Beijing may not be responsible for a hurricane in Boston. Simply stated, if one butterfly changes its flapping, this might cause a change in whether or when a hurricane occurs. This effect could, however, be countered by another butterfly changing its flapping patterns. In other words, there is no direct and unique link between a particular butterfly and a particular hurricane. Thus, we need to distinguish between sensitivity and causation. Hurricane behavior may be sensitive to the flapping of a butterfly's wings, but is not caused by it. In our effort to understand water management problems in the Zone of Complexity, we are looking at the interactions of a butterfly (or set of butterflies) with its environment and surrounding initial conditions that will be conducive to creating a hurricane that will hit Boston. This sensitivity of the initial configurations of a network to its future evolution—as opposed to cause–effect relationships between two states of a water system at different times—is critical to understanding, characterizing, and managing water networks.

Water management problems are the responsibility of multiple actors operating at multiple scales and levels in complex water networks. Some management interventions need only involve a limited number of actors at a single level. They may also involve only a few societal–natural interactions that can be counted on to produce predictable outcomes. When it is unclear which actors need to be involved in a water management effort, and the societal–natural interactions involve almost all the variables in the NSPD framework, this creates a Zone of Complexity. An intervention in this zone is likely to lead to unpredictable outcomes. Under appropriate structural conditions, however, water managers can find the Zone of Complexity (sometimes called a phase transition, or the "edge of chaos" in complexity theory and the chaos literature). It is in this Zone of Complexity that networks exhibit sensitivity to initial configurations and evolve due to associated interactions and feedback among key variables. In such situations, networks establish patterns that emerge and evolve over time. These patterns can be managed by designated actors (at multiple levels) working collaboratively to achieve desired outcomes through negotiated interventions.

We recognize that given the nonlinear coupling of natural and societal variables (within the political domain), the water professionals involved will probably not be able to understand and quantify every detail of the sub-system interactions their intervention will cause. It turns out, however, that this is not necessary. The actors in the management network need to focus their efforts to achieve a desired outcome given the configuration of the network and related interactions. Remember the butterfly effect: we are not looking for causal connections, but for sensitivity to initial configurations that will lead to desired outcomes in the future. Instead of trying to understand the intricacies of cause–effect relationships filled with uncertainty, ambiguity, nonlinearity, and feedback, water professionals should focus on the "emergent properties" of the network (as it seeks to re-organize its structure and interactions among various nodes). Each node in a management network tries to juggle conflicting constraints and find the best way of interacting with and adapting to all the other nodes to achieve the best possible outcome.

One way of mapping such emergence is to think of the conditions surrounding a water network as "an adaptive landscape" (Kauffmann 1993). Every node and associated interaction in a water management network aims to reach to the peak of fitness and avoid the troughs. Of course, what constitutes "fitness" is likely to change over time. In our fictional world of Indopotamia, for example, the decisions of Beta and Gamma to join forces and build a dam made a significant difference to Alpha's "fitness landscape." If the dam is built, Alpha won't have the water it needs to grow enough rice to meet the needs of its expanding population. Consequently, Alpha must adapt and find a water management strategy that fits the anticipated structural changes that it faces. Alpha's action will also change the structure of the network, and affect what Beta and Gamma need to do in the future as well.

Implementation in the Zone of Complexity

In Chapter 3, we talked about the four attributes of complex evolving systems: (1) collective behavior is emergent; (2) complex systems do not function in predictable ways; (3) boundary-crossing is pervasive; and (4) co-evolution is inherent. We can examine water management strategies in light of these attributes. Indeed, when we overlay these attributes on three types of water problems (simple, complicated, and complex), we can see the kinds of decision-making that is required (Figure 4.2).

The challenge for reflective water professionals is to have access to an array of situation-specific approaches and to be aware of when to use each one. Stacey (2007) proposed a matrix to help clarify the types of decision-making approaches appropriate to the three types of water management problems. Decisions are *close to certainty and close to agreement* when cause and effect linkages can be easily determined, there is agreement on means and ends, and one can extrapolate from past experience to predict the outcome of an action with relative certainty. This type of simple water management problem can be solved using "rational decision-making." At the other end of the certainty continuum, management decisions *far from certainty and far from agreement* arise when cause and effect linkages are not clear and extrapolation from past experience is not a reliable way to predict the future. Below, we describe five different types of decision-making scenarios based on the configurations, interactions, and feedback within a water management network.

Rational Decision-making for Simple Water Management Problems

Much of the water management literature has focused on simple water management problems. In these situations, cause–effect relationships are well understood, systems and sub-systems are clearly bounded, and agreement among

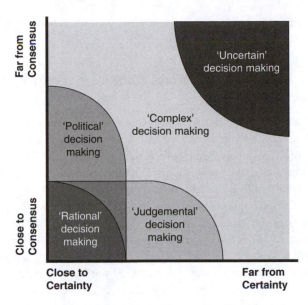

FIGURE 4.2 Degree of certainty and consensus: A lens to describe different types of decision making for different types of water problems described in Figure 4.1
Source: (adapted from Stacey 2007)

relevant stakeholders on means and ends is not hard to establish. Historical records are adequate to predict the future. Water professionals can plan and execute specific actions to achieve desired outcomes. Metrics for gauging success are readily-agreed upon. The tools of optimization can be used to increase efficiency.

Political Decision-making for Complicated Water Management Problems

Analyses of historical data are likely to yield useful predictions in this context. Political factors become increasingly important in these situations. Coalition-building, negotiation, and joint problem-solving are needed to manage interactions and achieve desired outcomes for this class of water management problems.

Judgmental Decision-making for Complicated Water Management Problems

Water professionals operating in these situations may have a high level of network agreement about what needs to be done, but very little agreement on how to proceed. Cause and effect linkages are likely to be unclear. In such cases, monitoring against a preset plan will not work. In these situations, steering the network towards an agreed-upon future state, even if the specific path for achieving it is not clear, may be the best way to proceed.

Complex Decision-making for Water Problems in the Zone of Complexity

There is a large zone in Figure 4.2 that lies between the region of disorder and regions of traditional management. The Zone of Complexity (also known as the "critical transition zone" or "edge of chaos" in the complexity literature). In this Zone, traditional water management approaches are not likely to be effective. This Zone shows significant sensitivity to initial configurations and interactions. New modes of operating and managing are likely to be required. In traditional water management, we spend much of our time working to address simple and complicated water problems by using tools and techniques relevant to rational decision making. In the Zone of Complexity, tools that work to address simple and complicated water management problems are not likely to be effective.

To manage complex problems in this Zone, water professionals need to be deeply cognizant of the network's initial configurations and open to new modes of operation. In such situations, they must recognize the highly sensitive nature of initial network configurations to any intervention. Causes may not be proportional to effects, and many interactions and most feedback is likely to be non-linear. They must recognize that the structure of the network and associated actions in this Zone—which may appear "messy" but are not random—are the result of interactions that are not yet fully understood. Once they understand them, as Kauffmann (1993) argues, water managers will see that networks in this Zone usually move toward the best possible solution space.

In this Zone, networks organize themselves in ways that are highly dependent on initial configurations. Consequently, unlike complex natural systems, we have some ability to change the initial conditions of a water management network. Specifically, we could determine which interactions and feedback in a water network would have to be changed to move the system to a position where the outcome is more predictable and desirable. The type and intensity of interactions among nodes are examples of initial conditions that we might be able to change. Or, we could analyze a network to determine which initial conditions and which variables have the most profound effect on our predictions and outcomes. As an example, recall that in our fictional world of Indopotamia, the decisions of Beta and Gamma to join forces and build a dam changed the configuration of the network. If the dam is built, Alpha won't have the water it needs to grow enough rice to meet the needs of its expanding population. Consequently, Alpha may try to change this initial configuration or adapt and find a water management strategy that fits the anticipated structural changes it will face in the future because of the new dam. Any action Alpha takes will also change the structure of the network and affect what Beta and Gamma need to do in the future as well. A reflective water professional must be cognizant of this type of co-evolution and feedback—and their impact—to manage a complex water network effectively.

Uncertain Decision-making for Disordered Water Problems

In water management situations in which sub-system interactions are highly uncertain, cause and effect linkages are ambiguous, and extrapolation from past experience is not a reliable predictor of the future, network managers are unlikely to be effective. Traditional methods of planning and negotiating are not likely to be very helpful.

We use two dimensions, shown in Figures 4.1 and 4.2: degree of certainty (i.e., the predictability of events and outcomes) and degree of consensus over means and ends. This framework allows us to describe five different types of management problems along associated approaches for decision-making. A key challenge for reflective water professionals is to identify the type of water management problem they are facing, based on a careful analysis of the networks within which they are operating. For example, managers need to realize that they cannot use predefined indicators to monitor unpredictable outcomes in the Zone of Complexity.

Adaptive Learning: A Key to Handling Evolving Complex Water Management Problems

Water managers can try to influence the environment of a complex water network by continuously observing, learning, and adapting to changing circumstances. Networks co-evolve with their environment and are "open" to flows of energy and information across boundaries. These flows do not usually follow fixed rules. Instead, they result from interactions among the nodes of the network. The macro structures that emerge in such situations constrain the choices of individual nodes and influence their interactions with other nodes in the network. Each node co-evolves with the structure resulting from the behavior and knowledge of all the others. In complex systems, surprise and uncertainty are part of the outcome.

Knowing that there are no certainties, but that there are opportunities for learning, can provide a direction for water managers. Any attempt to achieve a stable equilibrium within a setting that is inherently unpredictable is bound to fail. Effective management of complex water networks, especially in the long term, will result from continuous interactions among different networks components and ongoing adjustments in adaptation. Adaptive learning (shown in Figure 4.3) occurs at the intersection of identify, innovation, and implementation. Adaptive learning starts with the identification of a water management network in the certainty-consensus space (shown in Figures 4.1 and 4.2). This step is critical. Then, possible changes in past practice need to be considered. An implementation strategy will emerge from the structure and actions of the network.

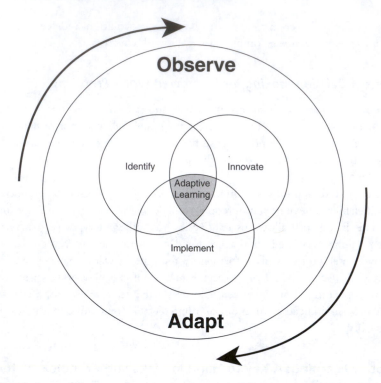

FIGURE 4.3 An evolving complex-systems perspective for water resource management through adaptive learning

Such an adaptive learning cycle reinforces the point that the future is unknowable for water problems operating in the Zone of Complexity. In such situations, long-term planning is more or less irrelevant; learning and continuous adjustment are essential. How to make this happen, that is, how to get agreement on what ought to be tried and what should be learned from each adaptive effort, are the challenges addressed in the next two chapters.

Selected Readings with Commentaries

M. Sivapalan et al "Water Cycle Dynamics in a Changing Environment: Improving Predictability through Synthesis," (2011a)

Introduction

This paper argues that many widely relied upon methods of hydrologic prediction are based on the assumption of stationarity. In a changing world, closer scrutiny of this assumption and its related implications are warranted. Several earlier attempts to reexamine the fundamental approaches used in hydrologic science are contrasted here: the physics–like "Newtonian" approach is compared with the ecology–like "Darwinian" approach. The authors suggest that contemporary challenges in hydrology and water–resource management require a synthesis of the two.

Hydrology and Water Resources in a Changing World (pp. 1–2)

Many current and widely relied upon hydrologic prediction approaches are founded on the assumption of stationarity [Milly et al., 2008], which permits extrapolation to the future using models that explain historical data. In a changing world, however, neither the structure (e.g., patterns of land use and land cover, connectivity between channels and riparian or wetland environments, or the extent of man-made structures), nor external drivers (e.g., temperature and precipitation forcing) of hydrologic response can be treated as fixed [Wagener et al., 2010]. Instead, changes in structure and drivers create the potential for new dynamics [Kumar, 2011] induced for example by hydrologic systems crossing unknown thresholds [Zehe and Sivapalan, 2009]. The potential for the emergence of such new dynamics poses significant challenges to predictability, especially on decadal or longer time scales.

Predictability Under Change

One way to cope with change is to take previously fixed or exogenous factors— such as climate, soil structure, river network topology, vegetation distributions or patterns of human land or water use—and treat them as an endogenous part of the predictive framework. This amounts to an expanded view of hydrology that considers the connections between the water cycle and climatic, ecological, social and earth surface systems. The behavior of the hydrologic system thus emerges from the co-evolution of the biotic, physical and anthropogenic systems that interact with it. Predictions under change are challenging because in this view, hydrologic predictability means understanding the interactions of multiple complex systems, including systems that are strongly driven by human decision making. These interactions must then be projected forward to make future predictions.

The co-evolution of the biotic and abiotic components in any particular ecosystem could (in theory) be simulated from detailed models that include all the relevant system feedbacks and couplings. Recent experience in hydrology has revealed the limits of the usefulness of such mechanistic models, even under the assumptions of stationarity [Blöschl and Montanari, 2010; Montanari, 2011]. The high dimensionality and process complexity of a coevolving system suggests that it may prove even more challenging to describe through detailed models [Strogatz, 1994]. Lower-dimensional approaches are needed [Dooge, 1992]. One possibility is to focus instead on the emergent outcomes of feedbacks and interactions between processes over time, which result in spatial and temporal organization of hydrologic systems: vegetation patterns, river networks and soil catena, or in the time domain, distributions of inter-event times and amplitudes of events. These "patterns"—loosely defined as consistent trends of commonality or difference between different places and/or times—contain information on the physical, biotic and socioeconomic mechanisms from which they emerged. Organized patterns not only reduce the dimensionality of the prediction problem but point toward the development of new kinds of understanding, relating for instance to underlying organizing principles or natural laws. Such understanding could lead to entirely new ways of modeling prediction under change [Kleidon and Schymanski, 2008; Schaefli et al., 2011].

Role of Synthesis

Given the potential importance of identifying trends, patterns and organization through time and space, classical hydrologic research faces a challenge. Hydrologic research tends to generate knowledge by collecting process- and place-specific data. This can lead to a body of understanding that is detailed and profound, yet fragmented in space and time and constrained by questions that motivated specific process studies. The arguments above suggest that overcoming this fragmentation is an urgent challenge for the field.

Hydrologic synthesis offers one approach toward overcoming this fragmentation of knowledge [Blöschl, 2006; Fogg and LaBolle, 2006; Hubbard and Hornberger, 2006]. The goal of synthesis is to make previously fragmented knowledge and understanding, and different disciplinary perspectives of the same phenomenon mutually intelligible across times, places, scales and disciplines [Blöschl, 2006], in the sense of Bronowski [1956]. Hydrologic synthesis aims to unify existing, diverse pieces of information (data, models and disciplinary theories), to discover unrecognized connections, and to develop scientific understanding that is valid across multiple places, scales and times [Blöschl, 2006]. This special section presents results from the NSF-funded University of Illinois (UIUC) Synthesis Project, further details of which are presented by Wilson et al. [2010] and Thompson et al. [2011a].

Focus on Emergent Patterns

An integrating framework for the synthesis approach presented in this series of papers is the focus on emergent patterns. Specifically, the research is aimed to examine if the patterns of hydrologic response found across multiple places correspond with existing hydrologic theory, and whether they could lead to new theories. Once emergent spatiotemporal patterns are identified, a range of questions can be formulated to investigate the nature, generation and consequences of these patterns:

Investigation of Emergent Patterns: Top-Down Questions

How do we measure and identify patterns and describe them? What can we learn from existing data sets? How should we design new observatories?

Theoretical Questions: "Deep Why-Type Questions"

Why do these patterns emerge? Under what circumstances do we expect them to occur? What are the underlying principles?

Bottom-Up Questions

What are the consequences of these patterns (their effects on processes of interest)? How do they scale up in time and space? How does understanding the pattern improve our capacity to make predictions?

Human Interactions

How do human activities respond to and modify these patterns in time and space? How are the patterns affected by human activities?

A Synthesis of Newtonian and Darwinian Approaches (p. 5)

There have been several calls to reexamine the fundamental approaches used in hydrologic science [Dooge, 1986, 1988; Gupta et al., 2000; Hooper, 2009; Torgersen, 2006]. Harte [2002] contrasted a physics-like "Newtonian" approach with the ecology-like "Darwinian" approach, and suggested that contemporary challenges in the earth sciences, such as dealing with environmental change, require a synthesis between the two. The Newtonian approach is exemplified in hydrology by detailed process-based models. This approach builds understanding from universal laws that govern the individual parts of the system. A Newtonian objective in hydrology is the mechanistic characterization of how water, energy and mass fluxes and transformations occur in the various parts of the landscape in the form of a boundary value problem. Even though the laws employed are taken to be universal, and not tied to a particular landscape, their solution depends strongly on the boundary and initial conditions, which must be characterized for a given landscape.

The Darwinian approach values holistic understanding of the behavior of the given landscape. It embraces the history of a given place, including those features that are relics of historical events, as central to understanding both its present and its future. The Darwinian approach gains predictive power by connecting a given site to several sites located along critical gradients. Laws in the Darwinian approach will seek to explain patterns of variability and commonality across several sites, as exemplified by the work of Kumar and Ruddell [2010] who used ecohydrologic data taken from several flux towers to discover underlying organizing principles. As previously argued by McDonnell et al. [2007] and Kumar [2011], the synthesis between the Newtonian and Darwinian approaches in hydrology thus offers the possibility for combining predictive understanding of the mechanisms of change with an explanatory understanding of the patterns that emerge when these mechanisms interact in real landscapes. This synthesis ensues when advances are made across the divide from both sides: when Newtonian process descriptions are used to develop explanatory hypotheses for variations between places, and when the particularities of many places, when viewed together at a certain distance, reveal commonalities that can help develop new process descriptions at large scales. The latter approach stands in contrast to the traditional reductionist approach where new process descriptions are developed through the treatment of phenomena at finer and finer scales.

The work represented in the papers appearing in this special section offers examples of both approaches. Newtonian process descriptions (necessarily simplified) were used in the works of Harman et al. [2011b], Thompson et al. [2011d, 2011b] … respectively, to develop parsimonious insights and explanations for the variations of vadose zone travel times, solute delivery and transformation, and differences in water balance between places with different climates and landscapes. In none of these cases would the models used be called "state of the art" in terms of the details of their Newtonian process descriptions, but each provided fundamental insights that a more sophisticated model run in one place for one time would not. In the other direction, the motivating works of Troch et al. [2009], Brooks et al. [2011], and Basu et al. [2010], each used large data sets to reveal intriguing patterns of commonality that cried out for explanation. These patterns were used by Voepel et al. [2011], Sivapalan et al. [2011b], and Harman et al. [2011a] to develop new predictive relationships for spatial and temporal variations in water balance and by Thompson et al. [2011d] and Basu et al. [2011a] to develop predictive models of reach and catchment-scale solute transformations. These process models rely on the emergence of ordered behavior at larger scales as a result of the evolutionary history of the systems, though the models do not (and need not) express that evolutionary process explicitly.

> In conclusion, the work reported in this special section can therefore be seen as tentative first steps toward a new approach to hydrologic science based on a synthesis of the Newtonian and Darwinian approaches. While significant breakthroughs are yet to be fully realized, it is our belief that if hydrologic synthesis as outlined in this paper is vigorously pursued it has the potential to generate transformative outcomes for hydrologic science.

Commentary

The Newtonian approach builds understanding from universal laws (that govern the individual parts of a system) using process-based models and a mechanistic characterization of how water, energy, and mass fluxes and transformations occur in the various parts of a water system. Even though the laws are universal, their solution depends strongly on boundary conditions that must be characterized in each situation. The Darwinian approach values holistic understanding of a given water management problem. The Darwinian approach relies on connecting information from a given location to several other locations along critical gradients.

A key assumption in the Darwinian approach is the need to find and explain patterns of variability and commonality. The authors argue that to understand and model a contemporary hydrologic problem, a synthesis of the Newtonian and Darwinian approaches is necessary. Such a synthesis has the potential to combine predictive and explanatory understandings in contrast to the traditional reductionist approach.

J.E. Kiang, et al Introduction to the Featured Collection on "Nonstationarity, Hydrologic Frequency Analysis, and Water Management," (2011)

Introduction

This article summarizes the findings and discussion from an important conference in 2010 that brought together senior personnel from a variety of federal agencies involved in water management in the United States. The group concluded that "decision-making under heightened uncertainty about future climate" (Kiang et al 2011) is highly likely, and that changes in water management techniques and strategies will be required.

Excerpt (pp. 433–435)

Water managers have long faced a nonstationary world. Changes in land use, declining groundwater levels, and urbanization are all drivers that can bring about nonstationarity in a watershed. With increasing demands on water resources and larger populations in vulnerable areas, floods and droughts have become more expensive and disruptive. Climate change is yet another potential driver of hydrologic change. Whatever the driver, nonstationarity can result in greater uncertainty about future floods, droughts, and their impacts, but particularly when the driver of the change is poorly understood. This featured collection explores methods for understanding future uncertainty and incorporating it into water management and planning, with particular attention to uncertainty created by the possible effects of climate change …

Scientific research to enhance understanding of nonstationarity can be categorized into two main approaches, as described by Hirsch [2011]: an empirical approach based on analysis of historical stream flow records and a modeling approach based on downscaled results of general circulation models. In his opinion piece, Hirsch asserts that although more research focus has been placed on modeling, the empirical approach is equally important, and that continued data collection and analysis is essential to furthering understanding of nonstationarity.

Many of the papers in this collection focus on the empirical approach. Villarini et al. [2011] analyze annual maximum peak discharges in the Midwest United States (U.S.) and Vogel et al. [2011] examine flood records for the entire U.S. Villarini et al [2011] characterize the flood-generating mechanisms in different parts of the Midwest and test for changes in flood magnitudes over time. Vogel et al. [2011] propose that a flood magnification factor and a recurrence reduction factor can be used to characterize changes in the magnitude and frequency of floods. While both papers present results indicating changes in some flood characteristics over time, both papers also note the difficulty in pinpointing the causes for any of these changes. Land use change, changes to agricultural practices, and operation of dams are likely causes of many of the changes. It is possible that climate change is also a contributing factor.

Lins and Cohn [2011] and Koutsoyiannis [2011] expand on the difficulties in interpreting results of statistical analyses to detect trends. As discussed by Lins and Cohn, hydrologic time series undergo periodic excursions from their long-term mean, a phenomenon Hurst (1951) first observed for the Nile River. This long-term persistence is a stationary process, but can cause natural fluctuations in hydrologic conditions that make it difficult to test whether or not anthropogenic climate change is affecting hydrologic time series. For example, the time horizon used for an analysis can have an enormous impact on whether a trend is identified. Koutsoyiannis [2011] continues with this theme. He discusses the use of the terms stationary and nonstationary and encourages water managers to adopt a Hurst–Kolmogorov framework to explicitly consider the long-term persistence in hydrologic time series. The Hurst–Kolmogorov framework admits more uncertainty into descriptions of hydrology, as may be desired when we are unsure of what the future may hold.

Stedinger and Griffis [2011] consider how to approach flood frequency analysis given the difficulty in interpreting trend analyses. They caution that while it is mathematically easy to formulate models to describe trends, it may not be productive to do so when the causes are not well-understood. Further, the considerable uncertainty of estimates of many statistics like the 100-year flood may further cloud the issue of whether climatic change needs to be incorporated into such analyses. While acknowledging these issues, Ouarda and El-Adlouni [2011] discuss methods for conducting hydrologic frequency analysis in the presence of nonstationarity, focusing on the Bayesian approach.

Statistical and empirical analyses are just one path toward better understanding of possible nonstationarities in the hydrologic system. Dettinger [2011] applies results from deterministic modeling to explore possible changes to flood-generating storms and river flow in California. Climate model results indicate that changes are possible to the frequency and intensity of flood events triggered by moisture brought into California by atmospheric rivers. While acknowledging the limitations of the models, Dettinger suggests that this is an example of how models can provide early insight into how flood risks may change in the future.

It is clear from the analysis by Villarini et al. [2011], Vogel et al. [2011], and many others (McCabe and Wolock, 2002; Lins and Slack, 2005; Hodgkins and Dudley, 2006) that some aspects of hydrology are changing. In many cases, it is less clear why the changes are taking place, whether climate change is playing a role, and whether the trends will persist into the future. Despite this uncertainty, water managers must continue to make decisions about how to operate existing resources and plan for the future. Brown et al. [2011] describe a framework for decision making under heightened uncertainty about future climate. Their approach seeks to identify robust water resources planning and management strategies by accepting that large uncertainties about the future exist, focusing on key vulnerabilities in water resource systems, and allowing for dynamic responses to evolving conditions. Waage and Kaatz [2011] outline five alternative planning methods that were originally described by the Water Utility Climate Alliance (Means et al., 2010). Different planning methods may be better suited for different water management agencies, and Waage and Kaatz also describe the criteria used by Denver Water to select a method for its planning process.

Arnell [2011] and Kundzewicz [2011] provide an international perspective on water management in the face of heightened uncertainty about the future. Arnell [2011] describes the evolution of regulations that require public water suppliers in England and Wales to consider possible impacts of climate change. Kundzewicz' [2011] discussion of conditions in Central Europe strikes a similar chord to the picture that emerges from this collection as a whole—nonstationarity in regional hydrology has been observed, but much of it can be attributed to nonclimatic causes. The degree to which anthropogenic climate change is a factor is unknown but the interplay of numerous factors affecting hydrology is important and numerous projects to study the issue further have been initiated.

Galloway's [2011] paper concludes this featured collection with an excellent summary of the January 2010 workshop. Different scientific communities are approaching the problem of understanding nonstationarity and water management in different ways. Galloway suggests that a common language and more collaborative efforts are needed to move the science forward so that water managers have more options to meet their immediate needs.

The workshop was one step in a continuing process of coordination and collaboration among the federal agencies that sponsored it. Efforts to enhance understanding of nonstationarity in hydrologic time series and to incorporate that knowledge into water management practices are ongoing.

Commentary

This is a summary of a collection of papers that were presented at the workshop on Nonstationarity, Hydrologic Frequency Analysis, and Water Management, held in Boulder, Colorado in 2010. Some scientists and technical specialists have analyzed changing characteristics of hydrologic signals and suggested that our ability to forecast or model likely changes in water supply in the future will be increasingly limited. Yet, there are water specialists who believe that continued improvements in modeling and forecasting tools and techniques will allow us to make reasonably good predictions even in the face of these uncertainties. Others argue that increased uncertainty, whatever its causes, calls for fundamental modifications in our approach to water management. They suggest the need to "accept that large uncertainties about the future exist, focus on key vulnerabilities in water resource systems, and allow for dynamic responses to evolving conditions" (Kiang et al 2011). This is consistent with our emphasis on collaborative adaptive management and negotiated approaches to managing complex water networks.

F. Berkes "From community-based resource management to complex systems," (2006)

Introduction

This article questions the emphasis and findings of most research in the area of common-pool resource management over the last several decades that relied on the simplicity of community-based resource management to develop theory. It raises questions with regard to scale and examines whether and to what extent the findings about small-scale, community-based commons can be scaled up to generalize about regional and global commons.

Issues of Scales and Levels in Common-pool Resource Management (p. 1)

There are ongoing debates in many areas of commons research. One of these concerns a scale related question: can findings from local-level commons be scaled-up? That is, can principles generated based on studies of micro-level systems be applied to meso-scale and macro-scale systems? Researchers have been dealing mainly with small scale, community-based systems. However, some of the experimental work on commons, using Prisoner's Dilemma models, has treated nation states as unitary actors in the analysis of global commons (e.g., Ostrom et al. 1994), with the implication that the same commons principles may apply across levels. More specifically, it is said "some experience from smaller systems transfers directly to global systems," but that "global commons introduce a range of new issues" (Ostrom et al. 1999:278). In this analysis, a number of factors are considered important, including the size and complexity of the system and the speed at which resources regenerate themselves. In addition to the scaling-up problem, Ostrom et al. (1999) indicate that there are several other challenges of global commons, concerning such factors as cultural diversity and inter linkages of commons. Other researchers have argued that the transferability of the small-scale commons experience is fraught with complications. The issue "is not fundamentally a matter of extreme size and complexity at the global level. Rather, the problem arises from differences between individuals and states, and from the separation between those who formulate the rules and those who are subject to them" (Young 2002:153). "Solving the tragedy of the commons at the local level is fundamentally a matter of self-regulation," but at the global level, "regulation is a two-step process" (Young 2002:152). Hence, "we should be particularly careful to avoid assuming unreflectively" that global issues can be treated in the same ways as local commons problems (Young 2002:149).

This debate may be approached by suggesting that it may be more useful to pose the issue as complexity management, rather than one of scaling-up. Commons management in many cases can be understood as the management of complex systems, with emphasis on scale, self-organization, uncertainty, and emergent properties such as resilience. Several authors have touched upon aspects of this (Levin 1999, Gunderson and Holling 2002, ... Adger et al. 2003, Berkes et al. 2003). There is general agreement that commons are often impacted by forces or drivers at various levels of organization ... There is also an agreement on the need to consider multiple levels of management (Ostrom et al. 1999, Young 2002, Adger et al 2003). Some of the commons literature explicitly deals with such multilevel management.

Co-management (Pinkerton 1989, Jentoft 1989) is by far the most widely discussed institutional form for dealing with commons management at two or more levels, but there is a diversity of institutional forms for dealing with multilevel commons (Berkes 2002). These other forms include epistemic communities (Haas 1990), policy networks (Carlsson 2000), boundary organizations (Cash and Moser 2000), polycentric systems (McGinnis 2000), and institutional interplay in which institutions may interact horizontally, i.e., across the same level, and vertically across all levels of organization (Young 2002, 2006).

These concepts have something in common: each provides an approach to understand cross-level linkages and to deal with complex adaptive systems. They all pertain to scale and to other aspects of complexity such as self-organization, uncertainty, and resilience. If so, one can argue that there is an evolution of thought toward dealing with commons management as complex systems problems. Such an approach provides an entry point to build a commons theory that proceeds to an analysis of commons as multilevel systems. The area of marine commons provides suitable examples to illustrate the phenomenon of resource management at all levels.

Using examples for marine commons, the objectives of the paper are to contribute to an understanding of commons as complex systems, and to the debate on scaling-up from the local to the global. The framework is provided by the three common scale challenges: (1) the failure to recognize important scale and level interactions, (2) the persistence of mismatches between levels and scales in human-environment systems, and (3) the failure to recognize heterogeneity in the way scale are perceived and valued by different actors.

Scale Related Complexities (p. 3)

The diversity and widespread prevalence of local level commons institutions indicate that they have been important for the survival of many societies and still relevant for contemporary resource management (Johannes 1998). However, there are certain limitations of the lessons learned from the study of local-level systems. Initially, research on commons issues often sought the simplicity of community-based resource management cases to develop theory. For example, Ostrom (1990:29) explains that her strategy has been to study small scale common property situations "because the process of self-organization and self-governance are easier to observe in this type of situation than in many others." This is not to say that such small scale systems are isolated from the rest of the world and immune to internal and external influences that affect self-governance. There are many such influences, and here I touch upon four scale-related issues that impact sustainability and resource management: (1) complexity at the level of the community itself; (2) the existence of external drivers of change; (3) the problem of mismatch of resource and institutional boundaries, i.e., the issue of fit; and (4) the necessity for community-based management to deal with cross-scale issues.

Scale Issues in Community Based Resource Management (p. 6)

The India case also illustrates some of the challenges related to scale in a community-based system that appear to have emerged as a response to certain external drivers. Southern India is home to a number of traditional community institutions for coastal resource management (Paul 2005). What have been called *padu* systems are found in Sri Lanka and the southern Indian states of Kerala and Tamil Nadu. These are lagoon and estuarine resource management systems, mainly for shrimp fisheries, characterized by the use of rotational fishing spots allocated by lottery. They are species- and gear specific, with rules to define fishing sites and rights holders, often according to social or caste groups (Lobe and Berkes 2004). Some *padu* systems in Sri Lanka go back to at least to the 18th and possibly the 15th centuries (Amarasinghe et al. 1997).

We investigated three community-based fisher associations, i.e., sanghams, in the Cochin estuary of Kerala, South India that use the *padu* system. The sanghams administered the rotational allocation of shrimp fishing spots, fished with stake nets that are rows of bag-like nets fixed to stakes driven into the ground. They operated under a set of well-defined rules serving livelihood, equity of access, and conflict resolution needs among their members.

As a commons institution, the *padu* system of the Cochin estuary only dates back from the late 1970s (Lobe and Berkes 2004). Tracing their origins showed that they arose out of two events. The first was the globalization of shrimp markets. Shrimp became "pink gold," as many small-scale fishers in South India abandoned other resources in pursuit of shrimp (Kurien 1992). The second factor was the centralization of fisheries management in Kerala. In 1967, the Kerala Fisheries Department started to institute a new licensing arrangement, replacing an older system of land and fishing site holdings. Beginning in 1974, state legislation required licenses for all fishers, but the state lacked the means to enforce the new law. Because shrimp fishing was lucrative and attracted new entrants, the resource effectively became open access, forcing the fishers to self-organize to consolidate what they considered to be their rights in a large and crowded estuary and lagoon system (Lobe and Berkes 2004).

Each *padu* association in the Cochin estuary dealt with the exclusion issue by limiting the access of nonmembers, and the subtractability issue through rules that provide for equity, social responsibility, and conflict management among its members. However, the Kerala State government does not recognize the three associations in the study area, nor does it license the fishers. They continue to fish only because of a 1978 court order establishing them "as fishers by profession" (Lobe and Berkes 2004), and ongoing state-level political action by their Dheevara caste organization to protect their rights (K. T. Thomson, *personal communication*).

The sanghams seem to be effective in dealing with the subtractability problem; they have well-defined and clear rules to regulate resource use among members. However, regarding the exclusion problem they are only partially effective. They control the stake nets that are in their rows of nets and have a say about who fishes them, including those that are leased out, but they have no control over the other fishers in the area. The three sanghams control only about one-half of the 289 stake nets owned locally, and that in turn is only a small fraction of some 13,000 stake nets used in the entire lagoon and estuary system.

In the heavily used estuary and lagoon system in Cochin, there appears to be no systematic data collection or stock assessment, but there is some enforcement of restrictive regulations. The lack of state recognition and mechanisms for cross-level coordination has limited the ability of the three sanghams in the Cochin estuary to contribute to management at the regional level. However, there is no effective regional-level management. Given the lack of resources in most developing countries, is it realistic to expect the management of such resources as used by *padu* systems of South Asia? There are, in fact, well-functioning *padu* systems with both local- and regional-level management, and they are found in the well-studied Negombo Lagoon of Western Sri Lanka (Amarasinghe et al. 1997, Amarasinghe et al. 2002)

There are differences between these two lagoon management cases that use variations of the same *padu* system. Both are species- and gear specific, with rules defining sites and rights holders, and both use a lottery-based, rotational use system for fishing sites. The differences are organizational. In Negombo, the fishers are organized at the community level through four rural fisheries societies (RFSs). They are subject to the rules made by each of the RFSs at the local level, and coordinated across the four RFSs at the regional level. This local control and regional coordination was made possible by the national government through the devolution of management authority to the RFSs and the Negombo (Kattudel) Fishing Regulations (Amarasinghe et al. 1997). By contrast, no effective cross-level linkages exist in the Cochin estuary and lagoon, even though the fishers are well organized at the community level and even though there is Kerala State legislation from 1995 that provides a directive to devolve resource management to municipal-level organizations.

The *padu* examples illustrate how external drivers related to economic development, i.e., international markets for shrimp, and resource management policies, i.e., state-level reorganization of fishing rights can affect community-based institutions. The *padu* systems of the Cochin area have their origins in ancient South Asian traditions of coastal resource use, but they are in fact the product of relatively recent economic and political transformations. Their existence is fragile because of a lack of state level policies and government recognition.

One of these two *padu* systems has effectively addressed most of the major challenges related to scale (Cash et al., *in press*), but the other has not. The Sri Lanka case has solved the scale mismatch issue though cross-level governance from the local to the national, thus engaging in management at multiple scales, both politically and geographically (Amarasinghe et al. 1997). By contrast, the Kerala case has no cross-level governance, no intermediate-level institutions, and no arrangements between the community and the government. Thus, the problem of scale mismatch remains unresolved, issues are defined only at the local association level, and cross-level linkages are developed only weakly at the state government level through lobbying by caste organizations.

Issues of Complex Adaptive Management for Common Pool Resources (pp. 10–11)

Commons thinking has been evolving to deal with the complexities of resource problems, turning to the examination of scale, self-organization, uncertainty, and resilience, all of which are concepts of complex adaptive systems (Gunderson and Holling 2002). Commons research has evolved through the critique of the "tragedy of the commons" model used "to paint a disempowering, pessimistic vision of the human prospect," and to rationalize central government control or privatization of all commons (Ostrom et al. 1999:278). Studies over the past 30 yr have documented, in considerable detail, the self organization and self-regulation capability of communities to solve the exclusion and subtractability problems of the commons.

However, research has also shown that community based resource management is vulnerable to external drivers and is often insufficient by itself to deal, for example, with problems of migratory marine resources. As the examples in this paper indicate, cross-level issues are pervasive in commons management. The marine commons cases considered here illustrate many of the scale challenges of ignorance, mismatch, and plurality (Cash et al., 2006). However, all three cases, Cambodia, Kerala, and Sri Lanka, and the ICCAT Caribbean, show various degrees of recognition of the scale problem.

Hence, the challenge of scale ignorance is not the dominant issue. The ICCAT does not respond to the livelihood problems of Gouyave fishers and the Government of Kerala seems unwilling or unable to set up regional level management, but nevertheless, most actors in the cases considered recognize scale and level interactions. Similarly, mismatch problems are addressed, although with mixed success. All three cases grapple with the tendency to define issues at only one scale. In Cambodia, Kerala, and Sri Lanka, the local level is the focus of management, whereas in the ICCAT Caribbean, focus is the level. The other levels are present but not effectively engaged, except in the Sri Lanka case. The Kerala case is unusual in that it lacks even an attempt at forging vertical institutional linkages. The challenge of plurality is pervasive, and resources are contested by multiple actors in each case. Kerala is the most crowded and contentious case, and it is possible that the lack of institutional solutions is related to the pessimism of the actors that win–win solutions are possible. Commons theory holds that solving the subtractability problem depends, among others, on the users having workable relations for monitoring, sanctioning, and conflict resolution (Ostrom 1990). The ICCAT Caribbean case provides a striking example of an asymmetrical relationship. Regional and international science is filtering down the levels, but fisher's knowledge and values are not being heard.

The mismatch in this case is about scale of management; the dispute is about equity and fairness. What is at odds is the management discourse of the politically powerful countries vs. the Caribbean nation states (Singh-Renton et al. 2003), reaffirming the socially constructed nature of knowledge (Lebel et al. 2005, Adger et al. 2003). There are a number of implications of these findings for the scaling-up debate (Ostrom et al. 1999, Young 2002). Can the findings of small-scale, community based commons be scaled-up to generalize about regional and global commons? It may be more illuminating to approach the debate by suggesting that commons management in many cases should be understood as the management of complex adaptive systems, rather than merely a question of the transferability of community-level findings to the global level.

Starting from this point of view, the results of this paper do not support Young's (2002:152–153) contention that self-regulation may suffice to solve the tragedy of the commons at the local level, but that at the global level, regulation is a two-step process involving those who formulate the rules and those who are subject to them. First, none of the cases show that self-regulation alone is sufficient to solve the commons problem at the local level. Second, there are no cases that show such a two step process; all cases include multiple levels and none of them shows a pure community-based case in which people who formulate the rules are those who are subject to them. These considerations indicate that, when multiple levels are involved, the emphasis of the inquiry should shift from the question of scaling-up to understanding linkages, their nature, and dynamics.

The study of cross-scale institutions such as co-management agencies, boundary organizations, and epistemic communities is important because these institutions provide a means to bridge the divide between processes taking place at different levels. In effect, they provide ways to deal with linkages in complex adaptive systems. Examining horizontal and vertical linkages, analyzing polycentric systems (McGinnis 2000), and dealing with policy networks (Carlsson 2000) are among the various means to understand the nature and dynamics of cross-level linkages and governance in general (Lebel et al. 2006).

Commons theory can provide insights into the solution of regional and global commons problems by looking beyond the community-based resource management paradigm, toward commons governance in complex systems (Dietz et al. 2003, Berkes et al. 2003). Communities, which themselves may be complex systems, are embedded politically and economically in larger complex systems. They respond to a range of drivers of change such as markets, central government policies, and international economic policies. The community level is important as the starting point for the solution of the tragedy of the commons (Berkes 1989, Ostrom et al. 2002). However, higher levels of organization are also important in providing monitoring, assessment, enforcement, and fostering local management. The importance of institutions that straddle levels and provide incentives for sustainability is increasingly recognized in marine resource management (Hilborn et al. 2005) and environmental management in general (Adger et al. 2003). All types of commons governance seem to start from the ground up and deal with cross-scale linkages in a complex systems context.

Commentary

A key question addressed in this paper is whether the success of self-governance in simple, site-specific instances can be scaled up to "higher" or "multiple" levels. The author's effort to introduce complexity theory into the management of common pool resources—characterizing a commons as a multi-level complex system is helpful. It does not, however, provide any guidance on how to do this at multiple levels at the same time.

C. Pahl-Wostl et al "Social Learning and Water Resources Management," (2007)

Introduction

This paper presents a framework for social learning and collaborative governance rooted in the more interpretive strands of the social sciences that emphasize the context dependence of knowledge. It argues that, historically, water management practices have largely been dominated by experts using technical means to design systems that can be readily controlled. In recent years, stakeholder involvement has gained increasing importance, and collaborative governance is increasingly considered more appropriate in the integrated and adaptive management regimes needed to cope with the uncertainty and complexity of social-ecological systems.

Adaptive Management and Social Learning (pp. 2–3)

There are different motives for increasing stakeholder involvement. One argument based on democratic legitimacy emphasizes that all those who are influenced by management decisions should be given the opportunity to actively participate in the decision-making process. Principles of equity and social fairness demand that the voices of the less powerful should also be heard (e.g., REC 1998, 1999, Renn et al. 1995 ...). A pragmatic approach is to build on the insight that complex issues and integrated management approaches cannot be tackled without taking into account stakeholders' information and perspectives and without their collaboration. Interdependence between government bodies and other stakeholders is increasing because of, for instance, decreasing government budgets that reduce the efficacy of the traditional command-and-control management style. Collective decisions are needed to implement effective management strategies, and the combination of top-down and bottom-up formation of institutional arrangements may lead to a greater acceptance by all the stakeholders involved.

This paper follows a pragmatic approach, albeit recognizing that effectiveness and legitimacy are related, and tries to provide evidence for the need for social learning processes that can cope with complex resource management problems. Social learning is analyzed as a means of developing and sustaining the capacity of different authorities, experts, interest groups, and the general public to manage their river basins effectively. This includes the capacity to deal effectively with differences in perspective, to solve conflicts, to make and implement collective decisions, and to learn from experience.

Uncertainty and Change

Some insightful examples for the importance of participatory governance come from the area of adaptive management. Adaptive management first focused on ecosystems but has increasingly embraced the importance of the human dimension (e.g., Berkes and Folke 1998, Lee 1999). Several authors emphasize the need for a shift toward adaptive co-management of social-ecological systems in which cooperation among a wide range of stakeholders and institutions is necessary. Hence, adaptive co-management combines the dynamic learning characteristic of adaptive management with the linkage characteristic of cooperative management ... Folke et al. (2003, 2005) explored the dimensions and the nature of governance that enable adaptive ecosystem-based management and identified the four critical factors for dealing with social-ecological dynamics during periods of rapid change and reorganization:

1. Learning to live with change and uncertainty.
2. Combining different types of knowledge for learning.
3. Creating opportunities for self-organization toward social-ecological resilience.
4. Nurturing sources of resilience for renewal and reorganization.

They emphasize the role of networks, sense making, leadership, diversity, and trust as well as the role of organizations capable of accumulating the experiences and collective memory they need to cope with surprise and turbulence. Bridging institutions play a major role in strengthening the generation of social capital and creating new opportunities and multilevel cooperation and learning. The question arises of how these characteristics are developed and sustained. Factors such as climate change, the rapid dynamics of socioeconomic development, and globalization are increasing the amount of uncertainty faced by managers from regional to global scales. This requires a more adaptive and flexible management approach that can speed up the learning cycle to allow for more rapid assessment and implementation of the consequences of new insights. This type of adaptive manager needs new skills and capabilities, informal and flexible management structures, and access to expert knowledge as well as local lay knowledge.

Folke et al. (2003) point out that social learning is essential for building up the experience needed to cope with uncertainty and change. They emphasize that "… knowledge generation in itself is not sufficient for building adaptive capacity […] to meet the challenge of navigating nature's dynamics … " and conclude that " … learning how to sustain social-ecological systems in a world of continuous change needs an institutional and social context within which to develop and act." Such findings support this paper's concept of social learning and knowledge as participation. Knowledge and the ability to act upon new insights are continuously questioned, applied, and regenerated or expressed alternatively in social processes. The social network of stakeholders is an invaluable asset for dealing with change. A similar argument was made by Tompkins and Adger (2004). They pointed out that community-based management enhances adaptive capacity in two ways: by building networks that are important for coping with extreme events and by retaining the resilience of the underpinning resources and ecological systems. Social learning increases adaptive capacity and leads to sustained processes of attitudinal and behavioral change by individuals in social environments through interaction and deliberation.

Social Learning: Implications for Water Governance and Management (p. 11)

This paper argues that processes of social learning and the presence of informal actor platforms are of major importance when it comes to implementing and supporting integrated and socially, environmentally, and economically sustainable resource management regimes over extended periods of time. However, it is of crucial importance to better understand the role of bridging organizations and the interplay between formal and informal institutions. We would like to highlight here three important implications for water governance and water management:

1.   The principle of integrated water resources management has been criticized for being unrealistic. For example, Biswas (2004) and several of those who responded to his article pointed out a number of barriers to implementation. The integration of sectors and issues would require more centralized policy development and implementation and thus larger, slower, and more bureaucratic authorities to handle all policy aspects. Furthermore, objectives such as

stakeholder participation and decentralization would be unlikely to promote integration. However, this paper portrays another perspective, i.e., that of a more dynamic actor landscape in which integration is not achieved by bureaucratic hierarchies but rather by processes of network governance. It also highlights the need and the requirements for processes of social learning to build the capacity to achieve joint solutions and to make thus stakeholder participation effective in terms of achieving the goals of water management.

2. Water management is facing increasing uncertainties because of climate change, fast-changing socioeconomic boundary conditions, and the goal of integration over a wider range of objectives. As a consequence, effective water governance must be adaptive. This paper highlights structural elements and processes in water governance regimes that make them more adaptive without compromising their stability.

3. In most countries, the structural conditions for integrated and adaptive water management have not yet been determined. Consequently, there is a need for major changes in which the kinds of processes highlighted in this paper will most likely play a major role.

Challenges still lie ahead, in particular, in collecting additional empirical evidence and in using available empirical studies for comparative analyses. Our analysis suggests that the development of such adaptive institutional settings involves continued processes of social learning in which stakeholders at different scales are connected in flexible networks and sufficient social capital and trust is developed to collaborate in a wide range of formal and informal relationships ranging from formal legal structures and contracts to informal voluntary agreements. The multiscale nature of institutional change is a quite fascinating and highly relevant area of research of which this paper could tackle only a few aspects.

Commentary

Pahl-Wostl et al (2007) emphasize that a significant change is underway in water resource management because of the increased recognition of the complexity of these systems and because of increased participation by stakeholders from different domains. The article endorses the idea that collaborative governance is required to cope with the complexity and uncertainty of socio-ecological systems. The authors argue that institutional dynamics are important to guaranteeing social learning. They focus on stakeholders (although they don't talk at all about the problem of representation), and the need for adaptive management. They build on the experience in Europe of participatory river basin management and emphasize the need to encourage distributed rather than multiple cognitions. In short, they view social learning as a multi-scale process, just as we do in our representation of water management network.

V. S. Saravanan "A Systems Approach to Unravel Complex Water Management Institutions," (2008)

Introduction

This article explores the complexity of water management institutions by analyzing the interactive nature of actors and rules. It highlights the importance of a comprehensive approach, as opposed to sectoral approaches to managing water. It emphasizes the importance of socio-political processes in which stakeholders, actors, and agents occupy and shift positions depending on the context they are in, and thereby affect the outcome in a dynamic way.

Complexity of Adaptive Water Management (pp. 202)

The study unravels the complexity of water management institutions by analysing the interactive nature of actors and rules to a particular water-related problem, using a systems approach in a hamlet in the Indian Himalayas. The approach builds on the strengths of institutional analysis development framework, but makes amendments to suit complex and adaptive water management institutions. It applies multiple research methods to collect both qualitative and quantitative information at different contextual levels. The information collected is applied in Bayesian belief network model to identify differential rules in influencing water management. Systems perspective in a problem context helped to comprehensively understand the socio-political process of water management by identifying broad array of actors and rules constraining water management, and at the same time identify actors and rules facilitating agents and their agency for a change in the water management process. In this socio-political process, the study reveals human entities—stakeholders, actors and agents—occupy different positions, which they actively shift in a problem context and when agents pursue "projects" by integrating diverse rules and resources to remain adaptive. It is this adaptive and dynamic behaviour that contemporary programmes and policies fail to acknowledge. In this dynamic behaviour of the transformative capacity or power is everywhere, but they are displayed, maintained and upheld, only when agents pursue their "project" by negotiating with other agents. The paper highlights the importance of comprehensive approach, in contrast to simplistic, linear and single package reforms to manage water. Such approach calls for conscious designing of rules and, at the same time, enabling actors to design rules. A conscious designing of rules is required to regulate water distribution, to build the capabilities of the poor, and to be adaptive to institutional and bio-physical crises. It calls for the development of infrastructures to further actors and agent's capabilities to design rules for informed water-related decisions. Such an approach will contribute towards sustainable water resource management.

Systems Approach to Unravel Complexity (pp. 203–205)

Water resources are intricately linked with a number of components that make it a socio-ecological system, in response to stimuli in various sub-systems ([Stephens and Hess, 1999] and [Anderies et al., 2004]). With many sub-systems operating at various levels, the failure of one of these units may be compensated by the functioning of the other, leading to adaptive change in the water management regime. It is this combination of adaptability and complexity that makes water resource management a complex adaptive system, characterised by openness, "ebb and flow" partnerships amongst multiple actors, and emergent properties ([Dorcey, 1986], [Stephens and Hess, 1999], and [Pahl-Wostl, 2007]). A systems approach is central to unravelling this complex adaptive system. Systems approach is often applied in contemporary literature in two forms; one calling for "broad" or comprehensive perspectives (Pahl-Wostl, 2007) to understand water management and the other calling for an integrative perspective (Bellamy et al., 2001), by identifying key variables influencing water management. Rather than considering these as opposing poles, Mitchell (2005) calls for their utilisation in a phased manner, thinking comprehensively at normative and strategic levels to identify and consider the broad array of variables influencing water management, whilst remaining integrated at a tactical and operational scale of water management. This helps to develop applied knowledge on water management practices, and focus on the levels of institutional action for designing water institutions.

The analytical framework builds on the strengths of the institutional analysis development (IAD) framework (Ostrom et al., 1994), which states that human entities and rules interact in decision-making arenas to manage water. However, various amendments are made by drawing on Dorcey (1986) and Gunderson and Holling (2002), to analyse complex adaptive water systems. The framework consists of three situational variables—human entities, prevailing rules and bio-physical resources, which interact in diverse decision-making arenas to reshape the same [Figure 4.4].

Implication for Integrated Water Management (pp. 212–213)

Using systems approach in a problem-context, the study unravels the complexity of water institutions in a hamlet. It identifies multiple actors negotiating a bundle of rules in a number of action arenas toward constraining and facilitating water management. These arenas are location-specific or generic, formal or informal, and are naturally evolved or deliberately created by strategically located actors and agents. The decision-making processes in these arenas do not represent communicative and consensual partnerships or strategic actions, but rather combines diverse social communicative skills over a period, making water management a socio-political process. Here integration of institutions does not have any tangible form, but are realised through linkages between pre-existing activities (Morrison, 2004). The scale of integration is complex that conceals the discrete distinction between local, national and

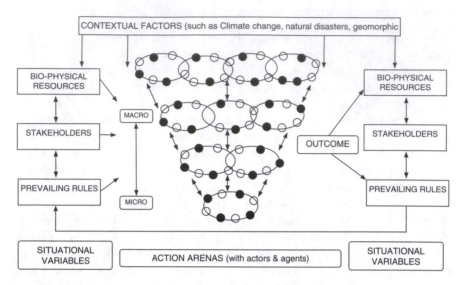

FIGURE 4.4 Framework for analysing institutional integration
Source: (from Saravanan 2008)

global, between state, market and community institutions, and among various sectors involved in water management. In spite of various institutions involved in framing the problem, the core belief remains that 'water is infinite' (and can be exploited) and prior water distribution cannot be meddled with. The hamlet witnessed noticeable change in the statutory actors (from the Ruler to the government of India, the GoHP, international agencies and local institutions), but even though the newly positioned actors ascertained that 'water is infinite', they did so without making any realistic assessment of water resources to meet the growing demands. What makes these actors adopt such a 'fire-fighting' approach (to address the food crisis, water conservation and in promoting democratic governance) is the absence of 'information' and 'scope' rules. Facilitating their decision-making process requires adequate opportunities that enable actors to voluntarily share and debate available information toward the formation of a consensual decision. This could be achieved through infrastructure facilities (road, telecommunications, mass media and others) that allow actors to interact and seek various options for desired outcomes using various forms of communicative skills.

The authority to distribute water changed from Princely Ruler, to the DoIPH and later to the BKIS. But the newly positioned actors still maintained the past "first-come-first-serve" basis of water distribution, thus making water distribution inefficient. In this case, there was no absence of information rules, rather its inadequacy (or misinformation that CBM were efficient), which resulted in the absence of any aggregation rule that resulted in the DoIPH (not monitoring and regulating water distribution) presuming that water distribution would be efficient being a community-controlled system now. This resulted in the wastage of water, inadequate crop planning and conflict, making distribution inefficient. This calls for strengthening the distributive form of governance, whereby sectoral statutory public actors monitor and regulate water distribution on the ground. Households' capability to access water is built over time by both statutory public actors (GoHP) and socially embedded informal actors (households), with the market playing a minor role. Availability of cattle, infrastructure facilities and a social network played a major role in influencing households' capabilities, which cannot be negotiated through communicative action, as presumed by contemporary development programmes. The differential capabilities enable them to adopt various actions to access water; resistance-, negotiation-, dissemination- and resignation-based types of action. These actions are multiple and are unpredictable. In these circumstances, the best option is to build their capability through various infrastructural measures and by directly targeting the disadvantaged households who do not have the capability to adapt to the institutional changes.

The interaction reveals the adaptive behaviour of actors by drawing on different rules to negotiate water policies, distribute water and in influencing households' capabilities. For instance, the households in Rajouri (upstream hamlet) who were granted the position of "employee" in Bagh exploited the contextual factors (location of their land) to demand irrigation rights from their Ruler. Similarly, granting of land ownership rights (position rule) to the households in Pipal enabled their legitimacy to demand irrigation rights, which was supported by the government of Himachal Pradesh, the World Bank, and the BKIS. The adaptive behaviour of actors is also notable with the introduction of infrastructure facilities that provided "positions" for households to build social networks to market milk, in contrast to a subsistence economy in the past. These actors (Princely ruler, community organisations, market institutions, government agencies, Supreme Court, multilateral and private agencies) are stakeholders having legitimate interest in influencing the water-related problem in context. The adaptability of these actors were dependent on the "position" granted by statutory rules, which were intelligently aggregated by socially embedded rules and contextual factors to evolve into new strategies and in the process constrained water management.

By constraining water management, actors facilitated the emergence of agents, who intelligently combined their practical (self-interest) with discursive consciousness (collective interest) to facilitate their agency. This is in contrast to agent-based studies (see [Saleth and Dinar, 2004] and [Janssen and Ostrom, 2006]), who presumes agents' as autonomous entity interacting with static rules in influencing institutional change. The transformative capacity or the power of these agents is gained incrementally and cumulatively with actors structuring them at various points of time. They use this capacity to integrate diverse actors and more importantly, change statutory rules in collaboration with other agents. For instance, PS emerged due to his experience as a "truck driver", which, in combination with his position as the President of BKIS and as a member of Punjabi community, gave him the authority for his actions. DC, who was a Scheduled Caste and had links with the underworld, exploited the failure of previous congress parties to gain immunity from the Indian democratic constitution as MLA. Similarly, SK gained his authority under the HP Administrative Service. Though socially embedded and statutory rules provided them with boundary and position, the authority to take decisions was provided by statutory rules. All of them use their individualistic goal to pursue a collective "project", similar to the study by Llewellyn (2007). The ability of agents to pursue dual goals makes them a "cunning players" (Randeria, 2003) in the socio-political process of water management. These agents are goal-oriented (PS), opportunistic (DC), and reactive (SK). The strength of these agents is located in informing other agents at the macro-level regarding their dissatisfaction with the existing institutional structures, and to then call for an adaptive approach by integrating diverse sets of rules. These agents do not always address issues related to poverty and environmental management or any philanthropic ideals of governance, due to limited knowledge and because of their cunning nature. It is important for statutory public actors to take notice of these external motives and to recognise this adaptive behaviour when evolving comprehensive strategies.

General Conclusion

Using a system approach in a water-related problem context, the study unravels the interactive nature of actors and rules in the hamlet Pipal by applying multi-methods. The information collected through these methods was applied in a Bayesian network to identify the relationship amongst variables and for its graphical representation in a network. This helped in identifying different types of rules and their interactions influencing water management. This demonstrated the reflexive behaviour of human actors and more importantly, in quantifying the probability of relationships. Though Bayesian network had several advantages compared to conventional analytical tools, they may be confronted with real problems when analysing biophysical reality in terms of a 100% probability. Furthermore, integrating both qualitative and quantitative information in the network may not be statistically representative.

Applying systems approach in a problem-context helped to comprehensively understand water management as a socio-political process. It is in this process that human entities with shared vision are triggered to make a well-informed strategic choice by drawing on diverse rules and resources. Examining the socio-political process in a problem-context helped to understand and analyse complexity at a manageable scale. This helped to identify broad array of human entities and rules constraining water management, and at the same time, identify actors and rules integrated by agents to bring about institutional change to address the problem. In this process, the stakeholders, actors and agents who are differentially positioned play an important role in facilitating the socio-political process. But they actively shift their positions depending on the problem context and by the "project" pursued by agents of institutional change, making the socio-political process of water management a dynamic and adaptive process. Here stakeholders are all natural resources users and managers, who have a passive role in influencing the socio-political process from outside the arena, but are drawn as actors depending on the context and by agents. Stakeholders as actors enter the arena depending on the legitimacy granted in the context and by the agents. These actors are organisations or groups of individuals, who not only have an incumbent role, as demonstrated by Archer, but also have a strategic role depending on the context. It is this latter role that facilitates the agents and their agency. Agents are people with "project" who develop out of actor-defined issues or problematic due to inadequacy existing rules and bio-physical resources. In this socio-political process, all human entities have transformative capacity or power, which is activated in a problem-context depending on the statutory and socially embedded rules facilitating them. The power is only revealed, displayed, maintained and upheld when agents negotiate their "project". Such negotiations do not always address issues related to poverty and environmental management or any philanthropic ideals of governance, due limited knowledge and their cunning nature.

> Given the dynamic and adaptive behaviour of human entities, the study high-lights the importance of a comprehensive approach to manage water resources, in contrast to contemporary highly simplistic, standardised linear policy and single package reform. Such an approach calls for statutory public actors to formulate strategic policies, but to benefit the poor, to enhance social justice and to bring about a sustainable future it calls for a combination of conscious designing of rules as well as enabling other actors to design rules is of utmost importance. Conscious designing of rules is required to monitor and regulate water distribution, to be adaptive to resource crises and more importantly, to target deprived households to build their own capabilities by modifying his-toric socio-cultural determinants. Enabling other actors to design rules requires infrastructure facilities (roads, transportation facilities, and mass media), to enable actors and agents to voluntarily share and debate available information on water resources and to build their capabilities toward self-organisation in integrating water-resource management.

Commentary

This paper employs a comprehensive approach to analyzing the interactive nature of water problems. It recognizes that water issues are the product of natural, societal and political forces interacting. These forces can affect each other at different scales (or levels) in ways that lead to reciprocal changes in network dynamics. They highlight power relations and their ability to impact micro–scale actions as well as macro–level outcomes.

References

Ackoff, R.L. 1979. The future of Operational Research is Past. *The Journal of the Operational Research Society*, 30(2):93–104.

Adger, W.N., Brown, K., Fairbrass, J., Jordan, A., Paavola, J., Rosendo, S., and Seyfang, G. 2003. Governance for sustainability: Towards a "thick" analysis of environmental decision-making, *Environment and Planning A*, 35(6):1095–1110.

Allen, P. M. 2000. Knowledge, learning, and ignorance, *Emergence, 2*(4): 78–103.

Amarasinghe, U.S., Chandrasekara, W.U., and Kithsiri, H.M.P. 1997. Traditional practices for resource sharing in an artisanal fishery of a Sri Lankan estuary, *Asian Fisheries Science*, 9:311–323.

Amarasinghe, U.S., Amarasinghe, M.D., and Nissanka, C. 2002. Investigation of the Negombo estuary (Sri Lanka) brush park fishery, with an emphasis on community-based management, *Fisheries Management and Ecology*, 9:41–56.

Anderies, M.J., Janssen, A.M., and Ostrom, E. 2004. A framework to analyse the robustness of social-ecological systems from an institutional perspective. *Ecology and Society*, 9: 18

Arnell, N.W. 2011. Incorporating climate change into water resources planning in England and Wales, *Journal of the American Water Resources Association*, 47(3): 541–549.

Bar-Yam, Y. 2004. Multiscale Variety in Complex Systems. *Complexity*, 9:37–45.

Basu, N.B., et al. 2010. Nutrient loads exported from managed catchments reveal emergent biogeochemical stationarity, *Geophysical Research Letters*, 37: L23404.

Basu, N.B., Rao, P.S.C., Thompson, S.E., Loukinova, N.V., Donner, S.D., Ye, S., and Sivapalan, M. 2011a. Spatiotemporal averaging of in-stream solute removal dynamics, *Water Resources Research*, 47: W00J06.

Bellamy, J.A., Walker, H.D., Mcdonald, T.G., and Syme, J.G. 2001. A systems approach to the evaluation of natural resource management initiatives. *Journal of Environmental Management, 63*: 407–423.

Bennis, W.M., Medin, D.L., and Bartels, D.M. 2010. The Cost and Benefits of Calculation and Moral Rules. *Perspectives on Psychological Science*, 5(2):187–202.

Berkes, F. (ed.). 1989. *Common Property Resources: Ecology and Community-Based Sustainable Development*. London: Belhaven.

Berkes, F.L. and Folke, C. (eds.). 1998. *Linking Social and Ecological Systems: Management Practices and Social Mechanisms for Building Resilience*. Cambridge: Cambridge University Press.

Berkes, F. 2002. Cross-scale institutional linkages for commons management: perspectives from the bottom up, in E. Ostrom, T. Dietz, N. Dolsak, P.C. Stern, S. Stonich, and E.U. Weber (eds.) *The Drama of the Commons* (pp. 293–321). Washington, D.C.: National Academy Press.

Berkes, F. 2006. From community-based resource management to complex systems. *Ecology and Society, 11*(1): 45.

Berkes, F., Colding, J., and Folke, C. (eds.). 2003. *Navigating Social-ecological systems: Building Resilience for Complexity and Change*. Cambridge: Cambridge University Press.

Biswas, A.K. 2004. Integrated water resources management: a reassessment, *Water International*, 29(2): 248–256.

Blöschl, G. 2006. Hydrologic synthesis: Across processes, places, and scales, *Water Resources Research, 42*: W03S02.

Blöschl, G. and A. Montanari. 2010. Erratum: Climate change impacts throwing the dice? *Hydrological Processes, 24*(8): 374–381.

Boland, J.J., and Baumann, D. 2009. Water resources planning and management, in C.S. Russell and D.D. Baumann (eds.) *The Evolution of Water Resources Planning and Decision Making*. Cheltenham: Edward Elgar, IWR Maass-White Series.

Bronowski, J. 1956. *Science and Human Values*. New York: Julian Messner Inc.

Brooks, P.D., Troch, P.A., Durcik, M., Gallo, E., Moravec, B., Schlegel, M., and Carlson, M. 2011. Predicting regional-scale ecosystem response to changes in precipitation: Not all rain is created equal, *Water Resources Research, 47*: W00J08.

Brown, C., Werick, W., Leger, W., and Fay, D. 2011. A decision-analytic approach to managing climate risks: Application to the Upper Great Lakes. *Journal of the American Water Resources Association, 47*(3): 524–534.

Carlsson, L. 2000. Policy networks as collective action, *Policy Studies Journal, 28*: 502–520.

Cash, D.W. and Moser, S.C. 2000. Linking global and local scales: Designing dynamic assessment and management processes, *Global Environmental Change, 10*:109–120.

Cash, D.W., Adger, W., Berkes, F., Garden, P., Lebel, L., Olsson, P., Pritchard, L., and Young, O. 2006. Scale and cross-scale dynamics: governance and information in a multi-level world, *Ecology and Society, 11*(2): 8.

Dettinger, M. 2011. Climate change, atmospheric rivers, and floods in California: A multi-model analysis of storm frequency and magnitude changes, *Journal of the American Water Resources Association, 47*(3): 514–523.

Dietz, T., Ostrom, E., and Stern, P.C. 2003. The struggle to govern the commons, *Science, 302*:1907–1912.

Dooge, J.C.I. 1986. Looking for hydrologic laws, *Water Resources Research, 22*(9), 46–58.

Dooge, J.C.I. 1988. Hydrology in perspective, *Hydrological Sciences Journal, 33*(1): 61–85.

Dooge, J.C.I. 1992. Sensitivity of runoff to climate change: A Hortonian approach, *Bulletin of the American Meteorological Society, 73*:2013–2024.

Dorcey, A.H.J. 1986. *Bargaining in the Governance of Pacific Coastal Resources: Research and Reform.* British Columbia: University of British Columbia Press.

Fogg, G.E. and LaBolle, E.M. 2006. Motivation of synthesis, with an example on groundwater quality sustainability, *Water Resources Research, 42*: W03S05.

Folke, C., Colding, J., and Berkes, F. 2003. Synthesis: building resilience and adaptive capacity in social-ecological systems, in F. Berkes, J. Colding and C. Folke (eds.) *Navigating Social-Ecological Systems: Building Resilience for Complexity and Change* (pp. 352–387). Cambridge: Cambridge University Press.

Folke, C., Hahn, T., Olsson, P., and Norberg, J. 2005. Adaptive governance of social-ecological systems, *Annual Review of Environmental Resources, 30*: 8.1–8.33.

Galloway, G.E. 2011. If stationarity is dead, what do we do now? *Journal of the American Water Resources Association, 47*(3):563–570.

Gunderson, L.H., and Holling, C.S. (eds). 2002. *Panarchy: Understanding Transformations in Human and Natural Systems.* Washington, DC: Island Press.

Gupta, V.K., et al. 2000. A framework for reassessment of basic research and educational priorities in hydrologic sciences, in *Report of a Hydrology Workshop, Albuquerque, NM, Jan. 31–Feb. 1, 1999, to the NSFGEO Directorate.* Albuquerque: CIRES.

Haas, P.M. 1990. *Saving the Mediterranean: The Politics of International Environmental Cooperation.* New York: Columbia University Press.

Harman, C.J., Troch, P.A., and Sivapalan, M. 2011a. Functional model of water balance variability at the catchment scale: 2. Elasticity of fast and slow runoff components to precipitation change in the continental United States, *Water Resources Research, 47*: W02523.

Harman, C.J., Rao, P.S.C., Basu, N.B., Kumar, P., and Sivapalan, M. 2011b. Climate, soil and vegetation controls on the temporal variability of vadose zone transport, *Water Resources Research, 47*: W00J13.

Harte, J. 2002. Toward a synthesis of Newtonian and Darwinian worldviews, *Physics Today, 55*: 29–34.

Hilborn, R., Orensanz, J.M., and Parma, A. 2005. Institutions, incentives and the future of fisheries, *Philosophical Transactions of the Royal Society Series B, 360*:47–57.

Hirsch, R. M. 2011. A perspective on nonstationarity and water management, *Journal of the American Water Resources Association, 47*(3): 436–446.

Hodgkins, G.A. and Dudley, R.W. 2006. Changes in the timing of winter-spring stream-flows in Eastern North America, 1913–2002. *Geophysical Research Letters, 33*: L06402.

Hooper, R.P. 2009. Toward an intellectual structure for hydrologic science, *Hydrological Processes, 23*: 353–355.

Hubbard, S. and Hornberger, G. 2006. Introduction to special section on Hydrologic Synthesis, *Water Resources Research, 42*: W03S01.

Hurst, H.E. 1951. Long term storage capacities of reservoirs, *Transactions of the American Society of Civil Engineers, 116*: 776–808.

Janssen, M.A. and Ostrom, E. 2006. Empirically based, agent-based models, *Ecology and Society, 11*(2): 37.

Jentoft, S. 1989. Fisheries co-management, *Marine Policy, 13*: 137–154.

Johannes, R.E. 1998. Government-supported, village-based management of marine resources in Vanuatu, *Ocean and Coastal Management, 40*:165–186.

Kauffmann, S.A., 1993. *The Origins of Order: Self-Organization and Selection in Evolution.* New York: Oxford University Press.

Kiang, J.E., Olsen, J.R., and Waskom, R.M. 2011. Introduction to the featured collection on "nonstationarity, hydrologic frequency analysis and water management," *Journal of the American Water Resources Association, 47*(3):433–435.

Kleidon, A. and Schymanski, S.J. 2008. Thermodynamics and optimality of the water budget on land: A review, *Geophysical Research Letters, 35*: L20404.

Koutsoyiannis, D. 2011. Hurst-Kolmogorov dynamics and uncertainty, *Journal of the American Water Resources Association, 47*(3):481–495.

Kumar, P. 2011. Typology of hydrologic predictability, *Water Resources Research, 47*: W00H05.

Kumar, P. and Ruddell, B.L. 2010. Information driven ecohydrologic self-oganization, *Entropy, 12*(10): 2085–2096.

Kundzewicz, Z.W. 2011. Nonstationarity in water resources–central European perspective, *Journal of the American Water Resources Association, 47*(3): 550–562.

Kurien, J. 1992. Ruining the commons and responses of the commoners: coastal overfishing and fishermen's actions in Kerala State, India, in D. Ghai and J. Vivian (eds.) *Grassroots Environmental Action: Peoples Participation in Sustainable Development* (pp. 221–258). London: Routledge.

Layzer, J.A. 2006. *The Environmental Case: Translating Values into Policy*, Second Edition. Washington D.C.: CQ Press.

Lebel, L., Garden, P., and Imamura, M. 2005. The politics of scale, position and place in the governance of water resources in the Mekong region, *Ecology and Society, 10*(2):18.

Lebel, L., Anderies, J.M., Campbell, B., Folke, C., Hatfield-Dodds, S., Hughes. T.P., and Wilson, J. 2006. Governance and the capacity to manage resilience in regional social-ecological systems, *Ecology and Society, 11*(1):19.

Lee, K.N. 1999. Appraising adaptive management, *Conservation Ecology, 3*(2):3.

Levin, S. A. 1999. *Fragile Dominion: Complexity and the Commons*. Reading, MA: Perseus Books.

Lins, H.F. and Cohn, T.A. 2011. Stationarity: Wanted dead or alive? *Journal of the American Water Resources Association, 47*(3): 475–480.

Lins, H.F. and Slack, J.R. 2005. Seasonal and regional characteristics of U.S. streamflow trends in the United States from 1940–1999, *Physical Geography, 26*: 489–501.

Liu, J., Dietz, T., Carpenter, S.R., Alberti, M., Folke, C., Moran, E., Pell, A.N., Deadman, P., Kratz, T., Lubchenco, J., Ostrom, E., Ouyang, Z., Provencher, W., Redman, C.L., Schneider, S.H., and Taylor, W.W. 2007. Complexity of coupled human and natural systems, *Science, 317*: 1513–1516.

Llewellyn, S. 2007. Introducing the agents, *Organization Studies*, *28*: 133–153.

Lobe, K., and Berkes, F. 2004. The padu system of community-based resource management: Change and local institutional innovation in South India, *Marine Policy*, *28*: 271–281.

McCabe, G.J. and Wolock, D.M. 2002. A step increase in streamflow in the coterminous United States, *Geophysical Research Letters*, *29*(24): 2185.

McDonnell, J. J. et al. 2007. Moving beyond heterogeneity and process complexity: A new vision for watershed hydrology, *Water Resources Research*, *43*: W07301.

McGinnis, M. D. (ed.). 2000. *Polycentric Games and Institutions*. Ann Arbor, MI: University of Michigan Press.

Means, E., Laugier, M., Daw, J., Kaatz, L.M., and Waage, M.D. 2010. *Decision Support Planning Methods: Incorporating Climate Change Uncertainties Into Water Planning*. http://www.wucaonline.org/assets/pdf/actions_whitepaper_012110.pdf (accessed April 2011).

Milly, P.C.D., Betancourt, J., Falkenmark, M., Hirsch, R.M., Kundzewicz, Z.W., Lettenmaier, D.P., and Stouffer, R.J. 2008. Stationarity is dead: Whither water management? *Science*, *319*: 573–574.

Mitchell, B. 2005. Integrated water resource management, institutional arrangements and land-use planning, *Environment and Planning A*, *37*: 1335–1352.

Mitchell, M. 2009. *Complexity: A Guided Tour*. Oxford: Oxford University Press.

Montanari, A. 2011. Uncertainty of hydrological predictions, in P.A. Wilderer (ed.) *Treatise on Water Science, vol. 2* (pp. 459–478). Amsterdam: Elsevier.

Morrison, T. 2004. *Institutional integration in complex environments: Pursuing rural sustainability at the regional level in Australia and the U.S.A.* Doctoral Thesis, School of Geographical Sciences and Planning, University of Queensland, Brisbane.

Orlove, B., and Caton, S. C. 2010. Water sustainability: Anthropological approaches and prospects. *Annual Review of Anthropology*, *39*: 401–415.

Ostrom, E. 1990. *Governing the Commons. The Evolution of Institutions for Collective Action*. Cambridge: Cambridge University Press.

Ostrom, E., Gardner, R., and Walker, J. 1994. *Rules, Games, and Common-Pool Resources*. Ann Arbor, MI: University of Michigan Press.

Ostrom, E., Burger, J., Field, C.B., Norgaard, R.B., and Policansky, D. 1999. Revisiting the commons: local lessons, global challenges, *Science*, *284*:278–282.

Ostrom, E., Dietz, T., Dolsak, N., Stern, P.C., Stonich, S., and Weber, E.U. (eds.). 2002. *The Drama of the Commons*. Washington, DC: National Academy Press.

Ouarda, T.B.M.J. and El-Adlouni, S. 2011. Bayesian nonstationary frequency analysis of hydrological variables, *Journal of the American Water Resources Association*, *47*(3): 496–505.

Pahl-Wostl, C. 2007. The implications of complexity for integrated resources management, *Environmental Modelling & Software*, *22*: 561–569.

Pahl-Wostl, C., Craps, M., Dewulf, A., Mostert, E., Tabara, D., and Taillieu, T. 2007. Social learning and water resources management, *Ecology and Society*, *12*(2): 5.

Paul, A. 2005. Rise, fall and persistence in Kadakkodi: an enquiry into the evolution of a community institution for fishery management in Kerala, India, *Environment and Development Economics*, *10*:33–51.

Pinkerton, E. (ed.). 1989. *Co-operative Management of Local Fisheries*. British Columbia: University of British Columbia Press.

Randeria, S. 2003. Glocalisation of law: Environmental justice, World Bank, NGOs and the cunning state in India, *Current Sociology*, *51*: 305–328.

Regional Environmental Centre (REC). 1998. *Doors to Democracy*. Szentendre, Hungary: REC.

Regional Environmental Centre (REC). 1999. *Healthy Decisions: Access to Information, Public Participation in Decision-Making and Access to Justice in Environment and Health Matters.* Szentendre, Hungary: REC.

Renn, O., Webler, T., and Wiedeman, P. (eds.). 1995. *Fairness and Competence in Citizen Participation: Evaluating Models for Environmental Discourse.* London: Kluwer Academic.

Rittel, H.W. and Webber, M.M. 1973. Dilemmas in a general theory of planning, *Policy Sciences, 4*: 155–169.

Rogers, P., MacDonnell, L., and Lydon, P. 2009. Political decision making: real decisions in real political contexts, in C.S. Russell and D.D. Baumann (eds.) *The Evolution of Water Resources Planning and Decision Making.* Cheltenham: Edward Elgar, IWR Maass-White Series.

Saleth, M.R. and Dinar, A. 2004. *The Institutional Economics of Water: A Cross-country Analysis of Institutions and Performance.* Cheltenham, UK: Edward Elgar and World Bank.

Saravanan, V.S. 2008. A systems approach to unravel complex water management institutions, *Ecological Complexity, 5*(3): 202–215.

Schaefli, B., Harman, C.J., Sivapalan, M., and Schymanski, S.J. 2011. Hydrologic predictions in a changing environment: Behavioral modeling, *Hydrology and Earth System Sciences, 15*, 635–646.

Sheer, D. 2010. Dysfunctional Water Management: Causes and Solutions. *Journal of Water Resources Planning and Management*, 136(1): 1–4.

Singh-Renton, S., Mahon, R., and McConney, P. 2003. Small Caribbean (CARICOM) states get involved in management of shared large pelagic species, *Marine Policy*, 27:39–46.

Sivapalan, M., Thompson, S.E., Hraman, C.J., Basu, N.B., and Kumar, P. 2011a. Water cycle dynamics in a changing environment: Improving predictability through synthesis, *Water Resources Research*, 47: W00J01.

Sivapalan, M., Yaeger, M.A., Harman, C.J., Xu, X., and Troch, P.A. 2011b. Functional model of water balance variability at the catchment scale: 1. Evidence of hydrologic similarity and space-time symmetry, *Water Resources Research*, 47: W02522.

Stacey, R.D. 1992. *Managing the Unknowable: Strategic Boundaries Between Order and Chaos in Organizations.* San Francisco, CA: The Jossey-Bass Management Series.

Stedinger, J.R. and Griffis, V.W. 2011. Getting from here to where? Flood frequency analysis and climate, *Journal of the American Water Resources Association*, 47(3): 506–513.

Stephens, W. and Hess, T. 1999. Systems approaches to water management research, *Agricultural Water Management*, 40: 3–13.

Stiglitz, J.E., Sen, A., and Fitoussi, J.-P. 2010. *Mis-measuring Our Lives: Why GDP Doesn't Add Up.* New York: New Press.

Stone, D. 2002. *Policy Paradox: The Art of Political Decision Making*, Revised Edition. New York: W.W. Norton and Company.

Strogatz, S.H. 1994. *Nonlinear Dynamics and Chaos, With Applications to Physics, Biology, Chemistry, and Engineering.* Reading, MA: Addison Wesley.

Thompson, S.E., et al. 2011a. Patterns, puzzles and people: Implementing hydrologic synthesis, *Hydrological Processes, 25*, 3256–3266.

Thompson, S.E., Harman, C.J., Troch, P.A., and Sivapalan, M. 2011b. Predicting evapotranspiration at multiple timescales: Comparative hydrology across AMERIFLUX sites, *Water Resources Research*, 47: W00J07.

Thompson, S.E., Basu, N.B., Lascurain Jr., J., Aubeneau, A., and Rao, P.S.C. 2011d. Hydrologic controls drive patterns of solute export in forested, mountainous watersheds, *Water Resources Research*, 47: W00J05.

Tompkins, EL and W.N. Adger. 2004. Does adaptive management of natural resources enhance resilience to climate change? Ecology and Society; 9(2): 10.

Torgersen, T. 2006. Observatories, think tanks, and community models in the hydrologic and environmental sciences: How does it affect me? *Water Resources Research, 42*: W06301.

Troch, P.A., Martinez, G.F., Pauwels, V.R.N., Durcik, M., Sivapalan, M., Harman, C., Brooks, P.D., Gupta, H., and Huxman, T. 2009. Climate and vegetation water use efficiency at catchment scales, *Hydrological Processes*, 23: 2409–2414.

Villarini, G., Smith, J.A., Baeck, M.L., and Krajewski, W.F. 2011. Examining flood frequency distributions in the Midwest U.S., *Journal of the American Water Resources Association*, 47(3):447–463.

Voepel, H., Ruddell, B.L., Schumer, R., Troch, P.A., Brooks, P.D., Neal, A., Durcik, M., and Sivapalan, M. 2011. Quantifying the role of climate and landscape characteristics on hydrologic partitioning and vegetation response, *Water Resources Research*, 47: W00J09.

Vogel, R.M., Yaindl, C., and Walter, M. 2011. Nonstationarity: Flood magnification and recurrence reduction factors in the United States, *Journal of the American Water Resources Association*, 47(3):464–474.

Waage, M.D. and Kaatz, L. 2011. Nonstationary water planning: an overview of several promising planning methods, *Journal of the American Water Resources Association*, 47(3):535–540.

Wagener, T., Sivapalan, M., Troch, P.A., McGlynn, B.L., Harman, C.J., Gupta, H.V., Kumar, P., Rao, P.S.C., Basu, N.B., and Wilson, J.S. 2010. The future of hydrology: An evolving science for a changing world, *Water Resources Research*, 46: W05301.

Wilson, J.S., Hermans, C., Sivapalan, M., and Vörösmarty, C.J. 2010. Blazing new paths for interdisciplinary hydrology, *Eos Transactions, American Geophysical Union*, 91(6): 53–54.

Young, O. 2002. *The Institutional Dimensions of Environmental Change: Fit, Interplay, and Scale.* Cambridge, MA: The MIT Press.

Young, O. 2006. Vertical interplay among scaledependent environmental and resource regimes, *Ecology and Society*, 11(1): 27.

Zehe, E. and Sivapalan, M. 2009. Threshold behavior in hydrological systems as (human) geo-ecosystems: Manifestations, controls and implications, *Hydrology and Earth System Sciences*, 13(7), 1273–1297.

5

A NON-ZERO-SUM APPROACH TO WATER NEGOTIATIONS

(with Peter Kamminga and Paola Cecchi-Dimeglio)

Historically, most negotiations over the use of water have taken a zero-sum approach, meaning that gains to one party have almost always been matched by losses to others (Bingham et al 1994; Tilmant et al 2007; Sgobbi and Carraro 2011). A zero-sum, or win–lose, approach implies that the parties have not succeeded in creating enough value so that the competing interests of the parties can be met simultaneously. If enough value is created, it is possible that the most important interests of all the parties can be met (Lax and Sebenius 1986; Arnold and Jewell 2003: 68). In this chapter, we review the dynamics of zero–sum water negotiations as well as the alternative: a non–zero–sum or informal problem-solving approach to water management.

In boundary-crossing water projects, zero–sum thinking has a self-fulfilling quality. That is, if water managers believe that supplies are limited, and that only some users will be able to use the water for what they want, then the potential "winners" act accordingly, and the result is inevitable. If the stakeholders in such situations set as their goal that they want to meet the objectives of almost all the relevant stakeholders simultaneously, can they always succeed? In theory, parties in water disputes ought to be able to find ingenious ways of using the same water in a variety of ways, or of helping each other reduce their demand for water, so that the interests of all parties can be met. There are instances in practice of such "joint gains" being achieved. For instance, the CALFED Bay-Delta program in California brought 25 federal, state, and non-governmental actors together to find a way to meet unmet water needs throughout the state. After decades of frustration, they were able to organize themselves in a way that led to a mutually advantageous outcome. Once they reframed their mission—shifting from a battle over who would make sacrifices to a search for ingenious ways of meeting multiple interests simultaneously—they were able to make progress. They created the Bay Delta Advisory Council to give stakeholders a venue to deliberate and a way to provide feedback on existing CALFED programs. They created subcommittees which included representatives of all the relevant constituencies and government agencies with related mandates. Using real-time water-use information, the participants were able to come up with ways of meeting both agricultural and urban interests that had eluded them previously (Fuller 2006; Innes and Booher 2010).

This is not necessarily easy to accomplish, even if the parties commit to a non-zero-sum approach to negotiation. The Danube River Basin Case in Europe is an example of a cross-boundary water negotiation that was unable to escape the trap of zero-sum thinking. The parties presumed they had no alternative but to choose between addressing economic development needs or environmental concerns. Nineteen countries share the Danube River Basin. For several decades, the stakeholders worked to formulate a legal basis for joint water management. In 1994, they produced the Danube River Protection Convention. The Convention focused solely on environmental questions at the expense of economic development concerns. As a result, a pre-existing economic plan for the river ended up in conflict with the environmental protection rules the same parties had negotiated. The 1994 Convention proposed activities directed almost entirely at floodplain restoration and protection of fish-spawning habitat. The previous agreement called for deepening and widening the riverbanks to enhance navigation, and thus allow the expansion of water-borne industry (Susskind and Ashcraft 2010: 60). The two issues were addressed separately instead of together, so it is not a surprise that opportunities for joint gain were missed. Rather than looking for ways of achieving both environmental and economic development objectives at the same time, the Danube negotiators presumed these interests were incompatible, and addressed them in sequence. Economic interests "won" in the first instance, while the environmentalists were "victorious" in the second. The results of the two sets of negotiations cancelled each other out as the parties failed to "trade across issues" as part of a value-creating strategy.

In an equally complex boundary-crossing water management case, the countries involved were able to meet multiple objectives simultaneously. Their success hinged on working out both new technological agreements and revising institutional arrangements. These negotiations go back to 1995 when the Mekong River Commission was established by Vietnam, Cambodia, Laos, and Thailand. Thailand's concern was that the other countries might try to veto its development plans and the related water allocations it had in mind. So, it suggested that each riparian should be allowed to use tributary waters within its territory without the approval of the others. At the same time, Vietnam, Cambodia, and Laos were very concerned about how they would maintain flow levels during the dry season. Vietnam suggested that any use of water from the river should be agreed upon by a joint technical committee before it could be implemented. They also made clear, though, that such consultations did not include a right to veto what other riparians had in mind. Ultimately, rules developed by the Commission provided assurances that all notifications of proposed uses would allow sufficient time before the onset of the dry season for all riparians to make appropriate adjustments. The parties also accepted the need for improved monitoring of intra-basin use and inter-basin diversions. These promises allowed them to reach an agreement acceptable to all the parties. A package was built that incorporated trades among the stakeholder groups emphasizing different locational priorities (tributary vs. mainstream), use priorities (inter- vs. intra-basin), and concerns about timing (wet vs. dry season). Thus, through an elaborate process of trading they generated a package that guaranteed each party that its top priority concern would be met. In the Danube case, because issues where treated separately, and no trades were possible, the parties failed to create value and resolve their differences. The Mekong case illustrates that for value creation to work, parties must cooperate, not just compete.

The Mekong River Commission countries also made the "common good" their concern, although there is no doubt that they were motivated primarily by self-interest (Innes and Connick 1999; Foster-Fishman et al 2001). To find the "common good," they did not need to substitute altruism for self-interest. Rather, they had only to realize that all sides would achieve a better outcome for themselves if each could find a "low-cost way" of meeting the interests of the others at the same time. This can only happen when parties in a complex water-management network trust each other enough to reveal their most important concerns, and if they work together to produce a package of trades in the spirit of reciprocity.

Basic rights, guaranteed by law, must also be taken into account in water negotiations, at whatever scale. Boundary-crossing agreements regarding the allocation of water must respect each governmental unit's right to use water, treat water, and release wastewater in ways guaranteed by law (Brooks and Trottier 2010: 110). On the other hand, while such agreements are guaranteed by law, they may still be difficult to implement. Resistance can best be avoided through informal problem-solving that encourages users and water network managers to move from zero-sum to non-zero-sum thinking (McKinney 1990; Scholz and Stiftel 2005; Fuller 2009; Kallis et al 2009).

In this chapter we describe how such problem-solving forums can be organized to increase the chances that value can and will be created, and that shared as well as separate interests of the parties will be met. It is our contention that opportunities to create value in boundary-crossing water disputes can best be achieved if the parties engage in joint fact-finding, formulate contingent agreements, and emphasize adaptive management. Such informal problem-solving, in our view, can work anywhere in the world, although appropriate place-specific modifications must be made in each setting (Boswell 2005; Scholz & Stiffel 2005; Yu 2008; Abukhater 2009; Kock 2010). We highlight the important role that trained mediators or facilitators (and we use these terms interchangeably) can play in assisting water network managers in shifting from competitive, zero-sum thinking to cooperative efforts to create value

More specifically, we address the following questions:

• What are the special attributes of boundary-crossing water negotiations that sometimes make value-creation difficult?

- If the zero-sum approach to negotiation often leads to sub-optimal outcomes, why do so many water professionals continue to rely on it?
- What are the keys to creating value in informal problem-solving forums?
- What are the most important tactics or techniques for generating mutually advantageous results through value creation?
- What is the added value of involving trained intermediaries in informal problem-solving efforts in the water sector?

From Competition to Value Creation

Reaching sustainable and equitable water agreements is not easy. In many places around the world, water is still viewed as a scarce rather than a flexible resource. Fear about water scarcity makes value creation difficult. In some of the cases described in the readings appended to this chapter, water disputes arise when: (1) a downstream party is worried that it won't have enough water during the dry season, or that it will experience flooding during the rainy season; (2) parties have opposing perspectives regarding the emphasis that should be attached to ecological, economic, or equity considerations in the allocation or use of water; (3) changes in the economic or demographic situation have led to rapidly growing demand for water; (4) existing water allocation agreements appear to be causing water scarcity for some, resulting in substantial price increases; (5) border disputes arise between government entities or parties that favor one use (e.g., agriculture) as compared to another (e.g., tourism); (6) threats to fishing downstream arise when pollution upstream creates water quality problems; and (7) political entities each feel they have the authority to set new rules regarding the use or re-use of water supplies.

In almost all these situations, water users think that their interests can only be met if some other group is denied what it wants. Deciding who gets which water, and for what reasons, is always a challenge. But, such challenges can either be framed as a choice about who wins or loses, or can be recast as problems that need to be solved jointly (Fuller 2006; Brooks and Trottier 2010).

There are almost always multiple stakeholders involved (directly or indirectly) in water management disputes. Sometimes, vast numbers of people—with what appear to be diametrically opposed interests—are affected by proposed changes in water quality standards or water allocation rules. Private parties, including agricultural, tourist or energy industries, see themselves competing for a share of a limited "water-pie." Even within the same sector, such as agriculture, there may be a dispute between two or more kinds of crop growers. The question remains, if particular water resources are used to meet the objectives of one group, does that mean these same resources cannot be used simultaneously to meet the interests of others?

The Green Acres Project in Orange County and Irvine Ranch Water District in Southern California is one instance in which the conflicting interests were resolved through a joint decision to try a new technology. In this region, sewage water is now recycled and added to drinking (sweet water) supplies. It is also used for landscape irrigation (parks, schools, and golf courses), food crop irrigation, and office cooling. Wastewater undergoes filtration and disinfection in addition to secondary treatment. This was made possible by using a new nano-filtration membrane that selectively removes organic material by molecular weight (Elkind 2011). So, reaching agreement on the use of a new technology can hold the key to meeting conflicting interests simultaneously.

Conflicts over water allocation can shift pretty regularly. Natural fluctuations in the available quantity of water, or shifting demand caused by changes in population levels or economic growth, can alter the nature of an allocation dispute or create a new one (McCarthy et al 2001; Guan and Hubacek 2007). Parties may be working in harmony at one moment, only to find themselves in opposition at another. Also, the focus of water disputes can evolve in response to new scientific or technical findings, or as a by-product of shifting ecological conditions (Guan and Hubacek 2007; Cascão 2009: 250–264; Brooks and Trottier 2010: 115).

A project in Southern Colorado illustrates how parties can come together to face an emerging challenge. One group of farmers was legally entitled to all the water it needed to grow crops; however, the way they produced their crops was causing increasing levels of salinity in a nearby river. At the same time, the overall amount of water available in the river was being reduced by reductions in the snow-melt. The River Basin Commission had to decide whether to invest in expensive water treatment facilities to deal with the salination problem, or to restrict the crop choices of the farmers. Neither option was appealing. In the end, the Commission was able to generate agreement among the many water users by implementing a trading program linked to a pricing strategy. They collected fines from agents who were polluting the river. This money was redistributed to agents not exceeding carefully specified pollution levels. Also, allowance holders were able to trade excess allowances with others, or purchase additional allowances to avoid violating maximum allowable discharges. The trading system produced greater efficiency in water use by enforcing quotas and distributing benefits (Kock 2010). The Commission avoided having to build costly water management infrastructure. The salination problem was solved. In a sense, more water was created through the efficiency achieved by a trading system.

When water disputes involve sovereign nations, negotiations aimed at creating value are more complicated. Diverse national interests, cultural imperatives, and internal political demands usually lead to efforts to protect sovereignty at all costs. Water resource allocation problems within or between countries occur episodically, regardless of whether or not they are handled effectively when they first arise (Wolf 1995; Megdal 2007). Subsequent negotiations can become increasingly difficult when stakeholders have a history of past negotiations that deadlocked. Perceived injustices can also create some of the most vexing difficulties in later problem-solving efforts (Furlong and Gleditsch 2003; Dixit and Gyawali 2010). Competition for water at every level and in every domain will inevitably lead to conflict. Water managers and stakeholders can approach such conflicts using traditional hard-bargaining techniques, or they can adopt a non-zero-sum approach that emphasizes problem-solving and value creation through technology innovation, reframing, and trades of all kinds.

Mediating Conflicting Claims

Generally, approaches to water management fall into two categories: a higher authority can *enforce* a decision, or the parties themselves can *negotiate* a settlement that formal authorities help to implement.

The option of assigning the authority to impose decisions on disputing parties to agencies, at national or international levels, has become increasingly unpopular (Jansky and Uitto 2005; Earle and Malzbender 2006). In theory, it might seem more efficient to have a "higher" authority impose a decision, but in practice reaching and, even more importantly, implementing such decisions is problematic. As soon as one party feels that its interests have not been addressed, it will search for a way to thwart implementation of whatever decision is imposed. The case of the restoration of South Florida's Everglades is a good example. Federal efforts to manage the Everglades left out key water users, including the sugar industry and environmental groups. These groups subsequently sued to block implementation of the government's plan for the ecosystem (Kiker et al 2001). The sugar industry felt the government was trying to hold it responsible for polluting the Everglades National Park. High levels of phosphorus—that were adversely affecting water quality—were found near the sugarcane fields. The reasoning of the state seemed to be that sugar companies should bear most of the cost of restoration since they caused the bulk of the pollution. At the same time, environmental groups argued that the State of Florida was responsible for setting and enforcing water quality standards in the Everglades National Park and that the state had failed to enforce water quality laws (Vileisis 1997: 343).

Litigation followed. It did not, however, create a forum in which value could be created (i.e., in which the interests of all sides could be met simultaneously through trades or inventions of various kinds). Efforts to settle the lawsuit led to a minimally acceptable outcome. The federal government bought large portions of the Everglades that had previously been used for sugar production. This provided compensation to the sugar industry while underwriting the cost of ecosystem restoration favored by the environmentalists. The resulting settlement, announced by Bruce Babbitt, U.S. Secretary of the Interior, required the agricultural industry to reduce its phosphorus concentrations, first to an intermediate level and then to a final limit five years later. Growers were allowed an extra ten years to meet previously-established standards, and were required to pay only a small fraction of the total cost. The settlement negotiations did not bring all the parties face-to-face in a problem-solving forum. Instead, their attorneys engaged in hard bargaining and reached a minimally acceptable agreement. Recalcitrant parties were not required to share information or invent new ways of working together to accomplish more than they could have on their own.

National governments usually have sufficient legal authority to impose decisions in regional or sub-national water disputes. Even if there is substantial political pushback, the government usually has the authority required to proceed. When water disputes cross national borders, negotiated solutions are the only option since sovereign nations cannot be forced to accept terms to which they object. There is no higher authority that can tell them what to do. In such situations, negotiated agreements either respond to the interests of all the agencies or political entities involved, or the goal of achieving "compliance without enforcement" cannot be achieved (Chayes and Chayes 1991, 1993).

Efforts to negotiate boundary-crossing water agreements (both within countries and between countries) often fail because those affected are unable to find a way to satisfy all the parties. For instance, as mentioned earlier, the Danube Basin countries failed to reach agreement on how the river should be used (Boswell 2005; Fuller 2006). In the Indus Basin, Pakistan and India are now hoping that arbitration will settle the question of whether the diversion of water from one tributary to another is allowed under the Indus Water Treaty (Mustafa 2002; Briscoe 2009, 2010). The parties have not been able to work this out on their own. The threat of arbitration, even though it is prescribed in a treaty that the two countries signed, is unlikely to settle things. It is more likely that negotiations will drag on or lead to an irresolvable impasse (Bingham et al 1994; Mustafa 2002). Unless and until the most important interests of all the parties are met, they are unlikely to agree to have a solution imposed on them.

One way of handling such situations is to involve a facilitator or mediator trained in guiding parties through a consensus-building process. Given the complexity of water disputes, and the difficulties the parties are likely to have formulating lasting agreements, a neutral facilitator or mediator can help in a number of ways. A mediator's role, initially, is to help the parties formulate a settlement process and then keep them on track. Subsequently, it may be to manage joint fact-finding, data gathering or other types of informal problem-solving. It might also include assisting the parties "away from the table" in dealing with internal conflicts within their own constituencies (Cash et al 2003). We have more to say about the selection of mediators below.

The Dangers of Zero-Sum Thinking

For negotiations to lead to sustainable solutions in the water sector, or in any other political domain, the parties must execute a well-designed problem-solving or negotiation process (Innes and Booher 2010). It is difficult, however, given the complexity of most water management situations to know how to design an appropriate problem-solving process (Susskind 1999; Margerum 2008). To build a sufficient basis for the negotiation, it is essential to start with pre-negotiation efforts to ensure that the right parties are at the table, are well prepared, know what their internal stakeholders are thinking, and specify a timetable, ground rules, agenda, and joint fact-finding procedures before they begin (Susskind 1999; Wolf 2002; Fuller 2006; Bingham and O'Leary 2008).

In practice, parties in water management conflicts often have very different levels of negotiation readiness. As a result, the least experienced and least confident parties are most likely to engage in "hard bargaining" since non–zero-sum negotiation requires greater preparation (Schelling 1960; Fisher et al 1981). In zero-sum negotiation the parties make exaggerated demands, because they know they will have to make subsequent concessions. They pursue only their own concerns and not the interests of their negotiating partners (Lax and Sebenius 1986). This is likely to happen in water negotiations when some of the negotiators just don't have sufficient experience to engage in joint problem-solving.

In hard bargaining, each party's primary objective is to claim the largest share of whatever limited water resources are available (and certainly more than the others). To achieve a result in these situations, parties deploy bargaining tactics typical of the bazaar, including deception, ultimatums, and threats (Lewicki et al 2003; Daoudy 2005). Underlying hard bargaining, of course, is a presumption that there is not enough water to go around. This zero-sum view presumes that water can only be used once, and only by one party. Not surprisingly, it yields outcomes in which one side "wins" and others "lose". For instance, a solution may favor only environmental protection (e.g., less pollution) or economic development (e.g., more water for industrial uses that create waste). This is what happened in the Piave River Basin simulation described in the published excerpt appended to this chapter (Sgobbi and Carraro 2011). The parties assumed they were in a zero-sum negotiating situation, and that water could be used either to support tourism or to enable agriculturalists to grow crops. The rules of the simulation did not permit the parties to explore the possibility of reframing their options in something other than zero-sum terms (Sgobbi and Carraro 2011). When there is no higher authority to enforce a fair decision, the distribution of gains and losses tends to favor those with the most bargaining power. Settlement negotiations in such situations are about which party can force the others to yield (Fisher 1983).

The results of hard-bargaining negotiations are generally not sustainable (Bernauer 2002). Unless agreements are viewed as fair (by those affected), efficient (by those who have to pay for them), and wise (by those with the expertise to judge), one or more parties, even if they reluctantly sign an agreement, will look for opportunities to reopen negotiations or to "get even" later (Susskind and Cruikshank 1987). Assuming that parties have to work together to implement water agreements—and that a water "war" is not the preferred outcome for anyone—the best result from a zero-sum perspective is compromise: a deal that guarantees most of the parties enough to keep them from pushing for more. Frequently, this means that a winning majority gets most of the benefits, while weaker parties get very little (Biswas 1993; Zeitoun and Mirumachi 2008). In the aftermath of an asymmetric negotiation, weaker parties often try to delay or disrupt implementation. For example, they can try to form a new coalition with a moderately powerful party to defeat the winning coalition that left them out. Or, they can initiate a campaign to undermine public acceptance of an agreement, even though they signed it (feeling they had no choice at the time). Sometimes weaker groups can appeal to the principle of fairness in an effort to gain public sympathy. For many years, the government of Nepal felt powerless in the face of India's insistence that Nepal provide water for India's growing population. Any hesitation was met with threats of military action. Eventually, in an effort to defend itself against India's demands, Nepal sought China's support. Once it was clear that China would support Nepal, India had to recalculate its negotiation strategy. This is how countries with less power negotiate with their more powerful neighbors (Pokharel 1996).

The way negotiations are framed when hard-bargaining tactics are involved makes negotiations even more difficult than they need to be. Hard bargainers withhold information, try to make their negotiating partners uncomfortable (so they'll want to end the back-and-forth as quickly as possible), offer few concessions, and rebuff requests to cooperate, claiming that this would be tantamount to "giving up." It is not hard to see why hard bargaining gets in the way of creative problem-solving.

Hard bargaining undermines relationships. Concession trading tends to undermine trust (Deutsch 1973; Putnam and Wilson 1989; Ostrom 2003). When agreements are reached, but trust has been lost, implementation, to say nothing of subsequent negotiations, becomes more difficult. Short-term victories that are the product of stronger parties pushing weaker parties around, or threatening them, make every subsequent interaction among those same players increasingly difficult. There is ample documentation to show that the implementation of "agreements" that parties reached because one side was bullied into capitulating, is more time-consuming and more costly (Coase 1960; Raiffa 1982; Putnam 1988; Zeitoun and Warner 2006). One example is what happened in the Ganges-Brahmaputra-Meghna River Basin. The 1996 Ganges River Water Sharing Treaty between India and Nepal regulates the sharing of river waters at Farakka for flow augmentation during the rainy season (Nishat 2001; Datta 2005; Priscoli and Wolf 2009). The treaty sought to provide mutual benefits for the people of both countries, enabling them to handle flood management, irrigation, and river basin development. A joint committee was appointed to oversee implementation and operation. The agreement, however, was not perceived as equitable. Some critics argued that it provided an unfair advantage to India, the stronger of the two riparians (Haftendorn 2000). Lack of public involvement in the negotiations heightened public skepticism, with some charging that the treaty only took account of political and not technical considerations. Secrecy around the negotiations and the confidentiality of the data that were used probably had something to do with why the public believed that the treaty favored India (Abukhater 2009).

In general, opting for a hard-bargaining approach to water management negotiations means that some parties will be left out, some will not have adequate time to prepare, some will not be invited to help design the negotiation process, and agendas will be set too narrowly. Agreements reached in this way are brittle (Sadoff and Grey 2002). When subsequent adjustments or improvements are needed there is insufficient good will and inadequate trust for negotiations to be re-opened (Gelfand et al 2007).

An Alternative Problem-Solving Approach

A more cooperative approach to negotiating water management agreements—in which parties work towards mutually advantageous outcomes—is likely to be more effective. A key obstacle, though, is our tendency to approach negotiations with a zero-sum mindset and focus only on self-interest without paying sufficient attention to the concerns of others. A shift to a problem-solving approach in water negotiation is possible when there is sufficient awareness of the benefits of such an approach, an understanding of how informal problem-solving works, and the professional skills required to achieve value creation.

The key difference between a value-creating approach to negotiation and the traditional hard-bargaining approach is that parties invest time in "trying to make the pie as large as they can" before allocating gains and losses (Fisher et al 1981). Creating value means inventing solutions that are substantially better for all sides than that which is likely to be left by not reaching an agreement. Value creation begins with efforts to increase the number of possible options or enrich various "packages" (Raiffa 1982). By taking this approach, parties can move away from the zero-sum assumption that there is a fixed amount of water that has to be allocated to one side or the other.

Here's a simple example of a value-creating approach to resolving what otherwise might have been a highly limited or constrained water allocation problem. All the riparians in this hypothetical case agree to make water available to the downstream users while guarantying a stable price for "extra water" for an extended period of time to upstream parties. This ensures that the upstream users earn enough over time to meet their needs while guaranteeing downstream users the water they require in an emergency in the short term. By emphasizing pricing and long-term trades, they can guarantee that the short-term gains to one side do not impose long-term losses on the other.

Taking a value-creating approach requires shifting the mindsets of the parties (Dinar 2008). Instead of starting with extreme positions and moving through a period of haggling to a barely acceptable outcome, the parties need to begin by making sure they understand each other's priority interests. Moving from a focus on "positions" or demands, to "interests," means clarifying the things that are important to each side (in rank order), instead of inviting inflated and indefensible claims that will ultimately have to be abandoned (Fisher et al 1981). When parties succeed in shifting from positions to interests, they are on their way to value creation.

Value creation also involves making trades or packaging across multiple issues simultaneously. If there is only one issue up for discussion, it is very likely that parties will get stuck in a zero-sum format. If there are multiple issues up for grabs, rather than considering items one at a time, the parties can explore packages that offer each more of what they consider most important (Sheer et al 1992; Susskind and Ashcraft 2010: 71). If Alpha has something that Beta wants very much, and the issue is not crucial to Alpha, then Beta ought to be willing to trade something to

Alpha that Alpha really does want. Such trades do not constitute compromise; rather they represent the creation of value. Thus, Alpha may be willing to adopt new industrial processes that conserve water while allowing Beta to draw increasing amounts of water from the river each year, as long as Alpha is guaranteed whatever it needs at a capped price in times of drought.

If the parties know each other's interests, opportunities for successful trading increase. Countries may agree on an allocation that favors a downstream country over an upstream country in the short term, as long as the upstream country can rely on the promise that food produced downstream will be made available to upstream riparians at a guaranteed price in the future (Susskind and Ashcraft 2010: 61). Such trades can radically alter attitudes toward the short-term allocation of water supplies. Unfortunately, unless the right kind of problem-solving forum is created, such linked agreements are unlikely to emerge.

A value-creating approach to water negotiation also requires sharing rather than withholding information (Sadoff and Grey 2005). While hard bargaining is all about keeping information secret, or only revealing portions of it in line with the arguments one wants to make, enlarging the "water pie" is only possible when information is shared among stakeholders. This does not mean that parties have to share strategically sensitive information that might leave them vulnerable to exploitation (Raiffa 1982). And it does not mean that all sides will interpret information in the same way, but when information is used as a weapon rather than a problem-solving tool, the credibility of that information is likely to be lost, making value-creation more difficult.

A Non-Zero-Sum Approach

In situations when stakeholders must negotiate a complex set of issues, it is important to prepare for and conduct negotiations in ways that support value creation (Yu 2008).

Approaches to water management that allow for value creation tend to emphasize technological innovation (like desalination) (Burkhard et al 2000; El-Sadek 2010), multiple uses of the same water, as well as attention to what is called "embedded" water or virtual water (Guan and Hubacek 2007; Velázquez 2007). The notion of virtual water or embedded water was first raised in Israel where exporting water embedded in crops was not considered sustainable (Fishelson 1994). Embedded water is defined as: "water used to produce food crops that are traded internationally" (Allan 1994). The idea is that water supplies in water-scarce countries should be used as efficiently as possible: meaning not to produce the most water-intensive products such as cement or paper. Instead water should be used for more productive purposes (Allen 1998). So, in value-creation terms: importing water-intensive products from water rich countries makes more sense than importing them from water-scarce countries. Shifts in this direction can open up opportunities in water-scarce countries that have been shipping embedded water beyond their borders.

In northern China, the majority of the water is used for agriculture and industrial production. Exporting water-intensive agricultural and industrial products such as crops and textiles to other regions means sending water out of the region embedded in these products. To determine whether there are opportunities for value creation, one could calculate the amount of virtual water that flows from the north of China to other countries or regions. This can be done by taking the total amount of virtual water transported to other regions and subtracting water-intensive products and services imported into northern China. To calculate the effective export of freshwater, the rainwater embedded in agricultural products needs to be subtracted since this water is not readily available for other forms of economic production if it is not used to produce crops. Thus, the export of virtual water involves only irrigated water. From a water conservation point of view, China should import water-intensive products rather than produce them. A water-rich area should try to export water-intensive products while importing water non-intensive products such as electronic equipment and social services.

Convening Problem-Solving Forums

Informal problem-solving forums typically involve representatives of stakeholder groups chosen by the groups themselves. These representatives are brought together, often at the invitation of government leaders, not only to pursue their own interests but also to learn more about the needs of other groups as well as the needs of the region or country as a whole. Such forums have been used all over the world to give stakeholders substantial input into decisions that affect their lives (Martinez and Susskind 2000). Informal face-to-face problem-solving cannot substitute ad hoc decision-making for formal decision-making by government agencies. Rather, problem-solving forums provide input into government decision-making by generating proposals (that have near-unanimous support) for elected and appointed officials to consider. Final decisions must be left to those with the formal authority to make them. Nevertheless, when informal problem-solving forums adopt a consensus-seeking strategy, their impact on formal decision-making is likely to be substantial (Lund and Palmer 1997; Yu 2008). This is because most public officials are interested in knowing what they can do that will yield across-the-board political support from all segments of society.

An example of an effective problem-solving forum is the CALFED Bay Delta Program (CALFED) mentioned earlier. CALFED, a consortium of 12 state and 13 federal agencies, was created in 1994 to address water management and ecosystem restoration issues in the California's Bay Delta region. After several impasses, a shift in the negotiation format allowed the parties to reach agreement. Professional facilitators helped the parties segue to a new problem-solving approach that ultimately changed attitudes and behaviors. Over a period of 24 months, after years of struggle, the participants moved from hard bargaining to collaborative problem-solving. They managed to reach agreement on a wide range of issues, from water quality and supply to ecosystem restoration and governance. The facilitators relied on joint technical inquiries to generate a common vocabulary and policy options that had not been considered before. The stakeholders first discussed general goals (e.g., the restoration of particular kinds of habitats), then moved to quantifiable objectives aimed at measuring how well those goals were being met (e.g., the maximum amount of salt that would be allowed to enter a habitat). From there, they moved to intervention strategies for achieving quantifiable objectives on a case-by-case basis. Their final report focused much more on reasons and methods for proceeding in particular ways than their earlier work had. Also, the facilitators helped the parties maintain relationships with their constituencies. This was crucial for grounding agreements in the political realities of the situation and building trust (Innes and Booher 2010; Fuller 2009). Ultimately, the shift in strategy—away from earlier rounds of hard-bargaining focused almost entirely on which areas would get how much water—won the political support required.

In cross-boundary water negotiations, such forums may be hard to create, but they have been used successfully all over the world. The dialogue that produced a joint management and dam construction plan in the Komati River Basin (a shared water-course between South Africa, Swaziland, and Mozambique) is described in an excerpt appended to this chapter (Dlamini 2006).

For problem-solving forums to succeed, the right parties need to be represented at the table. They need to agree on a negotiation agenda that incorporates the most important items of concern to all the parties, design and implement a joint fact-finding process (see Figure 5.1), and formulate mutually beneficial agreements. For these things to happen, the participants must "own" the design of the process. In some instances, staff from a lead agency can canvass potential participants ahead of time to help suggest an agenda. Preferably, though, the parties will employ a professional neutral, a facilitator or mediator, to assist.

Most water problems involve numerous parties concerned about a variety of issues. The more parties and issues (i.e., domains and levels) involved, the more difficult the negotiations tend to be (Susskind and Crump 2009). So, to keep a process manageable, the number of stakeholder representatives involved needs to be limited. Convening stakeholder groups and involving them in selecting the spokespeople they prefer usually begins with a Stakeholder Assessment (Susskind et al 1999).

This is the first of the five steps in the Consensus Building Process (see Figure 5.2). The others are: clarifying the responsibilities of the parties; deliberating, with a focus on value creation; generating a written agreement supported by an overwhelming majority of the stakeholders; and implementing that agreement, usually by asking elected decision-makers with formal authority to act on the proposals generated by the group (Susskind et al 1999).

The way a stakeholder assessment works is that the convener makes an inventory of all the categories of stakeholders with concerns about whatever water management issue needs to be addressed. Then, the convener selects a facilitator to interview all the relevant individuals or groups in each category. Based on the responses, the facilitator makes a judgment about whether it makes sense to

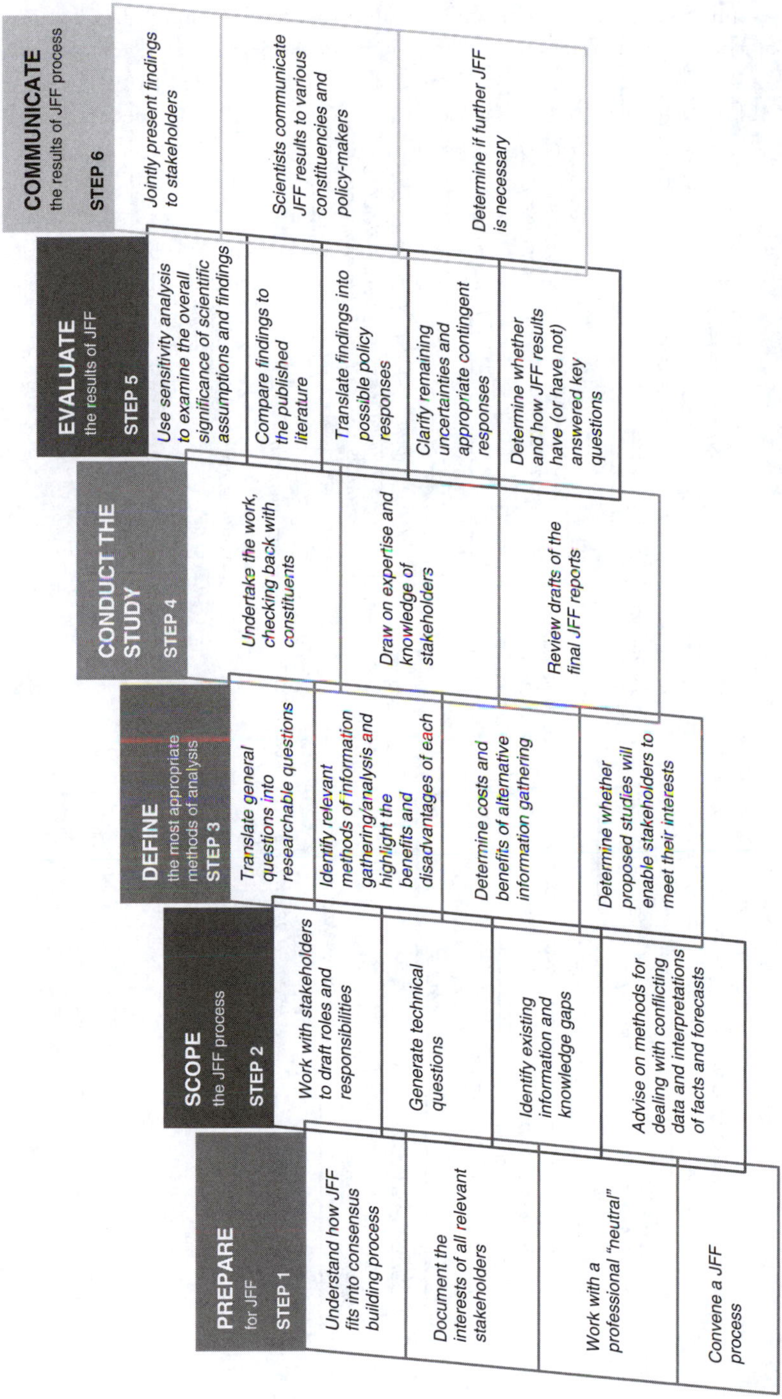

FIGURE 5.1 Joint fact-finding: Key steps in the process

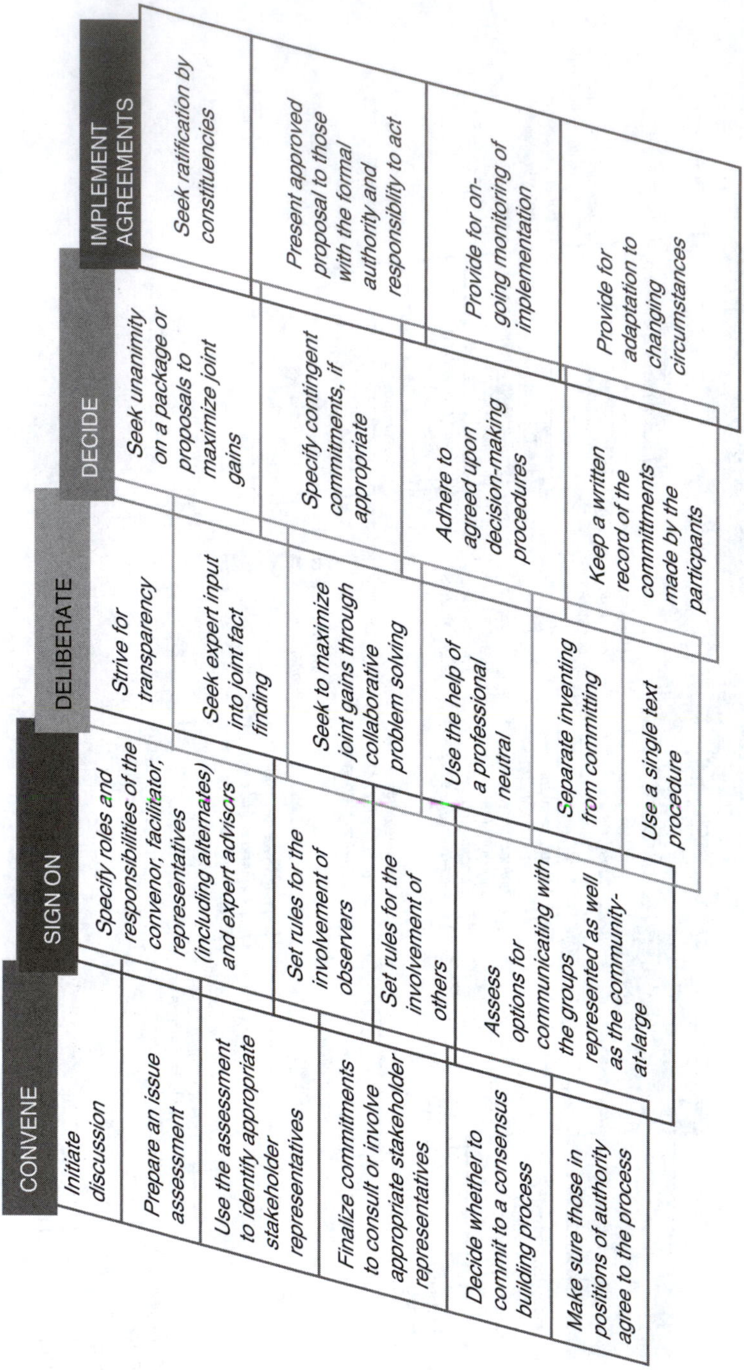

CONVENE	SIGN ON	DELIBERATE	DECIDE	IMPLEMENT AGREEMENTS
Initiate discussion	Specify roles and responsibilities of the convenor, facilitator, representatives (including alternates) and expert advisors	Strive for transparency	Seek unanimity on a package or proposals to maximize joint gains	Seek ratification by constituencies
Prepare an issue assessment	Set rules for the involvement of observers	Seek expert input into joint fact finding	Specify contingent commitments, if appropriate	Present approved proposal to those with the formal authority and responsibility to act
Use the assessment to identify appropriate stakeholder representatives	Set rules for the involvement of others	Seek to maximize joint gains through collaborative problem solving	Adhere to agreed upon decision-making procedures	Provide for on-going monitoring of implementation
Finalize commitments to consult or involve appropriate stakeholder representatives	Assess options for communicating with the groups represented as well as the community-at-large	Use the help of a professional neutral	Keep a written record of the commitments made by the participants	Provide for adaptation to changing circumstances
Decide whether to commit to a consensus building process		Separate inventing from committing		
Make sure those in positions of authority agree to the process		Use a single text procedure		

FIGURE 5.2 The consensus–building process

proceed given potential financial, institutional, and other constraints. If it does, the facilitator determines under what circumstances the stakeholders will agree to participate.

In the Komati River Basin case, the Water Authority was able to implement resource management measures (including the construction of several dams) by using a stakeholder assessment to identify appropriate participants and involve them in the decisions that needed to be made. The Authority asked the stakeholders to propose initiatives that would be both sustainable and politically plausible. The group set as a precondition that all affected groups must be better off with whatever projects they generated than they were prior to project implementation (Dlamini 2006).

Not only do stakeholder representatives have to be identified, but also experts who can ensure that the parties have the scientific, technical, and legal information they need. Generally, the facilitator talks to advocacy groups, community representatives, business leaders, and independent scientific experts. Based on the outcome of these conversations, the "neutral" proposes a work plan, timetable, ground rules, and budget. These are put in written form and sent to everyone interviewed to get their reactions and seek their endorsement. The strategy for joint fact-finding, and the identification of appropriate technical advisors is one of the products of the facilitator's conversations with all the parties.

In the CALFED process, the agencies decided to convene what became an Independent Review Panel of Agricultural Water Conservation Potential. This included technical advisors acceptable to both agricultural and environmental stakeholders. The Panel provided a framework that bridged different interests and views among the stakeholders (Fuller 2009).

Since the objective is to account for the full range of concerns that must be addressed, the participants should be expected to speak on behalf of the groups they represent (Susskind and Ashcraft 2010: 68–69).

Sometimes, it may be difficult to identify representatives who can speak for highly diffused or disorganized interests. For example, participants may feel it is appropriate to find a spokesperson for the interests of adjacent landowners who are not riparians. There probably will not be a standing organization of people or companies in this category. In the Mekong Basin negotiations, two of the six adjacent countries were not formally included (because they were not signatories to the treaty). But, the four parties at the table intended that the agreement be applicable to non-participating riparians. It would have been better to have identified surrogates or unofficial representatives of the missing countries. Given that the result of informal problem-solving is a proposal and not a binding decision, it is better to have proxies than no representatives at all (Susskind and Ashcraft 2010: 66).

Beyond preparing the assessment, a professional facilitator can assist a problem-solving forum in a number of ways as the parties try to decide whether and how to proceed. Furthermore, a neutral can be anyone who has no personal stake in the outcome and is acceptable to all the parties. Professional mediators, however, have training that allows them to interpret what potential stakeholders have to say. They are also adept at synthesizing the results of a great many confidential interviews. If necessary, a professional mediator can caucus a category of stakeholders to help them choose someone to speak for them and identify proxies or surrogates to speak for hard-to-represent interests. The facilitator's main responsibilities are outlined below:

Convening: Identifying Stakeholders and Generating a Commitment to Informal Problem-solving

- Meeting with potential stakeholders to hear their concerns and convince them that an informal problem-solving approach can work. Facilitators may use examples of situations where similar negotiations have produced better outcomes in the past.
- Explaining and seeking support for a set of ground rules.
- A facilitator can help parties work together even as they maintain a less-than-fully cooperative public stance to satisfy some of their constituents.
- A facilitator can be a link to the media for all the participants and speak on behalf of the process.

Clarifying Responsibilities

Once the parties are at the table, they need to agree on their roles and responsibilities. They also need to ratify the selection of a facilitator along with the agenda, work plan, budget, and joint fact-finding procedures. They need to decide who will prepare meeting summaries, maintain a web presence (if that is appropriate), moderate meetings, and make sure that all the group representatives keep their constituents informed about the progress of the forum.

The facilitator is usually the one who makes sure that all the parties understand how the process will unfold and enforces the ground rules, and may also help to build capacity in groups that have limited negotiation experience. In brief, in clarifying responsibilities a facilitator will:

- Develop discussion protocols and clarify the agenda.
- Remind the parties of the procedural commitments they have made.
- Keep parties on track and nudge the discussions if they become bogged down.
- Help parties shift from hard bargaining to value creation.
- Work with the group to propose and revise the agenda for each meeting.
- Enforce the ground rules. Parties often agree but then come unprepared, or lose track of the group's objectives.
- Make sure that the parties own the design of the process.

Sharing Information and Revealing Interests

This is the stage at which parties need to switch from being adversaries to becoming collaborative problem-solvers. It requires close attention to the needs and interests of all the other parties. Negotiation theory suggests that a substantial amount of time should be devoted to "inventing without committing" (Fisher et al 1981).

Checking the facts is wise in highly complex water management situations. A mechanism for collecting and sharing scientific or technical information is required. Together, the stakeholder representatives should decide what types of data they need, how data should be collected, what kinds of expert advisors would be helpful, and what possible interpretations of the data are plausible.

In the mid-1980s the Danube countries worked together at a technical level. In the Nepal dam case, pro and anti-dam proponents undertook a series of joint studies to assess Nepal's hydropower experience as an initial step to developing a set of country-specific guidelines based on the World Commission on Dam's report (Susskind and Ashcraft 2010).

In the Florida Everglades case, a neutral facilitator helped the group review the interests of each of the stakeholder groups involved. Then stakeholders decided what data they needed. They worked together to specify the forecasts they thought would be helpful (Fuller 2006). Local or indigenous knowledge from people living in the area was also gathered. This is the type of information that outside experts might miss entirely (Menkel-Meadow 2008).

Water-use planning in British Columbia, Canada, illustrates how information-sharing can lead to a fuller appreciation of interests, and ultimately to agreements that create value. Using full-group discussions as well as caucuses, the facilitator prepared a text for the group to review. A wide range of stakeholders including local residents, aboriginal representatives, and environmental leaders, worked to identify questions that needed to be addressed through joint fact-finding. Together they created a plan to examine claims, explore trade-offs, and search for mutually acceptable alternatives (Susskind and Ashcraft 2010). Web-based tools helped participants keep their constituencies informed.

This is also the stage at which like-minded stakeholders are likely to bond. Parties have a tendency to form coalitions if they think this will give them leverage in a negotiation. This gives them more influence when it is time to decide which package of options the group will support (Susskind and Crump 2009).

Coalitions can be either productive or counterproductive (Susskind and Crump 2009). Parties can coalesce in an effort to counteract the ideas that others, even the neutral, put forward. Or, they can serve as advocates of the winning proposal. Less politically powerful groups, in particular, need to pay attention to coalitional dynamics, especially if they are worried that others might be preparing to gang up on them. It is important to invest time in building coalitions early in a negotiation. This may involve thinking carefully about when to meet one-on-one with other parties, and trying to obtain solid commitments from others before committing to be part of an emerging coalition (Sebenius 1983).

In choosing coalition partners, it is important to be aware of what will happen if there is no agreement. In preparing for any water management negotiation, it is important to estimate one's best alternative to a negotiated agreement (BATNA) (Fisher et al 1981). If there are only two parties involved in a negotiation, there are only two BATNAs to consider in determining whether there is a likely zone of possible agreement (ZOPA) (Raiffa 1982). In most water negotiations, however, because there are usually multiple parties involved, figuring out whether there is a ZOPA is much more difficult. It requires putting oneself in the shoes of the numerous other parties, and gathering the information needed to make a multidimensional estimate of the overlap among the likely interests of all the stakeholders.

A professional facilitator is trained to sense when parties do not feel that their interests are being addressed. He or she may assist the stakeholder representatives in articulating their interests more clearly.

Deliberating

A facilitator will help the parties articulate their interests, undertake joint fact-finding, and engage in value creation:

- Persuade the parties to share information about their interests and brainstorm possible ways of meeting the interests of all the participants.
- Help to organize joint fact-finding.
- Prevent "group think" by encouraging the parties to speak their minds and press their views, even if they are unpopular.
- Suggest possible trades if the parties are stuck.
- Draft preliminary agreements.
- Help the parties consider what is being proposed in light of their BATNAs and the likely ZOPA, if there is one.

Helping the Parties Reach Agreement

Decision-making in complex water-negotiations may take many months, even years. The stakeholder groups involved typically set a long-term agenda as well as a timetable for regular meetings. Thus, they work towards a final meeting at which they can sign an agreed-upon text. It is important to leave enough time for parties to review all the possible trades invented along the way.

In the Mekong River Basin negotiations, the parties formulated a package that incorporated a variety of trades. These exploited differences in locational priorities, water use priorities, and concerns about timing. The participants also generated flexible agreements that would allow them to modify commitments as their interests shifted over time (Radosevich and Olson 1999).

When it is time to make decisions, it is important to seek unanimity but be willing to settle for overwhelming agreement on a package that will maximize joint gains, ensuring that no party is worse off than they would be had no agreement been reached. This usually includes contingent commitments as a way to deal with uncertainty or disagreements about the future (Susskind and Cruikshank 1987).

Once the parties produce a draft agreement, each participant is usually asked to check back with their constituents to ensure that they can, in fact, live with the terms of the emerging agreement. After such consultations have been completed, the parties come together one last time to ratify the agreement. Sometimes this involves a formal signing ceremony at which stakeholder representatives commit to supporting implementation of the agreement, as long as the relevant officials accept what has been proposed. When binding international agreements are involved, there are even more elaborate procedures required to ensure official ratification on all sides (Barrett 2003).

At this stage, parties sometimes spell out all the steps needed to implement the negotiated agreement. The objective is to design "nearly self-enforcing" agreements (Susskind 1994). To deal with political and scientific uncertainty, the parties may specify what will happen if various futures events occur. Final agreements should also include dispute resolution provisions indicating how parties who fear that something has gone wrong are expected to proceed.

During this stage, facilitators are expected to make sure that parties adhere to the ground rules. They may test the level of support or opposition to various packages that are on the table. The facilitator does so until the full text of an agreement is written down and representatives have something to show their constituents. This is done by making enquiries as to who cannot "live with" a package that seems to have widespread support. Those objecting are asked to suggest "improvements" that will make a package acceptable to them without leaving others worse off.

Deciding

Facilitators will help participants make choices, specify commitments, produce written accords. They will:

- Prepare a written summary of key points of agreement instead of allowing each party to write their own minutes, which can lead to confusing interpretations of what was agreed. The facilitator should produce a "single text." This will increase the chances of accurate reporting back to constituencies.
- Clarify the steps toward implementation.
- Make sure the parties are signing an agreement that spells out how their proposal will be linked to the formal actions needed to implement the negotiated agreement.

Holding the Parties to their Commitments

Stakeholder representatives must present their proposals to those with the formal authority to act. At this stage, it is also essential to specify how ongoing monitoring will be handled, and to provide for reconsideration of the agreement in light of changing circumstances.

Participants should make plans to stay in touch, often with the assistance of the facilitator. It is probably wise to agree on a schedule for the dissemination of regular reports about implementation progress. Periodic reviews will build participant confidence that others are meeting their obligations. In some cases, a permanent monitoring body may be proposed as part of a negotiated agreement. This would prescribe how reconvening will occur and spell out methods for reconsidering the agreement in light of new information.

In Nigeria, a Water Charter was developed for the Komadugu Yobe Basin using a participatory process that included plans for "ongoing communication, cooperation and coordination." The charter specifically addressed the responsibilities of the different stakeholders in implementing the agreement, as well as future mechanisms for cooperation among them. These mechanisms included regular meetings, details about strategies for cooperation and obligations for monitoring. It anticipated the need for dispute resolution, specifying that the signatories (the six Nigerian states and the Nigerian government) would try to resolve any differences amicably among themselves. Failing that, they agreed ahead of time to refer disputes to the National Council of States or the Supreme Court of Nigeria (Issa 2002).

Most efforts to negotiate water agreements end at this point, but in some situations ongoing oversight by the stakeholders may be necessary to ensure sustainable and equitable water management. Permanent advisory committees can meet on a continuing basis along the lines spelled out in an implementation plan (Margerum 2011). Such a committee can be empowered to recommend small changes in the agreement, in response, for example, to extreme seasonal changes that might result in a serious drop in water quantity.

In the Mekong agreement, parties anticipated the need for such ongoing negotiations. They were not satisfied with their agreement for dividing water resources unless it took account of changes over time. Thus, they created a process that enabled the parties to make proposals about how to handle changing conditions (i.e., rules for water utilization change if drought, flooding, or water surpluses occurred).

During this final step, the facilitator can be called upon to assist the stakeholder representatives in reporting back to their constituencies. It is a lot easier for a facilitator to explain how well a negotiator performed, than for that individual to make such a claim about themselves. Also, a facilitator will be in a better position to summarize in a completely dispassionate way the commitments that all the other parties have made.

Implement Agreements: the facilitator should ratify agreements, present them to officials with the power to act, put monitoring systems in place, and provide for adaptation:

- Make sure parties understand that any agreement will only be useful for a limited period.
- Suggest an adaptive management approach.

- Help parties build the institutional capacity required to monitor implementation.
- Make sure dispute resolution procedures are in place.
- Encourage the participants to think retrospectively about what they have learned.

Selecting a Neutral Facilitator

It should be clear by now that neutrals can play a key role in bringing consensus building efforts to a successful conclusion. When many agencies and stakeholder groups commit to working together to address water management problems, there is a chance that one or more will violate the most important principles of collaborative problem-solving. Being an intermediary in such situations is a difficult job. Individuals assigned to chair meetings often turn out to be unskilled in the techniques of facilitation. Many capable individuals have a hard time separating their own personal opinions about what ought to happen from the need for a discussion leader to take a strictly neutral stand. Moreover, if they actually have a stake in the outcome they really should not be the manager of the dialogue. Professional neutrals, on the other hand, bring no personal agenda to water negotiations. Group leaders can chair a session even as professional facilitators assist in managing problem-solving dialogue.

A key obstacle to engaging a professional neutral is that one or more public agencies may feel that agreeing to do so might be considered a sign of weakness or incompetence on their part. An initial suggestion to use a professional facilitator may be rebuffed by public officials who fear losing control. But, after experimenting with facilitation support, they often feel less threatened.

Another reason why some parties may be hesitant to involve a facilitator is that they are not used to dealing with anyone who claims to be neutral. How can they be sure a neutral is unbiased? Veto power is one way around this. Any participant who feels that a facilitator has forsaken their neutrality can ask the group to disqualify that individual. After giving the facilitator a chance to respond to such a charge, the group as a whole can dismiss that individual at any point during the process. In a sense, this gives all the participants a veto over the selection of the facilitator. Background and past affiliations of the intermediary are often fundamental considerations as well. The initial selection of the neutral should be reviewed and approved by all the participants the first time they meet as a group.

Neutrals have distinctive styles. In efforts to reach agreement on water management issues, the form of assistance provided is usually referred to as facilitation: helping to organize and manage the negotiation process in an even-handed way. But more "active" forms of assisted negotiation, including work with the parties away from the table, are called mediation. Mediation intensifies the involvement of the neutral without removing control over the outcome from the parties. Mediators have a greater role in confidential interactions among the parties compared to what facilitators do. One important function is carrying private messages among parties. Both facilitators and mediators can play this role. In either case, the neutral should have sufficient background in water disputes and be able to help the group secure additional expert advice as needed.

Sometimes there is poor chemistry between the neutral and some of the members of the group. It may be that the personal style of the neutral does not match the group's preferences. For example, the mediator may be too forceful or not forceful enough. Using an intermediary may not work because the parties have too little experience with informal problem-solving.

Important Lessons

1. Getting from hard bargaining to value creation in boundary-crossing water negotiations depends on the willingness of parties to use an approach that is focused on identifying interests, collecting all relevant information using joint fact-finding, taking time to brainstorm packages that meet the most important concerns of all the stakeholders, and relying on the process management skills of a professional neutral.

2. Many water professionals still rely on a zero-sum approach to negotiation. Involving more parties may seem unattractive because the process is likely to be more complex. Some fear the loss of authority implied by the invitation to other stakeholders to join in informal problem-solving. These same individuals often assume, mistakenly, that ad hoc assemblies will be given the power to make decisions when, in fact, they are being asked to produce proposals that must then be considered by those with the formal authority to make final decisions.

3. Using a process that enables joint problem-solving means involving all the stakeholders in the formulation of ground rules and holding them to these procedural mandates.

4. A professional facilitator should be better able to navigate the pushes and pulls of a complex multiparty negotiation than the parties operating on their own. Involving a facilitator early can help negotiators move through the five steps in the consensus building processes in a constructive and timely fashion.

Selected Readings with Commentaries

A. Sgobbi and C. Carraro "A Stochastic Multiple Players Multi-Issues Bargaining Model for the Piave River Basin," (2011)

Introduction

This article focuses on river and reservoir management in the Piave River Basin, located in the North East of Italy. The authors rely on game theory to analyze water allocation negotiations, seeking to predict how multiple parties involved in water negotiations are likely to react to uncertainty in a non-cooperative bargaining situation. Based on their mathematical efforts, the authors prescribe how public policy–makers ought to handle water allocations under various circumstances. However, they make a number of assumptions that depart from what happens in real boundary–crossing water negotiations.

We selected this article since it nicely illustrates the traditional zero–sum approach to water negotiations, and reflects what most people think of when they hear that "negotiation" is being suggested as a way to handle a water management dispute. It shows the limitations of thinking about water negotiations in zero–sum terms and assuming away many of the political and societal dynamics that are actually in play.

The Framework (p. 3)

In this framework, a finite number of players have to select a policy for sharing water resources from some collection of possible alternatives. If the players fail to reach an agreement by an exogenously specified deadline, a disagreement policy is imposed. The disagreement policy is known to all players: it could be an allocation that is enforced by a managing authority; it could be the loss of the possibility to benefit from even part of the negotiated variable; or it could be the continuation of the status quo, which is often characterised as inefficient. The constitution of the game as a finite horizon negotiation is empirically justifiable—as consultations over which policies to implement cannot continue forever, but policy makers have the power (if not the interest) to override stakeholders' positions and impose a policy if negotiators fail to agree. In finite horizon strategic negotiation models, it is unavoidable that "11th hour" effects play an important role in determining the equilibrium solution. In fact, last minute agreements are often reported in negotiation. In our model, unanimity is required for an agreement to be reached. This may seem restrictive in some cases, such as government formation, where simple or qualified majority rules may be more realistic. However, unanimity is empirically justifiable when no cooperation is the status quo, when there is no possibility for binding agreements, or when the enforcement of an agreement is problematic. Unanimity may also be appropriate when a compromise among different perspectives is sought. Finally, in our formulation of the game (part of) the players' utility is not known with certainty, as it depends on stochastic realisations of a negotiated variable. Players' strategies will then depend on the expected realisation of future states of the world.

The Background (pp. 4–7)

The potential usefulness of our multilateral, multiple players, stochastic bargaining model is illustrated in this section. The specific case study on which the model is tested is the Piave River Basin (PRB), which is among the five most important rivers in the North of Italy. Traditionally, in this area water management is primarily aimed at favouring irrigated agriculture and hydroelectric power production. However, the increase of other, non-consumptive, uses of water—such as recreation and tourism in the Dolomite valleys—and the rise in environmental awareness, coupled with variation in the water flow, have led to increasing conflict. Tensions over water management become fierce in the summer season, when the combination of dry months and peaks in demand often lead to local water scarcity situations. It is now widely accepted that the current exploitation regime for the Piave River Basin is not sustainable, as it has significant negative impacts on the river balances and ecological functioning, with consequent risks for the safety of local communities and economic activities (Franzin et al., 2000). Dalla Valle and Saccardo estimate an average water deficit of 3.4 million m^3 and, in dry years, this deficit can reach 75.5 million m^3, as happened in 1996 (Dalla Valle and Saccardo, 1996). The highly political and strategic nature of the problem has led to what is called "the battle of the Piave," with the problem of water resource exploitation at the centre of the debate, especially in relation to the requirements of the tourism industry in the Dolomite valleys, and the needs for agricultural water uptake downstream (Baruffi et al., 2002). The current situation with respect to water users and management plans in the Piave River Basin represents a good test case for the proposed model: the planning authority (the River Basin Authority of Alto Adriatico) intends to consider the interests and needs of all major stakeholders in planning for water use, yet its initial attempt has encountered their opposition.

Identification of the Issues, Players and their Utilities (p. 9)

The existing conflicts of interests among the Province of Belluno and the municipalities in the lower part of the basin on the one hand, and the agricultural water users in the Province of Treviso on the other, was singled out as the most important aspect of controversy of the Piave River. Irrigation needs condition the management of lakes and reservoirs to guarantee enough water for irrigation. As a consequence, much of the river flow in the middle part of the basin is reduced significantly for a large part of the year, and many of the river inlets, and the main river bed itself, are completely dry for long periods of time. On the other hand, upstream water use for tourism is important: with the current level of water abstraction permits, only the release of water stored in reservoirs can guarantee meeting the demand for water downstream, with important negative impacts on the socio-economic development of the upstream area, where the water reservoirs are located.

The downstream water users for irrigation are represented by the two major Land Reclamation and Irrigation Boards (LRB), the LRB of Destra Piave (LRB-Low), in the middle-lower part of the river basin and the LRB of Pedemontano Brentella (LRB-High), in the upper part. These are the institutions mandated with managing irrigation infrastructure and water distribution. The Province of Belluno (BELL), on the other hand, defends the interests of the mountain communities, related to the socio-economic development of the area (tourism industry) and the protection of the Dolomite National Park. The interest of the Province of Belluno is to maintain the reservoirs relatively full, given their landscape and tourist uses. Reservoir management, however, also affects the lower stretches of the river, where there are currently long periods of drought. Therefore, a municipality representative of all the municipalities in the lower part of the Piave River is also included (MUN), to defend environmental interests of the river downstream of Nervesa. In these conflicts, the role played by ENEL (an electricity company)—is important in determining water availability in the area. We only consider ENEL's water diversions at a specific point in the power production system directly linked to one of the major water reservoirs in the mountainous parts of the river. Virtually all the water diverted at this point is not returned to the Piave River, but transferred to the neighbouring Livenza River. Water use for power production at this point of the system can thus be effectively considered as consumptive use (about 40 m³/s). Finally, the River Basin Authority is not included explicitly as a player in the game. The Authority is interested in exploring the conflicts and potential solutions to tailor its management plan accordingly, thus identifying policies and allocation patterns which may represent a good compromise among competing water users. For our purposes, the River Basin Authority is assumed to establish the rules of the negotiation, as well as the default policy to be implemented in the case of no agreement.

Summary of Players and Preferences (p. 9)

TABLE 5.1 Summary of Players and Preferences

Actor	Player	Key preferences
Land Reclamation and Irrigation Board of Destra Piave	LRB-low	These are the key institutional players representing farmers' interests. This assumption is reasonable, since Land Reclamation and Irrigation Boards are actually composed of, and managed by, farmers in the area under the Boards' administration.
Land Reclamation and Irrigation Board of Pedemontano-Brentella	LRB-high	
Province of Belluno	BELL	This Province represents the interests of the mountain communities, with its elected representatives aiming at maintaining and further strengthening the tourism industry developed around lakes and reservoirs within their territory, as well as protecting the environmental values of the Dolomite Natural Park.
ENEL	ENEL	This hydroelectric power producing company takes part in the negotiation in as far as it manages the main reservoirs in the Piave River basin, as well as the network of channels and water diversions for hydroelectric power generation purposes.
Riverside municipalities	MUN	The riverside municipalities have a strong interest in ensuring an adequate river flow in the downstream of Nervesa. Their interests are partly aligned to those of the Province of Belluno, yet the maintenance of a minimum water flow in lower stretches may come at the expense of the mountainous reservoirs.

Findings and Analysis (p. 27)

The value added of the proposed approach to explore the water allocation problem lies in its ability to provide useful information to policy makers. In this section, we will explore the individual and social implications of two different water sharing rules, in the face of water scarcity. We compute the utilities of individual players and the overall welfare (in a utilitarian sense, this will be computed as the sum of players' utilities) under the two different sharing rules: fixed (downstream) and proportional allocation, when players account for uncertainty in water supply. A fixed rule allocates a fixed quantity of water to players, in an exogenously specified order: thus, the needs of the priority user are satisfied first, and the residual water is allocated to other uses. Fixed upstream distribution gives priority to the upstream users, while fixed downstream distribution prioritises downstream users. On the other hand, the proportional rule allocates a share of the resource to the users, which however does not need to be the same. We use our model to mimic a negotiation process in which the five players bargain over how to allocate a fixed quantity of resource ... We then assume, once an agreement is obtained, three realisations of water availability (very scarce, average, and abundant), and we compute the utilities that each player would derive from the equilibrium allocation agreement under the three water availability scenarios. For conditions of a severe and medium drought, the equilibrium quantities under the fixed sharing rule exceed the total quantity of water available. We therefore reduce exogenously the water quantities allocated to players, following current practices by the River Basin Authority and existing legislation (and as detailed in the River Basin Authority's Decision 4/2001). Key to our result is the assumption of fixed downstream allocation: it is in fact the case that, in cases of water shortage, all the water reservoirs are managed by ENEL in such a way as to ensure that downstream (agricultural) needs are satisfied. We thus assume that the Province of Belluno loses all of its allocation. Furthermore, we reduce the allocation to the downstream municipalities (the residual flow) to an emergency minimum water flow, defined by legislation for periods of water scarcity (ADB, 2001). The results indicate that, under fixed allocation, the players who have priority—namely two Land Reclamation Boards and ENEL—are able to obtain all their agreed allocation and, therefore, attain a high level of welfare. This is of course obviously embedded in the allocation rule, but it highlights the fairness implications of different water sharing rules. An insight from these results is interesting from a policy perspective: the utility gains enjoyed by the upstream water users are not sufficient to offset the loss of utility suffered by the Province of Belluno and the riverside municipalities. That is, overall welfare is higher, under uncertain water supply, under a proportional vs. fixed quantity allocation rule, as shown by the last two bars in the figure. These results reflect the fact that, under a proportional allocation, the risks of water shortage are shared equally among players. They are robust for a wide range of parameters for the underlying probability distribution, both in terms of changing means and spreads.

Commentary

This article illustrates why water negotiations become harder than they need to be when they are framed in zero-sum (win–lose) terms. It posits a static negotiating situation rather than a dynamic one that allows for value creation.

The authors assume that the parties will respond in predictable ways to uncertainty. They also assume that one party—the River Basin Authority (RBA)—will be empowered to unilaterally establish the ground rules for the negotiation and that its decision will be imposed if no agreement is reached. Both assumptions are unlikely in real life. If no agreement is reached by a certain date, it is more likely that nothing will happen.

Furthermore, all the parties are viewed as monolithic, with a single objective function characterizing what they want and what they value. In other words, the notion that parties are in fact complex, non-monolithic entities with internal factions pulling in different directions is not allowed in the model the authors have specified.

Moreover, in this analysis, the parties never meet face-to-face. They generate negotiation results by cycling through various "offers" made by one party at a time in the face of objective functions that the authors have set for each stakeholder group. No trades are ever discussed. In other words, there are no interpersonal relations among the players that might affect what the parties view as their options. The parties have no opportunity to reframe the problems they are dealing with, so no joint learning or adaptive management is possible.

Also, the parties are not allowed to form coalitions or consider contingent agreements, particularly in the face of uncertainty. The notion that more water can be created by changing operational rules or enhancing relations among the parties is not considered. In other words, they assume away the existence of a multifaceted water-management network that might engage in creative problem-solving.

In sum, this article illustrates how people respond to "wicked" water problems by assuming that a zero-sum bargaining approach is the only possible way to resolve water disputes. It shows why simplifying the dynamics of water negotiations not only overlooks key complexities, but also rules out opportunities to create value.

R. Burkhard, A. Deletic, and A. Craig "Techniques for Water and Wastewater Management: A Review of Techniques and Their Integration in Planning Review,"(2000)

Introduction

This article discusses various wastewater management techniques, and proposes alternative ways of integrating them into water planning. In addition, the authors illustrate how technologies for rainwater management, domestic wastewater management, and water and waste re-use can "change the equation" in water negotiations.

The article is of particular relevance to us because the authors highlight the importance of new technology as a tradable commodity in water negotiations. They illustrate how using new techniques might serve a variety of interests by encouraging greater energy efficiency, creating new jobs, and shaping social behavior. Thus, new technology can "open the door" to value creation when stakeholders consider gains and losses associated with using new technologies and techniques to deal with water scarcity.

Discussion and Conclusions (pp. 217–218)

6.1. Techniques

With respect to planning for water and wastewater management, there is a wealth of techniques available and yet most of them are hardly known by the main-stream engineer. In civil engineering undergraduate courses in higher education, ecological treatment systems are not included in water modules and students study conventional techniques. Only in specialised courses, do students have a chance to learn about new techniques. This fact makes the application of ecologically sound solutions more difficult. Similar observations can be made for other people who influence water and wastewater management, namely planners and site developers. Their attention is slowly drawn to these issues, but there is still a long way to go until there is greater awareness of water problems. That said, the UK Royal Academy of Engineering has recently decided to place "sustainable design" at the heart of its courses, to address social and environmental problems (THES, 1999). Similarly, a course was introduced almost three decades ago by the University of Connecticut's Civil Engineering Department (Laak, 1982) taking the above aspects into consideration.

6.2. Efficiency

Most ecological techniques match the performance of the well-established techniques. In some cases, they require more land, which can become a financial issue. Looking at the efficiency factor from a natural sustainability angle, the ecological techniques tend to fare better because they are designed to use less embodied energy in both building and running processes. Often, ecological techniques enhance the environment, create new habitats, replenish groundwater and can have a social function. Hence the efficiency aspect could easily be expanded into these fields and may have to be considered.

6.3. Economic Aspects

Further investigations into cost and life cycle analysis regarding wastewater management options are necessary. This is the case for conventional treatment systems where costs are not entirely broken down. This is also the case for novel techniques where costs are often difficult to obtain. The construction costs may change considerably over the next few years as the implications of the greenhouse effects may start to be felt. This may result in rising prices for construction and transportation, in case carbon dioxide taxes on fossil fuels are levied in future. Hence construction of wastewater treatment works, which consume a large amount of embodied energy, may turn out to be more costly than at present and this may tip the balance to less energy intensive ways of treating wastewater or even reducing domestic wastewater.

Some techniques may even generate revenue and jobs. Looking at the economic issue from this angle, ecological techniques may be seen in a better light, especially when there is the chance to partially or wholly substitute non-renewable energy and fertilisers. Jobs may be generated in building and maintenance of these techniques (with money saved from investing in conventional techniques), in further development of techniques, and in the marketing of eventual by-products.

6.4. Social Aspects

Most conventional techniques have little or no influence on the behaviour of users. For each of the areas discussed in this paper (rainwater management, domestic wastewater, and water-and-waste re-use), the more ecological techniques have a greater influence on the end-user. While it is clear that ecological rainwater management techniques are less likely to have an impact than for example compost toilets (apart from the earlier mentioned safety fears), the relationship between awareness and acceptance is not clear. Social acceptance and social sustainability are two areas in which further research is needed to ascertain the relative importance of the different criteria for the various techniques. The theory of participatory democracy holds that "individuals and their institutions cannot be considered in isolation from one another" (Pateman, 1970). Thus, a diverse range of opinions should be accommodated and considered within any planning decision. These opinions should include those which advocate the use of innovative ecological techniques, and also the multitude of views held in the public sphere. Moreover, public opinion should not only be sought out, but an attempt at understanding it should also be made, rather than simply dismissing it as "irrational public opinion", as is so often the case. There is also a clear need to develop effective educational strategies to develop environmental awareness and action competence (having the understanding and the tools necessary for environmentally friendly behaviour) at all levels, be it at home, in a primary school, in higher education, in planning departments, or developers' offices. Such attempts are actively made in Los Angeles, where demonstration gardens were set up to show residents that it is possible to have a nice garden without using excess water (Harasick, 1990).

6.5. Planning

Alongside this need for greater public participation, consideration should also be given to the planning system as a whole if "holistic planning" is to be achieved.

Thus, land use planning and control legislation which are sensitive to the requirements of such pollution prevention techniques such as those discussed herein
will go a long way towards achieving sustainability, howsoever it is defined.

Planning of integrated water and waste management requires thoughtful
consideration, because of the complexity of the process. Asano (1991) listed
elements of water re-use planning. However, to integrate all the above issues,
a new planning approach has to be developed in the form of participatory
integrated assessment, as suggested by Schlumpf, Behringer, Durrenberger,
and Pahl-Wostl (1999). Smerdon et al. (1997) show ways to plan and implement decentralised water supplies and sewage. They assess the application of
ecological techniques to single units as well as clusters of houses, villages and
towns and recommend the most appropriate technique for each size. To integrate all the above techniques and aspects, it is important to develop a tool that
is capable of incorporating all aspects of water and wastewater management.

Commentary

Most professionals agree that value creation sounds like a good idea. But, when
confronted with first-hand opportunities to do so, they are usually too committed
to hard bargaining to make the switch. In reality, some form of value creation
can almost always be achieved. This means that water negotiators need to figure
out what science and engineering have to offer when the parties are locked in
narrowly-framed economic and political debates.

The authors sketch the factors that should probably be considered when
choosing among technological options. More specifically, they urge negotiators to
analyze the costs and benefits of reusing grey-water, harvesting rainwater, recycling
solid waste, and re-using liquid wastes. They stipulate the need for a balanced
response to competing technical, economic, environmental, and social goals,
while satisfying the interests of the various parties involved. They also call for a
more "holistic" multi-level approach, suggesting that increased public participation
will make these tasks easier, not harder. Parties may be surprised to find how differently they value various technological options. This is good, as it creates more
opportunity for value-creating trades.

The article does not talk about these new techniques in the value-creating
way we would suggest. Nevertheless, it does offer hints about value-creating opportunities. If water allocation and network management are seen through the lens of
an infinitely expandable pie rather than a fixed one, parties are more likely to be
able to create value.

A value creation strategy requires looking beyond just the financial costs associated
with new technological solutions. Rather than arguing that water re-use is too
expensive, we would suggest that the parties expand their discussion to focus on
a more complete range of benefits, including job creation and the expansion of
"virtual" water supplies.

A. El-Sadek "Water Desalination: An Imperative Measure for Water Security in Egypt," (2010)

Introduction

This paper sketches Egypt's policy regarding the use of desalination. Desalination techniques offer a way to cope with water scarcity and are applicable in a range of developing countries. The article offers an excellent illustration of how international cooperation can be fostered relative to the management of water resources.

The following excerpts present the challenges associated with producing large amounts of water through desalination to meet water demands for agriculture and tourism. These passages illustrate how desalination can compete with more traditional water management strategies.

Desalination and More Traditional Water Management Strategies (pp. 876–878)

1. Introduction

The national water policy in Egypt has three aims: 1) to promote water quality protection, 2) to use the available resources optimally and efficiently, and 3) to foster international cooperation to safeguard and increase the supply in the Nile Basin. The optimal use of available water resources in Egypt is targeted with a multi-facetted package of policies that range from encouraging technical solutions, such as the use of closed pipelines to transfer water to the new lands, to economic instruments, such as cost recovery through greater farmer participation ... Policies aim at an increased efficiency in irrigation, cost recovery of infrastructure, the use of water efficient crops (e.g. less rice and sugar cane), groundwater use, the reuse of agricultural drainage water and sewage water, the desalination of brackish water and the harvesting of rainfall and flash floods.

Challenges, however, still exist to produce desalinated water for relatively large communities, for their continuous growth, development, and health, and for modern efficient agriculture, at affordable costs. Two main directions survived the crucial evolution of desalination technology, namely evaporation and membrane techniques. Many countries are now considering desalination as an important source of water supply. The desalination experience in Egypt is relatively new compared to other countries, especially in the Gulf States, but its importance began to grow recently as conventional water resources become fully exhausted. The difference in cost between desalted water, or a blend with desalted water, and that of conventional supplies has narrowed substantially in the past 10 years, especially if withdrawing the diminishing conventional groundwater. The public generally has the misunderstanding that costs for desalting are never competitive which inhibits not only the realization of this alternative water supply but also the research and development in this field, particularly in developing countries. Therefore, this attitude has been changed in order to comply with the fast decline in Egypt's conventional water resources, and the remarkable advances in desalination technology ... The objective of this paper is to study and investigate water desalination as a solution for water scarcity in Egypt. Moreover, the present work demonstrates the significance of seawater desalination for national development in Egypt.

2. Demand and supply forecast

With regards to increasing water supply quantity and quality, Egypt's Integrated Water Resources Plan (IWRP) 2017 ... and water expert judgments have reached several primary means to increase water resources in Egypt which include:

- Improving water storage facilities in order to decrease water losses; e.g. by using artificial aquifer storage recovery.
- Minimizing water losses on the River Nile's banks

Achieving this goal requires committing to high levels of co-operation with neighboring NB countries (possibly through the Nile Basin Initiative—NBI—in the form of joint projects) with the aim of minimizing Nile water losses on the length of the river's banks. However, an increase in Egypt's share of the River Nile as a result of these water-saving projects is not expected in the near future due to the rapid increase in Nile Basin countries' populations and the highly unstable, volatile political situations that govern such countries at present, and which increases their demands of water from the River Nile and even threatens Egypt's current share.

- Advocating Water Conservation among the Public

This requires increasing public awareness on the importance of water conservation especially in rural areas, which are characterized by high levels of unawareness regarding water problems that Egypt faces today and in the future. It is likely that increasing awareness through public campaigns and direct interactions with water users, especially farmers, will have, to a great extent, an impact on the amount of water that could be saved. However, changing water practices requires more than increasing awareness and is a very time-consuming process.

- Maximum Prevention of Water Pollution

Measures to prevent water pollution aim at eliminating harmful substances from products and wastes, to ensure that the pollutant does not contaminate the environment. The emphasis on these measures is to be posed on the agricultural and industrial sectors, due to the high degree of pollution they cause while using water resources for their activities. Water pollution can be minimized in the future through the production and utilization of clean products, and the improvement of in-plant processes.

- Water Pricing

The recovery of water costs through the collection of water service fees, represents an important measure in promoting the rational use of water resources. However, having to pay for a resource that used to be provided free of charge will not be accepted easily, especially, given Egypt's deeply rooted concept of viewing water as a basic human right. Thus, measures to introduce cost-sharing have been included for agricultural, industrial and domestic water supply.

- Applying Desalination Technologies

Given that seawater is available in unlimited quantities in coastal areas, it is expected that desalination plants for drinking water and industrial use in areas— where no other cheaper resources are available—will be developed to meet increasing demand. However, if brackish water is available in sufficient amounts, it might be a more preferred source for desalination due to its lower levels of salinity …

3. Desalination profile in Egypt

Desalination can be used as a sustainable water resource for domestic use in many parts in Egypt, since it has about 2,400 km of shorelines on both the Red Sea and the Mediterranean Sea. Desalination is currently being practiced in the Red Sea coastal area to supply tourist villages and resorts with adequate domestic water, since the economic value of a unit of water in these areas is high enough to cover the desalination cost …

At present, Egypt is encouraging not only the public sector but also the private sector to apply modern desalination technologies. Historically, the application of modern desalination technologies in Egypt started with distillation followed by ED and ending with the use of Reverse Osmosis (RO). The great achievements in desalination technology worldwide have now moved the costs for desalting in many applications from being "expensive" to being "competitive."

Challenges to Desalination (pp. 879–880)

5.1. Private Desalination Plants Raise Debate

The debate continues over water-related trade and investment agreements. If private water companies are permitted to operate desalination plants along Egypt's coast, the state could lose the right to regulate those plants. According to experiences of other countries, once multinational water companies are allowed to operate desalination plants, they could try to use international law to bypass state and local regulations. It is imperative to get the legislature to enact comprehensive policy regarding desalination plants, including a requirement that they be operated by public agencies. The effects of desalination on marine life and ocean water quality represent its potential inducement for growth and the huge amount of energy it takes to operate a plant.

Sclerotic legal and institutional arrangements have put paid to attempts to involve the private sector in desalination projects. Plans listed in Desalination Markets 2005–2015 for a plant with a capacity of up to 227,300 m^3/d (50 MIGD) at Sharm El Sheikh have still not been realised. In the long term, this state of inertia may change as the cost of brackish water desalination falls below the cost of transporting and distributing treated water from the Nile to the Red Sea coast region. Costs are already comparable. In the short- and medium-term, it is likely that desalination will continue to be used for industrial and tourism projects. This will largely be related to private investments, making it difficult to forecast with any degree of accuracy Egypt's future desalination requirements.

5.2 Energy Policies Needed for Desalination

The coupling of renewable energy and desalination systems holds great promise for increasing water supplies in water scarce regions. An effective integration of these technologies will allow Egypt to address water shortage problems with a domestic energy source that does not produce air pollution or contribute to the global problem of climate change. Meanwhile the costs of desalination and renewable energy systems are steadily decreasing ... while fuel prices are rising and fuel supplies are decreasing. In addition, the desalination units powered by renewable energy systems are uniquely suited to provide water and electricity in remote areas where water and electricity infrastructure is currently lacking ...

Commentary

The author of this article clearly points out the challenges involved in "productizing" large amounts of water for agriculture and tourism at affordable prices. He points out the need, and Egypt's awareness of it, to foster international cooperation as part of any strategy for dealing with water management.

Even though he touches on the social and economic challenges of introducing water pricing, he does not highlight the fact that the decreasing cost of desalination provides new opportunities for value creation. He points out the importance of building the knowledge of different stakeholders involved in water conservation, pollution prevention, and integrated water-resource management. Unless all parties have a clear sense of the problems and opportunities involved, negotiations are likely to produce hard-to-implement results. A high level of cooperation with neighboring countries is required to increase the quantity and quality of water supplied through the introduction of new technologies or production methods. Boundary-crossing projects are a way forward, but joint fact-finding should precede all decisions. The author calls for joint projects, but says very little about how agreement might be reached on what to do and how to do it.

Desalination is gaining increasing attention around the world, but it poses political challenges, given the tension between privatization of desalination plant construction and the need for the government to maintain its right to regulate such projects. The author unduly narrows the options for cooperation and value creation by only imagining public agencies relying on "command–and–control" forms of market regulation.

D. Guan and K. Hubacek "Assessment of Regional Trade and Virtual Water flows in China," (2007)

Introduction

The authors of this article describe water scarcity in one part of China at the same time as there is abundance in other areas. They use the concept of virtual water to assess the amount of water that can be extracted from the environment to produce products of various kinds. These products are exported to other regions that have rich water-resources. The authors evaluate inter-regional trade patterns in China and their effect on water consumption and pollution in the form of "virtual water flows."

This excerpt illustrates how virtual water provides a way of looking at the use and allocation of water—in this case for agriculture and industrial purposes in China—that may generate opportunities to create value. The analysis of water trading between the North and South of China can be applied to a wide range of cross-boundary situations. Taking a virtual water perspective generates opportunities for more efficient water allocation when water is a (costly) production factor. This should be taken into account when looking for opportunities for value creation in all water negotiations.

Virtual Water (p. 160)

1. The 'Economic Miracle' and Virtual Water Flows

Due to considerable regional differences in water supply and demand, and the need to assess regional trade flows, it is necessary to model water consumption on a regional level. Therefore we divide China into eight hydro-economic regions to establish water accounts for each region based on watersheds and provincial level administrative boundaries (see Hubacek and Sun, 2001). In this paper, we calculate and analyze the virtual water flows for two of China's regions: North China, which is characterized as water scarce, and South China which is abundant of water resources. The dataset for this study consists of two categories: detailed economic data (input–output tables)—to investigate the flow of goods and services between producers and consumers and the linkages between all production sectors; and hydrological data—comprising four sub-categories: water availability, fresh water utilization and fresh water consumption coefficients and wastewater discharge coefficients for each of the economic sectors.

Interregional Virtual Water Flows in China (pp. 164–168)

3.1 Virtual Freshwater Flows

Based on our calculations we find that North China imported a number of water intensive products and services. For example, North China spent 35.89 billion Yuan to purchase extra electricity from other regions in 1997, which means a virtual import of 147.9 million cubic meters of water which is withdrawn and used up in production processes in other regions. Another example is agriculture: North China received 44.67 billion Yuan through the export of agricultural products, and with it 7339.3 million cubic meters of virtual water have been transported to other regions. However, we have to consider that much of the agricultural land is rainfed in North China, which produces about 42 percent of total agricultural outputs. The amount of rainwater embedded in agricultural products would not be readily available for any other economic production even if crops were not grown on this land. Therefore, the effective export of virtual water in the agricultural sector only consists of irrigated water, which is 4284.2 million cubic meters. Annually, 4545.0 million cubic meters of fresh water virtually flow out of North China (which is used in the production of exports) excluding rainwater in the agricultural production. On the other hand, the import of virtual water was only 319.6 million cubic meters, which reduces the net flow to other regions to 4225.4 million cubic meters. From a water conservation point of view, North China, characterized as water-scarce, should import water-intensive products rather than produce them. According to this analysis, North China used up more than 5 percent of its total water resources for producing exports to other regions, mainly through the trade of water-intensive commodities such as agricultural crops, processed food, textiles and chemical products. By contrast, Guangdong is endowed with rich water resources, but virtually imported 444.8 million m³ of freshwater, 79 percent of which are through the trade of water-intensive products (e.g. irrigated agricultural products). On the other hand, Guangdong exports relatively water non-intensive commodities such as electric equipment and many commercial and social services.

By summarizing the virtual freshwater flows of both North and South China, we find that the trade patterns are apparently inconsistent with our original hypothesis: water-scarce regions in China produce and export water-intensive products but import water non-intensive commodities. Meanwhile, water-abundant South China imports water-intensive goods. One of the possible explanations could be that water has not been recognized as an important factor of production in China's economy as there are very low costs associated with the utilization of water resources for most production processes. Another reason could lie in the fact that North China has suitable climatic condition, soil and land for many agricultural crops (Heilig et al., 2000). A third reason refers to the design of economic policies: Guangdong is subject to more favorable policies and better circumstances for investments in industry and services sectors than other regions. Since the economic reform in 1978, many locations in South China (including Guangdong) have been established as "Special Economic Development Zones," which brought many commercial opportunities and triggered a regional economic boom. This is also reflected in changing water consumption patterns. These economic incentives led to a restructuring of the regional economy to higher value added products with relatively lower levels of resource inputs. Thus Guangdong imports and exports of virtual water reflect the economic structure of the more developed special economic zones. On the other hand, North China has a relatively lower economic growth rate and stronger focus on low value added and high water intensive production without these special policies.

If we consider multiple factors relevant for the existing production and trade structure such as environmental endowment (e.g. soil quality), land prices and other socio-economic or political factors we see that North China has a "comparative advantage" for producing and exporting agricultural products. In terms of water conservation it is important to effectively balance these factors. North China may sustain the export of rain-fed agricultural goods as rainwater cannot be effectively used by other production sectors. On the other hand, North China might want to reconsider the level of exports of irrigated agricultural products in order to make the scarce water resources (e.g. surface or ground water) available for other purposes which can contribute more to the economy and society in terms of value added and jobs.

From a water efficiency point of view, North China with limited water resources, should produce and export the commodities which have high value added per unit of water. North China has a comparative advantage in the production sectors of coal mining and processing, production of sawmill and furniture, machinery equipment, and many service sectors. Meanwhile, Guangdong has the advantage on producing agriculture, textiles, and metal products. Obviously this statement needs to be qualified by looking at other factors such as the availability of skilled labor and other essential factors of production, but the focus on water can provide a useful starting point.

Commentary

The authors take social-economic and environmental factors into account in their analysis of agricultural and industrial uses of water. Opportunities to create value emerge when the allocation of water is considered from a national perspective (i.e., a shortage in the North can be addressed with rich resources in the South). The main problem rests in the contradiction between export-oriented policies and water-saving policies. The tension between them makes it hard to find sustainable water agreements.

The authors fail to see trading opportunities that go beyond building infrastructure to transport water. Trade agreements involving virtual water don't necessarily require infrastructure. Virtual water flow accounting can give a boost to non-zero-sum approaches to water negotiation.

The authors also fail to identify opportunities to find negotiated solutions by involving stakeholders directly. In our view, if this is not done, important local or indigenous knowledge will be lost. A professional facilitator may well be the key to organizing and conducting discussions of the role of virtual water trades in traditional allocation and water quality disputes.

E. Velázquez "Water Trade in Andalusia, Virtual Water:
An Alternative Way to Manage Water Use," (2007)

Introduction

The author of this article describes the water shortage in Andalusia. She points out that water is not only a resource, but also part of the social and economic fabric of the region. This excerpt describes the social–economic, agricultural and industrial uses of virtual water. Virtual water is a means of reducing the pressure on water resources. It needs to be discussed as both a social and economic asset.

Water Scarcity (p. 202)

1.1. Water Scarcity in Andalusia: Physical or Economic Scarcity?

In the dry regions of the planet, where the shortage is not only physical, but also social and economic (Aguilera-Klink, 1993), and where it is difficult to allocate this resource to its appropriate uses, it is absolutely necessary to discover new ways to alleviate the pressure on water resources. It is important to try to identify the type of water scarcity in Andalusia and to determine if that scarcity is only physical or social and economic as well. There are two physical elements which define the water situation in the region: its geographical configuration and its climatic conditions. Considering the first element, Andalusia is shaped as a triangular depression opened to the Atlantic Ocean and bordered by two mountain ranges (Sierra Morena and Cordillera Bética). This configuration conditions the region's main climatic feature, namely the unequal distribution of precipitations in space and time. On the other hand, we can identify two hydrographic areas in Andalusia: the Guadalquivir river basin, that covers more than half the region's territory, and the coastal basins. These are the Atlantic basins (Guadiana and Guadalete–Barbate) and the Mediterranean basins (Sur and Segura). Most of the water resources in the region originate in superficial water (61 percent); only 21 percent of them is based on underground water (Consejería de Obras Públicas y Transportes, 1996).

Besides the physical factors, it is important to focus on the use of water in the region. The agriculture sector in Andalusia absorbs almost 80 percent of the water resources, urban water supply amounts to around 12 percent and the industrial use exceeds 8 percent (Consejería de Medio Ambiente, 2002). Irrigated agriculture is the most important specialized subsector, which has to compete with urban supply in the summer period. This situation results in Andalusia clearly being a water-deficient area, and causes not only serious economic problems but also environmental ones. To sum up, Andalusia, despite being characterized by a considerable physical water shortage, has gradually specialized in sectors whose water consumption level is extremely high (Velázquez, 2006). It is also important to point out that Andalusia, not only is specialized in intensive water use sectors, but it is also a net water exporter economy (Dietzenbacher and Velázquez, 2006).

There are two other elements that should be pointed out: the water price and water concessions. Regarding the water price, water in the Andalusian agriculture sector is not paid according to the amount consumed but depending on the number of irrigated hectares. The consequence of it is that there is no clear conscience in the region of the resource's real cost so that it becomes undervalued. At the same time, water concessions are granted by the different Water Regional Federations for extremely long periods (an average of 50 years), introducing a distorting element in the resource's management. So we can say that the physical water shortage in Andalusia has been aggravated by the institutional and economic scarcity due to the productive specialization.

Before this apparently irrational situation, the need arises to reflect upon the possibility of actually managing the region's water resources in a more sustainable manner. On the one hand, the transfer of massive quantities of water is difficult, costly, and most of the time, absolutely unsustainable. On the other hand, the building of hydraulic infrastructures is always expensive and problematical in social and environmental terms. In addition, in spite of the rise in the offer, the demand always ends up unsatisfied. This article supports the voices that defend virtual water as a means to mitigate the pressure on water resources. The most reasonable thing seems to be the import of water-intensive products to areas where the cost of water is very high and the export of other products which do not require so much water.

1.2. Objectives and Structure of this Study

We propose in this paper an in-depth analysis of the relationship of the agrarian production and its commercial dealings with the amount of water that has been consumed. In view of that, our first goal will be to find out, by means of the estimation of virtual water, the exported crops which have the highest water consumption. Similarly, we analyse the crops that are imported and therefore, might contribute to save water. Our second objective is to put forward new ways to save water by means of the virtual water trade and to focus on the sectors where we should work to improve this water situation. We are aware that the scope of our paper is far too ambitious and we know that it would be possible only with a more comprehensive work. Obviously, our recommendations for policies involving socio-economic and environmental aspects will be biased by economic, social, environmental, technological and institutional factors. Nevertheless, this paper tries to go one step forward in this field by presenting new questions and breaking fresh ground for future research.

Commentary

The author of this article discusses the utility of virtual water in the agricultural and industrial sectors. She explains why it is hard to find sustainable solutions to water conflicts. Physical shortages are aggravated by institutional and economic limitations. The social–economic, agricultural, and industrial use of virtual water can help to overcome some of these obstacles. The author also talks about water prices and water concessions. When water negotiators are not fully conscious of the real costs of water and the impact of granting long-term concessions, they can undercut value–creating possibilities.

G. Kallis, M. Kiparsky, and R.B. Norgaard "Adaptive Governance and Collaborative Water Policy: California's CALFED Bay-Delta Program," (2009)

Introduction

This article focuses on California's water system and more specifically on the CALFED Bay-Delta Program. The CALFED Bay-Delta program is one of the best–documented instances of water negotiation. The policy-makers involved tried to integrate collaborative efforts with a commitment to adaptive management. The authors discuss how this unfolded and what appear to be the pre-conditions for success. They also describe the institutional arrangements that have been put in place. They present their views on the limitations of collaborative adaptive management as an approach to resolving water disputes.

The excerpts locate collaborative adaptive management (CAM) at CALFED in its larger institutional context. They argue that CAM can only work if the right market conditions and relationships among the parties are already in place.

Collaboration (pp.636–637)

Collaboration is at the heart of adaptive governance. Collaboration means to co-labor, to work together (O'Leary et al., 2006). It is not merely power-brokerage, i.e. trading among predefined interests to find an optimal point of agreement (Fuller, 2009). Engagement and interaction may create new value and mutual social learning. Collaboration among partners in CALFED is said to have reframed a struggle over water users' entitlements to the collective question of "what do we want this watershed to do?" (Freeman and Farber, 2005, p. 3). Such reframing allows new ideas to emerge that were not part of a polarized solution spectrum. An oft-mentioned example in this respect is the Environmental Water Account (EWA) (Innes et al., 2007; Freeman and Farber, 2005; Ingram and Fraser, 2006; Lejano and Ingram, 2009). In the EWA environmental and water agencies trade water for fish with water for drinking and agriculture in real time. Innovative ideas like the EWA, some scholars have argued, are most readily conceived through informal interaction between agencies and stakeholders, such as those in the CALFED working groups (Innes et al., 2007; Ingram and Fraser, 2006). Such interaction not only produces innovation but also creates a "cascade of changes in attitudes, behaviors and actions" and "social and political capital" with long-term positive effects (Connick and Innes, 2003).

However in this Issue we want to go one step further, not only in terms of understanding how collaboration works, but also engaging with its "dark side" and shortcomings (McGuire, 2006). Because collaboration is new, or because it produces new results, it does not follow that in and of itself it must be desirable (McGuire, 2006).

Whereas distinctions between collaborative, adaptive governance and hierarchical state regulation and competitive markets are often emphasized, in the real world these three forms necessarily coexist and depend on one another (Jessop, 1998). CALFED for example did not replace but incorporated conventional regulatory agency programs, in the process allowing them to develop new connections and innovations. Governance needs State forms of governing. Court decisions for example formed the background entitlements with which CALFED partners sit at the negotiating table (Freeman and Farber, 2005). State support, financial and symbolic, was crucial and so were state assurances that agreements will be implemented. Furthermore, the State offers a governance process the democratic legitimacy that it otherwise lacks given the ad hoc selection of participants. The downfall of this is that inversely, governance suffers from the shortcomings of State administration and it is vulnerable to external, political changes (Thompson and Perry, 2006). Many of the shortcomings of CALFED, for example, have been related to general problems of public administration in the participating agencies and CBDA, such as understaffing, budget management processes, competition between state and federal agencies or entrenched agency mentalities (Lurie, 2004). Political changes such as the election of the George W. Bush administration after the signing of the ROD and California's budget crisis shortly thereafter also undermined CALFED (LHC, 2005). However, such changes cannot be viewed as unexpected aberrations that derailed an otherwise successful governance program; in the real world, governments change and crises happen. Nor can government interventions be seen as "messing" with an otherwise innocuous governance process; governance needs the State. Governance processes therefore have to be studied within their real-world institutional context, and in their real, messy, hybrid form.

4. When and How does Collaborative Governance and Adaptive Management Work?

Networked, collaborative governance arrangements are crucial for a culture and practice of adaptive experimentation (Folke et al., 2005; Gunderson and Light, 2006). Favorable conditions for their emergence include: an impasse which makes warring factions ready to negotiate alternatives (i.e. "fail their way into collaboration", Bryson et al., 2006); a relative balance of legal, economic, and/or political power (Duane, 1997); pre-existing social capital and networks; stakeholders with the resources and expertise necessary to generate new solutions; political mandate, pressure and support; and the presence of—or prospect of access to—external financial resources that would not otherwise be available to participants (Freeman and Farber, 2005; Bryson et al., 2006). These conditions were largely met in CALFED (Innes et al., 2007; Freeman and Farber, 2005). The ballot defeat of the Peripheral Canal and the series of legal decisions empowered environmental groups and created legal and political impasses. Expertise and scientific knowledge were distributed beyond agencies and Universities. Stakeholders had already started networking in the San Francisco Estuary Project and the Three-way process (Connick, 2003). And federal and state leaders, most notably Interior Secretary Bruce Babbitt of the Clinton administration, pushed the process and lubricated it with federal subsidies and state bonds.

If these conditions bring stakeholders to the table for negotiation, they alone are not sufficient to create successful collaborations and partnerships (Fuller, 2009). Innes and Booher (1999) and Bryson et al. (2006) identify several important procedural attributes for effective collaboration such as: the presence of shared practical tasks; initial agreements; a reliance on self-organization rather than an externally imposed structure; the use of high-quality, agreed-upon information sources; proceeding with agreements when there is overwhelming support; external legitimacy of the process; resources and commitment to equalize power differences between participants; continuous trust-building activities, and genuine engagement in productive dialogue. The contributions in this Issue elaborate further how and when collaboration works.

4.1. Looking Inside Collaborative Processes

Contributors heed Agranoff's (2006, p. 56) call to "go beyond heralding the importance of collaborations to look inside their operations." They engage with the question of how and when innovative agreements result, looking at processes and working groups within CALFED that produced breakthrough results, and others that clearly failed (Lejano and Ingram, 2009; Fuller, 2009). Contributions in this issue delve deeper into the question of how collaborative processes work, training a magnifying lens on the mechanics of sub-processes within the larger CALFED program. Their common starting point is that "it is the shared learning process that is critical" (Norgaard et al., 2009).

A critical institutional avenue towards encouraging shared learning is the creation of boundary organizations. These refer to the institutionalized forums where different knowledge and stakeholders work together to bridge the gaps between disparate frames and viewpoints. The Science Program (Taylor and Short, 2009; Norgaard et al., 2009) or the Environmental Water Account (Lejano and Ingram, 2009) served as boundary organizations. They provided opportunities for direct, personal and sustained engagement of scientists and stakeholders, facilitating shifts in concepts and the emergence of new language to talk about problems and solutions (Taylor and Short, 2009).

Within such boundary organizations, boundary objects are used to develop a shared language—an "inter-language" in Fuller's (2009) terms. Boundary objects are "artifacts that individuals work with … that cross disciplinary or cultural barriers" (Carlile, 2002:446, as cited in Fuller, 2009), such as models, maps, reports, spreadsheets or power point presentations, or even the very conferences and workshops that create a space for shared interaction. Boundary objects offer stakeholders a new vocabulary to talk about problems and a platform for modifying and re-organizing concepts in a way that is acceptable from all perspectives (Fuller, 2009; Lejano and Ingram, 2009). For example, in the EWA, games and modeling simulations of pumping scenarios and their impacts on fish, allowed stakeholders to get a grasp of what water trade meant and offered a base for negotiation and agreement (Hudgik and Arch, 2003; Innes et al., 2007). Identifying commonalities between CALFED and global scientific assessments, Norgaard et al. (2009) underscore how shared language can take the form of a new meta-model or the complementary use of multiple analytical models with different scales or functions as a way to allow participants to communicate across disciplinary perspectives.

However, as Lejano and Ingram (2009) show, the narratives and perspectives of different stakeholders are not reconciled and integrated just through the creation and use of a master frame. It is in the conversation, translation and exchange of different knowledges, i.e. the dialectic juxtaposition of concepts, that "magic occurs", not in the mere combination of the knowledge stored by each camp (Lejano and Ingram, 2009, p. 4).

Commentary

The authors suggest the importance of institutional factors in the success of collaborative adaptive management efforts has been underestimated. They illustrate by parsing what they feel are important distinctions between collaborative and adaptive approaches to state regulation and the creation of competitive markets. They fear that by emphasizing the success of collaborative processes, policy analysts may be neglecting the importance of the courts and market conditions.

They seem to have a rather stilted view of consensus building, focusing only on the extent to which the parties are satisfied with the substance of agreements, neglecting the relationships that may have been built or the increasing levels of trust that were established.

Consensus-building involves more than just engaging stakeholders to "get a deal." It requires assessing the real world constraints on implementation and seeing if relationships can be improved to the point where previously immovable constraints can be surmounted. The authors recognize the importance of creating value. They see this as a social process that depends on "reframing" to generate shared commitments. They also acknowledge the importance of informal interactions among the stakeholders as a means of changing attitudes and behaviors.

They point out the limitations of collaborative processes and the importance of what they call traditional governance, noting the important role played by "boundary organizations," separate groups that give opportunities for direct, personal, and sustained engagement of scientists and stakeholders. These can be key to informal problem-solving. The authors deconstruct the elements of a consensus-building process, including the role of the neutral, without seeing how all the pieces fit together. They assume that collaboration in adaptive governance is merely power-brokering, which they refer to as "trading among predefined interests to find an optimal point of agreement." They downplay how face–to–face interaction and coalitional dynamics can alter each party's ranking of its own interests and its responsiveness to the interests of others.

B. W. Fuller "Surprising Cooperation Despite Apparently Irreconcilable Differences: Agricultural Water Use Efficiency and CALFED," (2009)

Introduction

This article describes the challenges and breakthroughs encountered by the decision makers involved in the CALFED Bay-Delta Program. The author describes how stakeholders, conveners, and facilitators reconciled "apparently irreconcilable differences." The article illustrates how consensus–building permits parties to move from impasse to agreement. It also introduces several concepts from negotiation theory—like inter-language—and shows how they have been applied. The following excerpts illustrate how consensus building and the use of a facilitator can produce agreement. As opposed to the previous author who discusses the same case, Fuller has a more in–depth view of the collaborative adaptive effort that took place at CALFED.

CALFED and Collaborative Adaptive Management (pp. 668–672)

The Steering Committee's original mission was to provide feedback and ideas to the CALFED Program Manager rather than seek consensus. However, the Babbitt–Dunn group, a high level federal-state policy group that was negotiating certain elements of the overall CALFED Program, offered the Steering Committee a challenge as it began deliberations: if the Steering Committee could craft and agree upon the outline of an Agricultural Water Use Efficiency Program (Program), the Babbitt–Dunn group would likely incorporate it into the Revised Phase II Report for CALFED. If not, the Babbitt–Dunn group would get someone else to do it. The Steering Committee's first instinct was to reject the offer; however, in the end they could not pass up this rare opportunity to have direct input into the policy crafting, especially (as we shall see just below) once they discovered a new framework for moving forward. Despite the daunting challenges and failures of before, and their lack of mandate to seek consensus, the Steering Committee decided to try. In a few intense working days, they surprised themselves and others by crafting the outlines of a very different kind of Agricultural Water Use Efficiency Program. Not only that, but over the next 2 years, they created an innovative and stakeholder supported Agricultural Water Use Efficiency Program.

How did the Steering Committee do it? The answer is more than the trading across interests found in negotiation theory, more than learning and reframing found in theories of intractable conflicts, and more than in the joint fact-finding advocated by consensus building theory.

4.1. Getting a New Conceptual Model for the Program

After the Work Group was disbanded, CALFED decided to convene what became the Independent Review Panel of Agricultural Water Conservation Potential (Panel) in the hopes that its findings would resolve the ongoing disagreements about the estimates of water conservation potential for agriculture and what efficient water management practices were most appropriate … CALFED started work on assembling this Panel at the same time it was convening the Steering Committee.

The stakeholders and facilitators played an active and key role in the Panel. For example, the Panel was designed to include "technical advisers" selected by agricultural and environmental stakeholders. The role of the technical advisers was to ask questions of the panelists, provide information as required, and to otherwise serve as the technical eyes of their constituency. In addition, at the request of the facilitators and non-government stakeholders, CALFED expanded the Panel deliberations to include a pre-deliberation Scoping Session.

The Scoping Session was held 1 week before the first meeting of the Steering Committee. The Scoping Session was convened to provide stakeholders and members of the public with information on the Panel and CALFED's reasons for convening it as well as to provide input into the structure and preparation of the Panel. However, its impact was much larger, in large part because the Panel provided an overarching framework that bridged the different interests and perspectives of the stakeholders (Susskind and Field, 1996).

In the Scoping Session, the Panel proposed that, instead of focusing on EWMPs, CALFED should set downstream goals (e.g. the restoration of a particular habitat), identify specific quantifiable objectives (QOs) that would help achieve those goals (e.g. a maximum amount of salt entering that habitat), and then seek to identify on a case-by-case basis interventions upstream through which those QOs could best be met. This approach quickly became one of the key frameworks used by the Steering Committee and they used it to craft the outline of a program for the Revised Phase II Report as requested by Babbitt–Dunn group.

The concepts suggested by the Panel integrated and provided leverage on several key sticking points between agricultural and environmental stakeholders. First, it recognized and specified environmental objectives as the key driver for the program, which the environmentalists had always wanted. Second, the Panel's suggestion of a case-by-case approach to meeting the QOs addressed two fundamental concerns of agricultural stakeholders: (1) that each water district be allowed to adopt water conservation methods suitable to its particular, local conditions (soil, water, climatic, economic, etc.) and (2) that before agriculture was asked to act, there must be some framework that clearly linked actions with downstream benefits, which in turn opened the door for the Steering Committee to explore more nuanced ways of allocating costs and benefits.

To appreciate the impact of the Panel's suggestions in the Scoping Session, it is important to understand the context. The Steering Committee had only 2 weeks, within which they organized four working days, to draft the Program description for the Agricultural Water Use Efficiency Program. Though the Steering Committee worried that the text would be misused and that there was not be enough time to consult their constituencies, they also realized that this was a rare opportunity to directly influence CALFED's decision-making. What tipped the balance in the end was the new framework, which all sides found they could accept as a starting point for future policy work.

It would be a mistake, though, to say that the Panel's ideas were influential because they were new. In fact, individual agricultural and environment stakeholders had suggested similar ideas during the Work Group sessions, though these ideas had little impact. So why did the Panel's suggestions become so influential?

From interviewing stakeholders, the answer lies at least partly in the character of the Panel. Overall, stakeholders perceived the Panel as wise (using the best information possible, and including both inside and outside experts) and impartial (the panelists were outsiders vetted by the constituencies and watched by the stakeholders' technical advisers).

However, there is another aspect here that would be easy to ignore. It is useful here to reverse the question above, namely: why did the same ideas not gain support in the Work Group? Part of the answer, as discussed earlier, was that stakeholders within the Work Group could and would not explore dangerous and unlikely cooperation.

4.2. Closing the Door and Creating Openings

> They told us that, "Well, [this idea] just isn't going to fly. I wish it could but it's just won't." We had developed enough trust to know that they were not just stonewalling; it really was a fact. If this method does not work, is there another method that would work? Why is this sensitive? What are the issues that these [stakeholders are] dealing with? Now that we have this, and now that we know why we are perceiving it this way, now ... are there things that we can get to? (Interview with Steering Committee member, Summer 2004.)

The quote above from one of the Steering Committee members highlights one of the key attributes of the Steering Committee: the openness of its members in expressing views and exploring ideas without fear of immediate retribution.

The first statement of the quote above truly stands out in this regard. Before, stakeholder representatives often were unable and unwilling to experiment. In contrast, one stakeholder quoted above agrees with an idea but claims that it will not be accepted by his constituency. Strikingly, his counterparts in the Steering Committee (a) seem to trust him; (b) are willing to work with him to understand the sources of resistance; and (c) come up with new ideas to address those concerns.

The willingness of stakeholders to trust one another in the Steering Committee is a remarkable change compared to the lack of sharing found in the Work Group. Why were stakeholders able to be more open and explore their disagreements and differences effectively? One important reason was their deliberations occurred without direct public observation after the facilitators and CALFED Program Manager decided to close the Steering Committee's meetings to outsiders. Away from the everyday observation of their colleagues, Steering Committee members were able to build relationships and explore innovative ideas that would not have been possible otherwise.

Legally, they were able to do so because of the existence of BDAC, which remained the final and official point of public participation and which would review whatever the Steering Committee produced. CALFED and the facilitators were also very careful in selecting participants, working with each side to identify well respected representatives that the constituencies felt could speak about their concerns effectively.

The Steering Committee members earned that trust from their counterparts by their actions during the deliberations. It was not that the differences among stakeholders vanished, they said. Instead, trust building was iterative. Each overture built upon a preceding one, and as trust grew the representatives grew more willing to treat their differences as constraints within which problem solving would take place, rather than insurmountable obstacles.

Like CALFED's decision above, choices about the purpose and transparency of a group are a growing concern in the consensus building literature. Closed door meetings provide breathing room for representatives, but questions of cooptation and accountability arise. Similar to CALFED's structure of BDAC and Steering Committee, Susskind et al. (2003) propose the separation of informal, brainstorming discussions from more formal, vetting and decision-making negotiations in order to create room for creativity in contentious negotiations.

These choices, however, were not the only factor in opening up problem solving. Consider, the following words from the CALFED Program Manager who convened the Steering Committee.

> [The facilitators] worked really hard to maintain some of the rules involving no ownership of ideas. That really becomes pervasive. Sometimes [the facilitators] synthesize [your idea] with stuff that other people have said, and this new direction starts to emerge. The group gets this feeling that they invented it—and they did, they honestly did ... In some cases we literally invented new words, or new buzzwords I probably should say. (Interview with the AgWUE Program Manager, Fall 2003.)

Recall how problematic words and style were before. In the Steering Committee, the process of exploring concerns and generating ideas and solutions occurred using a set of new or re-defined words and concepts created and employed under a set of norms that became firmly entrenched in the group's behavior. Some words, concepts, and rules were formal; other informal ones emerged as well during the process, such as listening respectfully as well as no ownership of ideas. The creation by the group of this "interlanguage" (Galison, 1997)—including both words and concepts as well as rules and norms—provided accepted means that the Work Group had lacked previously for handling both everyday and difficult situations—for example, to describe the situation on the ground, to vent, or to identify and manage an impasse on a particular issue.

This interlanguage did not exist at first. Words and concepts had to be painstakingly created or re-defined through intensive discussion and then vetting with constituencies. The ground rules were created by the facilitators and modified and vetted by stakeholders at the beginning of the process. The informal norms and habits, on the other hand, were developed through experience during the Steering Committee's deliberations. Many Steering Committee members noted that the facilitators of the group modeled behaviors—e.g. giving respect—that the rest of the group adopted over time. In some other cases, Steering Committee members or CALFED staff were the initiators. Whatever their source, however, these norms and rules allowed the parties to deliberate more effectively and avoid escalating the conflict when discussions became difficult.

4.3. Crafting Solutions from Pieces

Creating a safe space for deliberation can increase parties' willingness to talk, but it does not guarantee that the group will have the ability to effectively define, analyze and solve the difficult problems in front of them. Negotiation and consensus building theory (Susskind, 1999) promote cooperation through the exploitation of trades across differently valued interests and pursuit of common interests. However, the economical trades envisioned in these theories seemed to less present in the Steering Committee deliberations, as the words of this next Steering Committee member helps us begin to explore:

> The facilitators took pains to always make sure that this brainstorming was occurring within the context of, and was captured in, a conceptual framework which made sense and which was understood and approved by the participants. They would make sure that the relationship between the objectives of the programs and the tools ... were always represented in a conceptual model. That the relationships between the different types of programs and the different types of funding were always part of an architecture. That the relationship between the participation of various entities in the Program and the institutional, legal, political implications of that participation—in terms of assurances and other issues—was always part of an architecture. That was extremely useful because, very often, groups like these will come up with some good concepts which don't necessarily survive as part of a coherent and congruent whole. (Interview with environmental representative, Fall 2003.)

We get the sense from the quote above that Steering Committee members identified, analyzed, and reconfigured a large number of "pieces"—including not only ideas and interests, but also the political, legal, programmatic, and other elements—as part of constructing their Program. Second, that the facilitators helped stakeholders construct an "architecture" in which the various pieces could be seen in relationship to one another as part of a "coherent and congruent whole."

The quote misses another important part of the architecture, though. As the Steering Committee developed their Program, it also worked with CALFED to develop the scientific framework needed for calculating the quantifiable objectives and the links between them and upstream proposed projects.

The science was also built piece-by-piece, with each element joined by a conceptual model of flow paths. Some pieces were in the form of new language, which was needed to make the new concepts understandable to people with little technical background. Spreadsheets and PowerPoint presentations also captured important pieces and relationships among them. For example, PowerPoint presentations provided diagrams and other concise ways of encapsulating the main ideas visually that stakeholders found key in building understanding and agreement. Spreadsheets, on the other hand, provided an architecture in which different elements were captured as variables and linked through equations. The architecture mentioned above and the spreadsheets were snapshots that portrayed the complex world in which CALFED worked and provided the means for them to be re-constructed into a more desirable outcome. Put simply, give someone Lego and they can often use it to model houses, cars, and other products they imagine they want, even if they do not share a common perspective. Innes and Booher (1999) have observed this process within other negotiations and called it "bricolage" and the pieces mentioned here are the "boundary objects" mentioned in Carlile (2002).

It would be a mistake, however, to imagine the architectures described above as static. Consider the words of the CALFED AgWUE Program Manager, for example, as he describes how the science evolved.

> [W]e literally had to invent the science behind this. Inventing that science and continuing to ask for review of that science in a group that was not technical— that was in essence a policy-level group that had a very wide spectrum of technical background, all the way from layperson to expert—was tough. … We continued to challenge ourselves to do the good work but present it in a way that was digestible by this policy group. … [W]e challenged the policy group to tell us, "Okay folks, you've got to keep grilling us until we're really sure we know why you don't like this." Then we would follow up with the challenge of, "Okay, now you've got to tell us the right way to present it" (Interview with the CALFED AgWUE Program Manager, Fall 2002).

The Steering Committee, CALFED, and several experts developed the scientific framework, and the other conceptual architecture too, as needed to support their deliberations about and design of the Program they were crafting. The experts, for example, continually presented the latest draft of the pieces—including spreadsheets, PowerPoint presentations, conceptual drawings, and more—to the Steering

Committee for their approval and feedback, and the Steering Committee deliberated about and suggested modifications for those same pieces so that the science remained aligned with their array of interests.

Furthermore, the quote above also tells us about another dimension of interaction, that between the constituencies and the CALFED and Steering Committee collaborative effort. As the Program Manager mentions above, "Okay, now you have got to tell us the right way to present it." The Steering Committee decided that CALFED would present the final draft of the Program to the constituencies, in part because they were worried that both the agricultural and environmental constituencies might reject the ideas without reflection if they thought the ideas came from members of the opposite side. The Steering Committee knew that the message had to be precisely targeted to the interests and perspectives of each constituency to make it clear why its members should support it. To accomplish this, the Steering Committee and CALFED developed the presentation together. Terms were modified and clarified, diagrams designed and vetted to capture key concepts, and every other element was carefully planned to strengthen the message using the perspectives and language of the particular constituency.

4.4. Connecting to Constituencies

Communicating the final proposed Program to constituencies was only one of many efforts to communicate ideas to and get feedback from the different communities. Earlier, I argued that CALFED's decision to make the Steering Committee's meetings closed door was crucial for creating a safe space for exploration. The success of that decision cannot be understood without looking at how the Steering Committee interacted with the different constituencies. Stakeholders agree that the Steering Committee would not have succeeded without the careful work it did to keep key members of each constituency, including both decision-makers and experts, engaged with the evolving solution and its surrounding architectures. And, as possible points of resistance were identified in one or more of the constituencies, the Steering Committee would find a way to make the Program acceptable by either packaging it correctly or finding another alternative to achieve their goals.

The communication with constituencies was ongoing throughout the Steering Committee's deliberations. Each of its member was very active in using email and phone calls, their "hot lines," to get feedback on and ideas for the emerging Program from specific people in their constituencies. As the Program evolved, the links between the table and the constituencies became more and more varied. For example, as the group started working on the specific procedures and calculations for their Program, they convened a group of Regional Liaisons who provided regional-specific expertise as required. Strikingly, the members of this group were primarily from the agricultural community, a sign of the growing trust among the parties.

This careful work with constituencies also shows the limits of transforming and reframing approaches suggested by Lewicki et al. (2003) and Schon and Rein (1994), which focus largely on the perceptions and understandings of individuals at the table and how these can be changed to encourage solutions. Recall a portion of an earlier quote:

> They told us that, "Well, [this idea] just isn't going to fly. I wish it could but it just won't."

Changing minds is not enough, or may not even be the right path when constituencies are being represented. Strategic thinking and communication is required, both in terms of vetting and refining ideas and also in mapping the limits of what solutions can be considered at all.

4.5. More than Incentives

One could point to the presence of significant incentives, positive and negative, as the main factor that encouraged this particular agreement. CALFED certainly had significant funds available for programming, all of the relevant state and federal agencies were present and represented, and the stakeholders were certainly tired of the decades-long conflict. And yet, each of these incentives was also present during the Work Group deliberations. Even Secretary Babbitt's challenge to the Steering Committee, while certainly important, is hard to single out. The non-government stakeholders' first response was to reject it. That they eventually provided Babbitt the asked for text is largely due to the fact that the process described here helped them find an attractive way forward out of their painful stalemate.

Thus, while the potential availability of funding and the presence of the key agencies certainly encouraged the parties to meet and seek a solution, they cannot explain by themselves why the Steering Committee reached an agreement, and the Work Group did not.

5. Lessons

What can conveners, facilitators, participants, and constituencies involved in collaborative policy-making learn from the experiences of the Steering Committee? This paper shows that stakeholders can make progress on difficult and contentious policy issues despite the existence of apparently irreconcilable differences among stakeholders. Decision-makers often convene collaborative policy-making efforts when they recognize that stakeholders have pronounced differences about how the problem is defined and solved. The findings here are crucial for the design and facilitation of future efforts, especially those in which the disagreements about stakeholders seem irreconcilable.

The paper puts forward several key and related lessons. First and most broadly, the presence of the right stakeholders and the right incentives may not be enough. Work Group members had very similar incentives as the Steering Committee, but failed to make progress. Second, while it is clear that stakeholder representatives do learn more, and sometime reframe their personal ideas conceptions about the problem, its solutions, and about each other when they deliberate, they need to do more. What the Steering Committee did was to create an interlanguage (words and concepts), as well as the habits, rules, and procedures to use them effectively. Those helped the stakeholder representatives understand each other and solve problems.

Third, rather than making individual trades across their interests, Steering Committee members developed their solution by identifying, modifying, and organizing concepts (using boundary objects) until they found a combination that they thought looked acceptable from all perspectives. Strikingly, the creation of the words, boundary objects, and habits to support the solution and the crafting of the solution happened in parallel.

Fourth, boundary objects—such as spreadsheets, Power-Point presentations, diagrams of concepts, and other material artifacts—can play an important role in bringing aspects of the world to the table and making them into puzzle pieces that can be manipulated and combined for solution crafting. These material objects support problem solving where the real phenomena being represented cannot. Until rivers and eco-systems can be brought to the table as pieces of a larger puzzle, stakeholders have no common means of talking about them. The findings here raise interesting questions about the role of objects in enabling problem solving and negotiation when stakeholders have conflicting and divergent perspectives.

Fifth, and especially in situations of apparently irreconcilable conflict, conveners and facilitators need to strategically design how constituencies will be represented in consensus building groups. More specifically, they need to balance the transparency of the group as they seek to create safe spaces for problem solving, and yet keep the results grounded in the political realities of the situation. In this case, the Steering Committee kept observers away from the meetings, but kept constituencies engaged throughout the deliberations through informal and formal communications. As such, the products of the members' creativity and learning became the means and focal points for reaching out to constituencies, and these products were in turn constantly tested and re-grounded in the realities of each and every constituency. Practically, this means that the pieces, architecture, and eventual solution need to make sense to each and every constituency separately away from the table as well as jointly at the table. In this case, the complex coordination of different worldviews benefited largely from the use of material objects carefully crafted by stake-holders as a means to reach out to various constituencies.

Finally, a note of caution. Like any element of a long and difficult policy process, the agreements made in consensus building processes are not the final step. Much can change in implementation, as it did in CALFED, especially when an administration changes. Despite the strong stakeholder support for the Program, the incoming Bush administration and Congress never provided the financial support promised by their Clinton-era predecessors. This paper shows some ways to improve the ability of consensus building processes to reach agreement supported by stakeholders. It does not provide a full answer to the problems posed by changing administrations and faltering funding, which plague all policy-making efforts.

Commentary

In this article the focus on trust-building is important. The ability of stakeholders to increase the level of trust among the Steering Committee was remarkable, especially compared to the trust problems that the earlier Work Group encountered. Trust building was continuous. Each overture built upon a preceding one, and as trust grew, the representatives grew more willing to treat their differences as constraints within which problem solving could take place.

The group managed to separate informal brainstorming from more formal vetting and decision-making to create space for problem-solving. They collected, analyzed, and reconfigured a large number of "pieces," including not only ideas and interests, but also the political, legal, programmatic, and other elements. Ultimately, exploring concerns, generating new ideas, and reframing became the new behavior of the group.

The process described by the author stresses the importance of constant communication between stakeholder representatives and their constituencies. This was ongoing, not just at the end of the Steering Committee's deliberations when they had provisional agreement. The group used a facilitator to synthesize ideas and invent new concepts that became the group's inter-language. The facilitator also emphasized the importance of a "safe space" in which stakeholders were freed from the symbolic roles they had to maintain whenever they were in public view. On the other hand, "closed doors" pose a trade-off: they offer a safe environment for joint problem-solving, but they risk alienating outsiders who worry whenever transparency is reduced.

D. Brooks and J. Trottier "Confronting Water in an Israeli–Palestinian Peace Agreement," (2010)

Introduction

In this article the authors view trans-boundary water agreements in terms of ongoing joint management that allows for continuous resolution among shifting water demands and uses. They specify five key principles for shared water management: water allocations that are not fixed, but vary over time; equality in rights and responsibilities; priority for demand management over supply management; continuous monitoring of water quality and quantity; and mediation among competing uses of fresh water. The paper focuses on the water shared between Israel and Palestine, but their approach has more general application. The authors' objective is to demonstrate that an agreement about shared waters is possible if all the political, economic, social, agriculture, and environmental contingencies are considered using a conflict resolution mechanism.

This excerpt illustrates how value creation, using mediation, can reconcile two or more parties in disagreement about how to divide trans-boundary waters.

Principles of Shared Water Management (pp. 109–113)

[P]rinciples must be adopted to guide the design of the institutional structure for joint management of shared water. Many such principles, as with equitable and reasonable use, are common to all forms of trans-boundary water management (see, for example, Rahaman, 2009). We focus here on five supplemental principles that are critically relevant to our proposal for joint management of water shared by Israelis and Palestinians; viz:

*   definition of water rights,
*   equality in rights and responsibilities,
*   priority to demand management,
*   acceptance of the historic standing of local forms of management,
*   continuous monitoring of quantity and quality in all shared water and mediation of conflicting uses, demands and practices.

Definition of Water Rights

In the Israeli–Palestinian context, water rights have been perceived until now as fixed and permanent quantities of water allocated to one side or the other. Such a definition of water rights may serve as a political slogan, but the variability of flows and the interconnection among sources make it problematic. A more appropriate definition of water rights relies on recognition of the mutual interdependence of both parties in sustaining the quality and quantity of all shared water. Water rights are therefore defined as a bundle of rights and responsibilities to manage water according to a set of mechanisms whereby each party has a right or duty to (a) access water, (b) use water, (c) treat water, and (d) release waste water, as well as to set the limits necessary for the access, use, treatment and release, in ways that will maintain the quantity and quality of flow in all shared water sources within limits set by (and perhaps changed by) natural conditions.

Parallel rights and obligations for the citizens and the institutions of the two parties imply that existing patterns and volumes of water use have some standing within the bundle. This principle does not extend so far as to give those patterns and volumes permanent status. However, they can be altered only after due consideration of impacts, and the changes must be implemented gradually to permit time for adjustment.

Equality in Rights and Responsibilities

Israelis and Palestinians must have equality in all rights and responsibilities related to the management, development and use of shared water. The Oslo Agreement created the Joint Water Committee (JWC), composed of equal numbers of Israelis and Palestinians, to function on the basis of consensus when making decisions about water. However, its role is truncated. The JWC only makes decisions concerning

Palestinian water management and development in the Occupied Territories. It has no role with respect to water management or development carried out by Israel. Clearly, the current institutional design fails to satisfy the principle of equality in rights and responsibilities. In contrast, the set of institutions in our proposal have been designed from the start to accommodate this principle. Just as with the objective of equity, this does not mean that each party can expect to receive an equal volume of water. It does mean that each party will have equal standing within each of the institutions for joint management of shared water bodies.

Priority to Demand Management

There is enough water in the region shared by Israelis and Palestinians to satisfy all of their needs and to provide a high quality of life, but far less than enough to satisfy all their desires. Therefore, the main focus of water management must shift from supply to demand. Demand Management is a very broad concept with dimensions of both quantity and quality, as well as timing of use (Brooks 2006; Brooks et al., 2007). In management terms, it means that, when a request is expressed for more water, attention must be paid first to determining whether the proposed used of water can occur without increasing the supply of water. All requests for funding of new supply must be considered against policy and program options that reduce the need for additional water, that reduce the quality of water required for the end use, or that shift the timing of use to off-peak periods. Moreover, though water use within a state, and pricing of water, are within the sovereign authority of that state, both parties must recognize that efforts to reduce the use of water through demand management are so fundamental to their respective futures that they are appropriate issues for negotiations between them.

The priority given to demand management is of course far different from the supply-management mentality that typically pervades government bodies across the entire region (Brooks and Wolfe, 2007). Whether the focus is drinking water for cities or irrigation water for farms, planners always look first to new supply, even when opportunities abound to reduce water use (Brooks, et al., 2007). Nowhere in the entire Middle East is there a government agency tasked primarily with water demand management and given the bureaucratic status and the budget to make its role effective (Brooks and Wolfe, 2007).

Acceptance of the Historic Standing of Local Forms of Management

Local forms of water management must receive formal standing. In effect, this principle conveys what are called "soft" or informal water rights. Local, communal forms of water management have been all but extinguished in Israel, but they remain common in Palestine. Despite the evidence that they can operate efficiently and effectively (Trottier, 1999), communal forms of water management are increasingly treated as vestiges of the past that have to give way to centralized institutions managed by the state. The Israeli pattern of top–down centralized management may appear convenient to officials in the Palestinian Water Authority as a replacement for the existing bottom–up pattern. However, evidence, both in the area and elsewhere in the world, shows that such a change usually impairs the goal of practical and implementable water arrangements (Buckles, 1999; Mabry, 1996).

Existing local and communal management institutions, whether formal or informal, should, at a minimum, be given a chance to prove themselves in a new State of Palestine. True, some minimal structure is needed to indicate which institutions do have standing, and which do not, something that can be contentious. Nevertheless, the process can be undertaken in a fairly rapid yet sensitive and transparent way, and it can stop far short of centralization.

Continuous Monitoring of Quantity and Quality in All Shared Water and Mediation among Competing Uses, Demands and Practices

Continuous monitoring and ongoing mediation processes must constitute the main management tools to ensure that the goals of equity, efficiency and sustainability are achieved. This principle is not just a technical detail; rather, it is the basis on which decisions will be reached concerning adjusting withdrawals from each well or reservoir, or modifying use of water from a spring. It has many implications, including the need for fair treatment of water users who find themselves requested to reduce their rates of extraction. For example, users of a well supplying household water might require immediate replacement with water from a different source. In contrast, users of a well supplying irrigation water might be asked to cut back at certain times of the year or to accept monetary compensation (along with technical advice) for shifting to rain-fed methods. Practices that are not directly linked to water, but affect water availability and quality, must also be considered. For example, urban developments that increase the area covered by impermeable surfaces or farming practices that allow polluted water to flow into aquifers must be challenged as undesirable land-use changes that require coordination between water management and land management officials.

It is important to emphasize that this principle is not solely continuous monitoring but rather continuous monitoring and mediation. Mediation implies discussions with the groups involved and solutions devised at the lowest possible level according to the principle of subsidiarity. For example, a farmer-managed institution using well water for irrigation would be the most appropriate body to propose a new schedule of extraction in order to minimise the loss of crops while respecting new, lower extraction rates. Ongoing mediation could also allow a similar group to request a halt to neighbouring urban development in order to protect a spring recharge area. Most importantly, the ongoing mediation means that all actors involved can appeal to the proposed Water Mediation Board, whether they are scientists who perceive themselves as the spokespersons for the environment, or the members of the institutions that manage water, whether these institutions are state institutions, private institutions or communal institutions. The Water Mediation Board therefore provides a transparency to the interactions between the scientific claims and the social and political claims. This breaks with the usual approach that places water scientists (or other "experts") in the role of an enlightened despot concerning water and the environment, effectively relying on the political and social values of the scientists alone in interacting with the scientific claims.

By virtue of the second principle of equality, continuous monitoring and mediation mechanisms will apply equally to both parties. It will also apply to withdrawals from any shared water, whether the system is private, communal, or public. However, mediation mechanisms will be more relevant to the existing Palestinian institutions than to Israeli ones because the latter are so centralized.

Institutional Structure

The four objectives and five principles described above guided our work in designing an institutional structure for joint Israeli–Palestinian management of shared water. The resulting structure divides power over water along several axes:

- between the Israeli and Palestinian governments,
- among several joint Israeli–Palestinian institutions,
- between scientific and political dimensions of management,
- whether local or national,
- among institutions working over several scalar levels.

This division of power is systematically accompanied by mechanisms for mediation to seek resolution of competing claims. More specifically, it is designed to allow both parties to de-securitize water and to treat water as a resource that is the object of competition among many different actors, acting over varying scalar levels, and involved in political, social and economic relations that are not determined solely by nationalist lines. In many ways, the success of the agreement on water will be measured less by the record of how many disputes are peacefully resolved than by the unrecorded number of disputes that are prevented and never come to mediation.

Bilateral Water Commission

The Bilateral Water Commission (BWC) will replace the existing Joint Water Committee, but will have responsibility for all shared water, not only Palestinian water (as is currently the case for the JWC). It will report directly to the Israeli and Palestinian governments with a mandate that is central and critical, but limited. Most importantly, it will have the mandate to:

- Establish limits for withdrawals, standards for treatment and targets for releases of water from aquifers on the basis of the recommendations set by the Senior Science Advisors.

- Grant permits for new drilling projects on the basis of the recommendations set by the Senior Science Advisors.
- Develop extraction rates for contained aquifers, which are inherently non-renewable resources, so that their use is balanced over time against the ability of those using the water to develop alternative sources or to reduce demands for water.

The BWC can reject recommendations it receives from the Senior Scientific Advisors, but it cannot issue an alternative decision on its own. Rather, it must explain its reasons for rejection to the Senior Scientific Advisors and wait for new recommendations. If, after two exchanges, the BWC finds it impossible to reach agreement with the Senior Scientific Advisors, it can refer the matter to the Water Mediation Board (see below), but in no case can it issue its own decisions concerning the scientific soundness of these limits and standards.

Just as for its relationship with the Senior Scientific Advisors, the BWC will have final authority to accept or reject, but not to adjust or modify, decisions from the Water Mediation Board. When rejecting a decision from the Water Mediation Board, the BWC would be requested to explain the reason for its rejection. The Water Mediation Board would then re-examine its decision in light of this explanation and propose a new decision to the BWC. In short, the responsibility of the Bilateral Water Commission is solely to confirm the above-mentioned limits, permits and standards and to ensure that they are implemented.

Residual competence would generally lie elsewhere than with the BWC except that it would have final authority to accept or reject, but not to adjust or modify, decisions from the Water Mediation Board. Any responsibility that is not specifically detailed in the Agreement belongs to residual competences. An institution that retains residual competence retains all of its responsibilities except for those that are specifically attributed to another institution.

The BWC would be comprised of seven members, three selected by the governments of each Party, plus one member elected by the other six from any state other than the two parties. Decisions of the BWC would be made by majority rule provided that at least two members from the three selected by each Party must be in favour of any decision.

Water Mediation Board

The Water Mediation Board (WMB) will receive the complaints of any community or institution that argues that it is being negatively affected by either a planned water project, or an ongoing practice within another community or institution—including cases when these practices, such as urban planning, are not directly linked with water management. It will also receive complaints related to inequitable distribution of water or to inadequate water quality. For all of these situations, the main role of the Board will be to approach the parties involved in the complaint in order to hear their respective cases, and then to attempt conciliation.

In cases when either the conciliation process fails or the impact alleged by the entity bringing up the complaint is not proved to be attributable to the entity or entities incriminated, the Water Mediation Board will be responsible for investigating the complaint independently. Its investigations will include economic and other social science analyses to consider the incommensurable losses as well as the commensurable losses that any community or group claims to suffer. Open forums or public hearings may be held, and various dispute resolution options tried. Records shall be kept and published of all public hearings, and all recommendations to and from the Water Mediation Board shall be public.

The final decision of the Water Mediation Board will be implemented in a binding fashion. As necessary, BWC will confer with those ministries or authorities outside the water sector that have competence and authority in such areas as urban planning to determine how best to implement decisions on, for example, land-use or urban development.

The Water Mediation Board will be constituted by two Israeli members and two Palestinian members, each nominated by their respective ministries of justice, and one member elected by the members of the Local Water Management Board (see below).

Office of Scientific Advisors

The Office of the Scientific Advisors will consist of two "Senior Science Advisors," one each seconded from appropriate agencies in their respective governments, plus supporting staff. Their office will have the responsibility for reporting to the BWC on relevant issues related to water quality and water quantity and of recommending appropriate abstraction licenses and drilling limitations to the BWC. In addition to the other roles, the two Senior Science Advisors will be expected to have access to and to provide the BWC with commentary on three broad sorts of information: Water quantity data (including mapping), water quality data, and ecological limits on water withdrawals and wastewater disposal. Their role is not to maintain an independent database but to ensure accessibility of the databases maintained by the two Parties. As requested, they will make reports to the BWC and to other bodies created under this agreement.

The Senior Scientific Advisors will also have responsibility for establishing and then monitoring ecological "red lines" that define the minimum flow volumes and minimum quality standards that are required to maintain the ecological health of watersheds carrying shared water. After official review of red lines and of flow regimes, the BWC can adjust previous determinations for acceptable water withdrawals and release. Where existing quotas for withdrawal already exceed water availability in average rainfall years, or where water quality is impaired by existing quota levels, a schedule for gradual retirement or adjustment of quotas with appropriate compensation shall be negotiated with those holding such quotas. Quotas were introduced on Palestinian wells following the Israeli occupation of the territory. They determined the maximum yearly extraction each well could proceed with. The quota that was imposed on a well corresponded to the metered abstraction of that well over the previous year. It was not determined through modelling of the impact of that abstraction on the aquifer. The novelty of the proposal here is twofold. First, it consists of having quotas that are determined as a function of the uses made and as a function of the impact on the environment, and, second, it applies to both Israeli and Palestinian wells via a joint process. At present, Palestinians have no say in Israeli abstractions. In this proposal, domestic uses are rated as having highest priority and a minimum domestic allocation, corresponding to a "human right for water" is to be guaranteed to every household, whether Israeli or Palestinian … All withdrawal limits and changes of appropriate flow regimes for shared water will be made public by the BWC.

Any community, local water institution, or non-governmental organization may protest excessive or inadequate limitations on water withdrawals or flow regimes to the Water Mediation Board.

Mountain Aquifer Authority

Because the Mountain Aquifer is both the most important and the most vulnerable of the shared water bodies, it requires special attention. It is therefore proposed that a Mountain Aquifer Authority be created to represent the BWC for the western and north-eastern blocks of the Mountain Aquifer and to provide advice for the eastern block, which, as indicated in Section "What water is shared and what is not," is not shared water. The basic goals of the Authority will be to protect the aquifer from excessive withdrawals and from pollution—in effect serving in the role of the BWC for the same key functions, but with particular attention to the integrity of the aquifer. In addition, the Mountain Aquifer Authority will work cooperatively with national agencies in the two governments to limit flows of polluted surface water or of inadequately treated effluents into the aquifer. All of these priorities shall be accomplished prior to secondary priorities for promoting local and national economic development.

Possible designs and structures for a Mountain Aquifer Authority were studied by Israeli, Palestinian, and international scientists over a period of about 6 years with grants from Canada's International Development Research Centre. Therefore, further discussion here is not needed. The results have been published in several formats (Feitelson and Haddad, 1998, 2000).

Local Water Management Board

A single Local Water Management Board will identify and register all bodies, families, communities or private entities that manage water resources locally and distribute the water to a community. The criteria used for this identification will be the existence of "rules-in-use" locally. "Rules-in-use" are the rules according to which a resource is actually managed by a group in specific situations. They often differ from formal rules that have been recognized in writing. They can remain oral, yet be scrupulously obeyed within a community. In effect, the process of registering local water institutions is to give them standing in subsequent interaction with the bodies described just above.

Whenever a complaint is brought to the Water Mediation Board, the Local Water Management Board will, as requested, assist the experts of the Water Mediation Board to identify the institutions responsible for management of the water sources in question, and it will ensure that these institutions are fully consulted within any investigation under the auspices of the Water Mediation Board. It will further ensure that the conclusions and recommendations reached by the Water Mediation Board and BWC are communicated to appropriate local people or bodies. When these conclusions involve a change affecting these institutions, such as a reduction in extraction flow, the Local Water Management Board will negotiate with these institutions to develop a time schedule for implementing changes, and also some form of compensation. Such compensation needs not involve money but rather should preferably aim to develop mechanisms whereby the negative consequences of these changes will be mitigated as much as possible.

> The Local Water Management Board will be comprised of four members. Initially, two members of the Board will be selected by the Palestinian Ministry of Local Governments, and two members by the Israeli Ministry of Social Affairs. Within 3 years of its creation, the registered local management bodies will each be given one vote, and their representatives will elect future members of the Local Water Management Board.

Commentary

The authors of this article pinpoint the importance of water rights in understanding the mutual interdependence of parties that must share water. Depending on how water rights are defined, parties may be able to maximize joint gains through collaborative problem solving. They also argue for transparency as the key to creating effective partnerships.

To some extent, inequalities of power can be addressed using mediation. Parties will have to de-securitize water and treat it as a resource that is the object of competition among parties vying for a share at different scales of decision-making.

References

Abukhater, A. 2009. Fostering citizen participation, *GIS Development (Neogeography), February, 2010.*

ADB. 2001. Annex A of Decision N. 4/2001: Adozione Delle Misure di Salvaguardia Relative al Piano Stralcio per la Gestione Delle Risorse Idriche. Autorità di Bacino dei Fiumi Isonzo, Tagliamento, Livenza, Piave, Brenta-Bacchiglione, Venice.

Agranoff, R. 2006. Inside Collaborative Networks: The Lessons for Public Managers, *Public Administration Review, (Special issue)*: 56–65.

Aguilera-Klink, F. 1993. El Problema de la Planificación Hidrológica: Una Perspectiva Diferente, *Revista de Economía Aplicada, 2*(1): 209–216.

Allan, J. 1994. Overall perspective on countries and regions, in P. Rogers and P. Lydon (eds.) *Water in the Arab World: Perspectives and Progress*. Cambridge, MA: Harvard University Press.

Allan, J. 1998. Virtual water: A strategic resource, global solutions to regional deficits, *Ground Water*, 36(4): 545–546.

Arnold, C.A. and Jewell, L. 2003. *Beyond Litigation: Case Studies in Water Rights Disputes*. Washington, DC: Environmental Law Institute.

Asano, T. 1991. Planning and implementation of water re-use projects, in R. Mujeriego and T. Asano (eds.) *Water Science and Technology*, 24(9): 1–10.

Barrett, S. 2003. *Environment and Statecraft: The Strategy of Environmental Treaty-Making*. Oxford: Oxford University Press.

Baruffi, F., Ferla, M., and Rusconi, A. 2002. "Autorità di bacino dei fiumi Isonzo, Tagliamento, Livenza, Piave, Brenta-Bacchiglione: Management of the Water Resources of the Piave River Amid Conflict and Planning." Second International Conference on New trends in Water and Environmental Engineering for Safety and Life: Eco-compatible Solutions for Aquatic Environments, Capri, Italy, 24–28 June 2002.

Bernauer, T. 2002. Explaining success and failure in international river management, *Aquatic Science, 64*: 1–19.

Bingham, G., Wolf, A., and Wohlgenant, T. 1994. *Resolving Water Disputes: Conflict and Cooperation in the United States, the Near East, and Asia*. Arlington, VA: ISPAN for USAID.

Bingham, L.B. and O'Leary, R. 2008. *Big Ideas in Collaborative Public Management*. Armonk, NY: M.E. Sharpe, Inc.

Biswas, A.K. 1993. Management of international waters: problems and perspective, *International Journal of Water Resources Development*, 9(2): 167–188.

Boswell, M.R. 2005. Everglades restoration and the south Florida ecosystem, in J.T. Scholz and B. Stiftel (eds) *Adaptive Governance and Water Conflict: New Institutions for Collaborative Planning* (pp. 89–99). Washington, DC: RFF.

Brooks, D.B. 2006. An operational definition of water demand management, *International Journal of Water Resources Development, 22*(4): 521–528.

Brooks, D.B., Thompson, L., and El Fattal, L. 2007. Water demand management in the Middle East and North Africa: Observations from the IDRC forums and lessons for the future, *Water International, 32*(2): 193–204.

Brooks, D. and Trottier, J. 2010. Confronting water in an Israeli–Palestinian peace agreement, *Journal of Hydrology, 382*(1–4): 103–114.

Brooks, D.B. and Wolfe, S.E. 2007. "Institutional Assessment for Effective WDM Implementation and Capacity Development. Water Demand Management Research Series," International Development Research Centre, Cairo, Egypt.

Briscoe, J. 2010. Perspectives: Troubled waters – Can a bridge be built over the Indus?, *Economic & Political Weekly, 45*(50): 28–32.

Briscoe, J. 2009. Water security. why it matters and what to do about it, *Innovations, 4*(3): 3–28.

Bryson, J.M., Crosby, B.C., and Stone, M.M. 2006. The design and implementation of cross-scale collaborations: Propositions from the literature, *Public Administration Review*: 44–55.

Buckles, D. (ed.).1999. *Cultivating Peace: Conflict and Collaboration in Natural Resource Management*. Ottawa: International Development Research Centre.

Burkhard, R., Deletic, A., and Craig, A. 2000. Techniques for water and wastewater management: A review of techniques and their integration in planning review, *Urban Water, 2*(3): 197–221.

Carlile, P.R., 2002. A pragmatic view of knowledge and boundaries: Boundary objects in new product development, *Organization Science, 13*(4): 442–455.

Cascão, A.E. 2009. Changing power relations in the Nile river basin: Unilateralism vs. cooperation?, *Water Alternatives, 2*(2): 245–268.

Cash, D.W., Clark, W.C., Alcock, F., Dickson, N.M., Eckley, N., Guston, D.H., Jager, J., and Mitchell, R. B. 2003. Knowledge systems for sustainable development, *Proceedings of the National Academy of Science, 100*(14): 8086–8091.

Chayes, A. and Chayes, A.H. 1991. Compliance without enforcement: State behavior under regulatory treaties, *Negotiation Journal, 7*(3): 311–330.

Chayes, A. and Chayes, A.H. 1993. On compliance, *International Organization, 47*(2): 175–205.

Coase, R. 1960. The problem of social cost, *Journal of Law and Economics, 3*: 1–44.

Connick, S. 2003. "The Use of Collaborative Processes in the Making of California Water Policy: The San Francisco Estuary Project, the CALFED Bay-Delta Program, and the Sacramento Area Water Forum." Environmental Science, Policy and Management. Ph.D. Dissertation. University of California, Berkeley.

Connick, S. and Innes, J. 2003. Outcomes of collaborative water policy making: Applying complexity thinking to evaluation, *Journal of Environmental Planning and Management, 46*(2): 177–197.

Consejería de Medio Ambiente. 2002. *Junta de Andalucía.* http://www. cma-juntaandalucia.es.

Consejería de Obras Públicas y Transportes, Junta de Andalucía. 1996. *El Agua en Andalucía. Doce años de gestión autonómica 1984–1995.* Junta de Andalucía.

Dalla Valle, F. and Saccardo, I. 1996. *Caratterizzazione Idrologica del Piave.* Venezia: ENEL.

Daoudy, M. 2005. *Le Partage des Eaux entre la Syrie, l'Irak et la Turquie: Négociation, Sécurité et Asymétrie des Pouvoirs, Moyen-Orient.* Paris: CNRS.

Datta, A. 2005. The Bangladesh-India treaty on sharing of the Ganges water: Potentials and challenges, in S.P. Subedi (ed.) *International Watercourses Law for the 21st Century: The Case of the River Ganges Basin* (pp. 63–104). Andover: Ashgate Publishing, Ltd.

Deutsch, M. 1973. *The Resolution of Conflict.* New Haven, CT: Yale University Press.

Dietzenbacher, E. and Velázquez, E. 2006. Analyzing Andalusian virtual water trade in an input–output framework, *Regional Studies, 41*(2): 185–196.

Dinar, S. 2008. *International Water Treaties: Negotiation and Cooperation along Transboundary Rivers.* London: Routledge.

Dixit, A. and Gyawali, D. 2010. Nepal's constructive dialogue on dams and development, *Water Alternatives, 3*: 106–123.

Dlamini, E. M. 2006. Decision support system for managing the water resources of the Komati River Basin. Presentation at the Enhancing equitable livelihood benefits of dams using decision support systems Working Conference held by CGIAR Challenge Program on Water and Food on January 23–26, 2006 in Adama/Nazareth, Ethiopia. Environmental Consulting Services (ECS) (undated) The Maguga Dam Project, ECS webpage available at www.ecs.co.sz/magugadam/index.htm

Duane, T.P. 1997. Community participation in ecosystem management, *Ecology Law Quarterly, 24*: 771–797.

Earle, A. and Malzbender, D. (eds.) 2006. *Stakeholder Participation in Transboundary Water Management.* South Africa: African Centre for Water Research.

Elkind, E.N. 2011. "Drops of Energy: Conserving Urban Water in California to Reduce Greenhouse Gas Emissions". Report from the Emmett Center on Climate Change and the Environment and the Environmental Law Center at UCLA School of Law, and the Center for Law, Energy and the Environment at the UC Berkeley School of Law, (available at http://www.law.berkeley.edu/files/Drops_of_Energy_May_2011_v1.pdf).

192 Non-Zero-Sum Approach to Water Negotiations

El-Sadek, A. 2010. Water desalination: An imperative measure for water security in Egypt, *Desalination, 250*(3): 876–884.

Feitelson, E. and Haddad, M. 1998. *Identification of Joint Management Structures for Shared Aquifers: A Cooperative Palestinian–Israeli Effort*. World Bank Technical Paper No. 415. Washington, DC: The World Bank.

Feitelson, E. and Haddad, M. (eds.) 2000. *Management of Shared Groundwater Resources: The Israeli–Palestinian Case with an International Perspective*. Ottawa: Kluwer Academic Publishers.

Fishelson, G. 1994. The allocation of marginal value product of water in Israel agriculture, in J. Isaac and H. Sshuval (eds.) *Water and Peace in the Middle East* (pp. 427–440). Amsterdam: Elsevier Science B.V.

Fisher, R. 1983. Negotiating power: Getting and using influence, *American Behavioral Scientist, 27*(2): 149–166.

Fisher, R., Ury, W.L., and Patton, B. 1981. *Getting to Yes: Negotiating Agreement without Giving In*. New York, NY: Penguin Books.

Folke, C., Hahn, T., Olsson, P., and Norberg, J. 2005. Adaptive governance of socio-ecological systems, *Annual Review of Environment and Resources, 30*: 441–473.

Foster-Fishman, P.G., Berkowitz, S.L., Lounsbury, D.W., Jacobson, S., and Allen, N.A. 2001. Building collaborative capacity in community coalitions: A review and integrative framework, *Journal of Community Psychology, 29*(2): 241–261.

Franzin, R., Fiori, M. and Reolon, S. 2000. "I Conflitti dell'acqua: Il Caso Piave", mimeo.

Freeman, J. and Farber, D.A. (2005). Modular environmental regulation, *Duke Law Journal, 54*: 795–912.

Fuller, B.W. 2009. Surprising cooperation despite apparently irreconcilable differences: Agricultural water use efficiency and CALFED, *Environmental Science and Policy, 12*(6): 663–673.

Fuller, B.W. 2006. "Trading Zones: Cooperating for Water Resource and Ecosystem Management when Stakeholders have Apparently Irreconcilable Differences". Dissertation, Department of Urban Studies and Planning, Massachusetts Institute of Technology.

Furlong, K. and Gleditsch, N.P. 2003. The boundary dataset: Description and discussion, *Conflict Management and Peace Science, 20*(1): 92–117.

Galison, P., 1997. *Image and Logic: A Material Culture of Microphysics*. Chicago: University of Chicago Press.

Gelfand, M.J., Major, V.S., Raver, J., Nishii, L., and O'Brien, K.M. 2007. Negotiating relationally: The dynamics of the relational self in negotiations, *Academy of Management Review, 31*(2): 427–451.

Guan, D. and Hubacek, K. 2007. Assessment of regional trade and virtual water flows in China, *Ecological Economics*, 61: 159–170.

Gunderson, L.H. and Light, S.S. 2006. Adaptive management and adaptive governance in the Everglades, *Policy Sciences, 39*(4): 323–334.

Haftendorn, H. 2000. Water and international conflict, *Third World Quarterly, 21*(1): 51–68.

Harasick, R.F. 1990. Water conservation in Los Angeles, in H.R. French (ed.) *Proceedings of the International Symposium on Hydraulics/Hydrology of Arid Lands*. New York: ASCE.

Heilig, G.K., Fischer, G., et al. 2000. Can China feed itself? An analysis of China's food prospects with special reference to water resources, *International Journal of Sustainable Development and World Ecology*, 7: 153–172.

Hubacek, K. and Sun, L. 2001. A scenario analysis of China's land use change: Incorporating biophysical information into input–output modeling, *Structural Change and Economic Dynamics, 12*(4): 367–397.

Hudgik, C.M. and Arch, M.A. 2003. "Evaluating the Effectiveness of Collaboration in Water Resources Planning in California: A Case Study of CALFED." IURD Working Paper Series, WP 2003–06.

Ingram, H. and Fraser, L. 2006. Path dependency and adroit innovation: The case of California water, in R. Repetto, (ed.) *Punctuated Equilibrium and the Dynamics of U.S. Environmental Policy* (pp. 78–109). New Haven, CT: Yale University Press.

Innes, J.E. and Booher, D. 1999. Consensus building and complex adaptive systems: A framework for evaluating collaborative planning, *Journal of the American Planning Association, 65*(3): 412–423.

Innes, J.E. and Booher, D. 2010. *Planning with Complexity*. New York: Routledge.

Innes, J.E. and Connick, S. 1999. San Francisco estuary project, in L. Susskind, S. McKearnan, and J. Thomas-Larmer (eds.) *The Consensus Building Handbook: A Comprehensive Guide to Reaching Agreement* (pp. 801–828). Thousand Oaks, CA: Sage Publications.

Innes, J.E., Connick, S., et al. 2007. Informality as a planning strategy, *Journal of the American Planning Association, 73*(2): 195–210.

Issa, S. 2002. Access to Lake Chad and Cameroon–Nigeria Border conflict: A historical perspective, in S. Castelein and A. Otte (eds.) *Conflict and Cooperation Related to International Water* (pp. 67–75). IHP-VI (Technical Documents in Hydrology (TDH) No. 62. Paris: UNESCO.

Jansky, L. and Uitto, J.I. 2005. *Enhancing Participation and Governance in Water Resources Management: Conventional Approaches and Information Technology*. Tokyo: United Nations University.

Jessop, B. 1998. The rise of governance and the risks of failure, *International Social Science Journal, 155*: 29–45.

Kallis, G., Kiparsky, M., and Norgaard, R.B. 2009. Adaptive governance and collaborative water policy: California's CALFED bay-delta program, *Environmental Science and Policy, 12*(6): 631–643.

Kiker, C. F, Milon, J.W., and Hodges, A.W. 2001. Adaptive learning for science-based policy: The Everglades restoration, *Ecological Economics, 37*: 403–416.

Kock, B. 2010. "Addressing Agricultural Salinity in the American West: Harnessing Behavioral Diversity to Institutional Design." Dissertation, Department of Urban Studies and Planning, Massachusetts Institute of Technology.

Laak, R. 1982. Integrating onsite system design into sanitary and environmental curricula, in L. Waldorf and J.L. Evans (eds.),*Proceedings of Eighth National Conference on Individual Onsite Wastewater Systems*, Ann Arbor, MI: University of Michigan Press.

Lax, D.A. and Sebenius, J.K. 1986. *The Manager as Negotiator: Bargaining for Cooperation and Competitive Gain*. New York: The Free Press.

Lejano, R.P. and Ingram, H. 2009. Collaborative networks and new ways of knowing, *Environmental Science and Policy, 12*(6): 653–662.

Lewicki, R.J., Gray, B., and Elliott, M. (eds.). 2003. *Making Sense of Intractable Environmental Conflict: Concepts and Cases*. Washington, DC: Island Press.

LHC (Little Hoover Commission) 2005. "Still Imperiled, Still Important: The Little Hoover Commission's Review of the CALFED Bay-Delta Program," Little Hoover Commission Report #183. http://www.lhc.ca.gov/lhcdir/report183.html.

Lund, J. and Palmer, R. 1997. Water resources system modeling for conflict resolution, *Water Resource, 108*: 70–82.

Lurie, S. 2004. "Interorganizational Dynamics in Large-scale Integrated Resources Management Networks: Insights from the CALFED Bay-Delta Program," Unpublished Ph.D. Thesis. University of Michigan.

Mabry, J.B. (ed.) 1996. *Canals and Communities Small-scale Irrigation Systems*. Tucson: University of Arizona Press.

Margerum, R.D. 2008. A typology of collaboration efforts in environmental management, *Environmental Management, 41*(4): 487–500.

Margerum, R.D. 2011. *Beyond Consensus: Improving Collaborative Planning and Management*. Cambridge, MA: The MIT Press.

Martinez, J. and Susskind, L. 2000. Parallel informal negotiation: An alternative to second track diplomacy, *International Negotiation, 5*(3): 569–586.

McCarthy, J., Canziani, O.F., Leary, N. A., Dokken D.J., and White, K.S. 2001. *Climate Change 2001: Impacts, Adaptation, and Vulnerability: Contribution of Working Group II to the Third Assessment Report of the Intergovernmental Panel on Climate Change*. Cambridge, UK: Cambridge University Press.

McGuire, M. 2006. Collaborative public management: Assessing what we know and how we know it, *Public Administration Review, 66*: 33–42.

McKinney, M.J. 1990. State water planning: A forum for proactively resolving water policy disputes, *Water Resources Bulletin, 26*(2): 323–331.

Megdal, S. 2007. Arizona's recharge and recovery policies and programs, in B.G. Colby and K.L. Jacobs (eds.) *Arizona Water Policy: Management Innovations in an Urbanizing, Arid Region*. Washington, DC: RFF Press.

Menkel-Meadow, C. 2008. Getting to "Let's Talk": Comments on collaborative environmental dispute resolution processes, *Nevada Law Journal, 8*: 835–852.

Mustafa, D. 2002. To each according to his power? Access to irrigation water and vulnerability to flood hazard in Pakistan, *Environment and Planning D: Society and Space, 20*(6): 737–752.

Nishat, A. 2001. Development and management of water resources in Bangladesh: Post-1996 treaty opportunities, in A.K. Biswas and J.I. Uitto (eds.) *Sustainable Development of the Ganges-Brahmaputra-Meghna Basins* (pp. 80–99). New York: United Nations University Press.

Nishat, A. and Pasha, M.F.K. 2001. "A review of the Ganges treaty of 1996". Presented to The Globalization and Water Resources Management: The Changing Value of Water AWRA/IWLRI– University of Dundee International Specialty Conference.

Norgaard, R.B., Kallis, G., Kiparsky, M., 2009. Collectively Engaging Complex Socio-Ecological Systems: Re-envisioning Science, Governance, and the California Delta. *Environmental Science and Policy, 12*(6): 644–652.

O'Leary, R., Gerard, C., and Blomgren Bingham, L. 2006. Introduction to the symposium on collaborative public management, *Public Administration Review, (Special issue), 66*: 6–9.

Ostrom, E. 2003. Toward a behavioral theory linking trust, reciprocity, and recognition, in E. Ostrom and J. Walker (eds.) *Trust and Reciprocity* (pp. 19–79). New York: Russell Sage.

Pateman, C. 1970. *Participation and Democratic Theory*. Cambridge: Cambridge University Press.

Pokharel, J.C. 1996. *Environmental Resources: Negotiation Between Unequal Powers*. Noida, India: Vikas Pub. House.

Priscoli, J.D. and Wolf, A.T. 2009. *Managing and Transforming Water Conflicts*. Cambridge, UK: Cambridge University Press.

Putnam, L.L. and Wilson, S. 1989. Argumentation and bargaining strategies as discriminators of integrative outcomes, in M.A. Rahim (ed.) *Managing Conflict: An Interdisciplinary Approach* (pp. 549–599). Newbury Park, CA: Sage.

Putnam, R.D. 1988. Diplomacy and domestic politics: The logic of two-level games, *International Organization, 42*: 427–460.

Radosevich, G.E. and Olson, D.C. (1999). "Existing and Emerging Basin Arrangements in Asia: Mekong River Commission Case Study." Third Workshop on River Basin Institution Development June 24, 1999. Washington, DC: The World Bank.

Rahaman, M.M. 2009. Shared water–shared opportunities: associated management principles, *International Water Resources Update*, 22(2): 14–18.

Raiffa, H. 1982. *The Art and Science of Negotiation*. Cambridge, MA: Harvard University Press, Belknap Press.

Sadoff, C. and Grey, D. 2002. Beyond the river: The benefits of cooperation on international rivers, *Water Policy, 4*(5): 389–403.

Sadoff, C. and Grey, D. 2005. Cooperation on international rivers: A continuum for securing and sharing benefits, *Water International, 30*(4): 420–427.

Schelling, T. 1960. *The Strategy of Conflict*. Cambridge, MA: Harvard University Press.

Scholz, J.T. and Stiftel, B. 2005. *Adaptive Governance and Water Conflict: New Institutions for Collaborative Planning*. Washington, DC: RFF Press.

Schlumpf, C., Behringer, J., Durrenberger, G., and Pahl-Wostl, C. 1999. The personal CO_2 calculator: A modelling tool for participatory integrated assessment methods, *Environmental Modelling and Assessment, 4*: 1–12.

Schon, D.A. and Rein, M. 1994. *Frame Reflection: Toward the Resolution of Intractable Policy Controversies*. New York: Basic Books.

Sebenius, J.K. 1983. Negotiation arithmetic: Adding and subtracting issues and parties, *International Organization, 37*(2): 281–316.

Sgobbi, A. and Carraro, C. 2011. A stochastic multiple players multi-issues bargaining model for the Piave river basin, *Strategic Behavior and the Environment, 1*(2): 119–150.

Sheer, D.P., Ulrich, T.J., and Houck, M.G. 1992. Managing the lower Colorado river, *Journal of Water Resources Planning and Management, 118*(3): 324–336.

Smerdon, T., Wagget, R., and Grey, R. 1997. *Sustainable Housing: Options for Independent Energy, Water Supply and Sewage*. Bracknell: BSRIA.

Susskind, L. and Ashcraft, C. 2010. Consensus building, in J. Dore, J. Robinson, and M. Smith (eds.) *Negotiate – Reaching Agreements over Water*. Gland, Switzerland: IUCN.

Susskind, L. and Crump, l. 2009. *Multiparty Negotiation: An Introduction to Theory and Practice*. Volumes I-IV, Thousand Oaks, CA: Sage Publications

Susskind, L., Fuller, B., Fairman, D., and Ferenz, M. 2003. The organization and usefulness of multi-stakeholder dialogues at the global scale, *International Negotiation: A Journal of Theory and Practice, 8*(2): 235–266.

Susskind, L, McKernan, S., and Thomas-Larmer, S. 1999. *The Consensus Building Handbook*. Thousand Oaks, CA: Sage Publications

Susskind, L. 1999. A short guide to consensus building, in L. Susskind, S. McKearnan, and Thomas-Larmer J. (eds.) *The Consensus Building Handbook: A Comprehensive Guide to Reaching Agreement*. Thousand Oaks, CA: Sage Publications.

Susskind, L. and Field, P. 1996. *Dealing with an Angry Public: The Mutual Gains Approach to Resolving Disputes*. New York: The Free Press.

Susskind, L. 1994. *Environmental Diplomacy: Negotiating More Effective Global Agreements*. New York: Oxford University Press.

Susskind, L. and Cruikshank J. L. 1987. *Breaking the Impasse: Consensual Approaches to Resolving Public Disputes*. New York: Basic Books.

Taylor, K. and Short, A. 2009. Integrating scientific knowledge into large-scale restoration programs – the CALFED Bay-delta program experience, *Environmental Science and Policy, 12*(6): 674–683.

Thompson, A.M. and Perry, J.L. 2006. Collaboration processes: Inside the black box, *Public Administration Review, (Special issue), 66*: 20–32.

Tilmant, A., Van der Zaag, P., and Fortemps, P. 2007. Modeling and analysis of collective management of water resources, *Hydrology and Earth System Sciences, 11*: 711–720.

Times Higher Education Supplement (THES). Analysis. 29 October 1999.

Trottier, J. 1999. *Hydropolitics in the West Bank and Gaza Strip*. Jerusalem: Palestinian Academic Society for the Study of International Affairs.

Velázquez, E. 2007. Water trade in Andalusia virtual water: An alternative way to manage water use, *Ecological Economics, 63*(1): 201–208.

Velázquez, E. 2006. An input–output model of water consumption: Analysing intersectoral water relationships in Andalusia, *Ecological Economics, 56*: 226–240.

Vileisis, A. 1997. *Discovering the Unknown Landscape: A History of America's Wetlands*. Washington DC: Island Press.

Wolf, A.T. 2002. *Conflict Prevention and Resolution in Water Systems*. Cheltenham, UK: Elgar.

Wolf, A.T. 1995. *Hydropolitics along the Jordan River: Scarce Water and Its Impact on the Arab-Israeli Conflict*. New York: United Nations University Press.

Yu, W. 2008. "Benefit Sharing in International Rivers: Findings from the Senegal River Basin, the Columbia River Basin, and the Lesotho Highlands Water Project." World Bank AFTWR Working Paper 1, November.

Zeitoun, M. and Mirumachi, N. 2008. Trans-boundary water interaction I: Reconsidering conflict and cooperation, *International Environmental Agreements, 8*: 297–316.

Zeitoun, M. and Warner, J. 2006. Hydro-hegemony: A framework for analysis of trans-boundary water conflicts, *Water Policy*, 8: 435–460.

6

THE PRACTICE OF WATER DIPLOMACY IN A NUTSHELL

(with Elizabeth Fierman)

There are six key tasks central to the practice of water diplomacy. Each has been described briefly in one of the previous chapters. In this chapter, we take a more hands-on look at what it will take to implement these ideas in various parts of the world.

The first task is to ensure that the appropriate stakeholders and network interests are identified and adequately represented in water management efforts, so that the full range of perspectives and all available local knowledge can be tapped. Second, these parties need to engage in joint fact-finding to generate a shared understanding of how the key variables in the NSPD interact in their particular settings. Factual disagreements, particularly those caused by uncertainty and complexity, need to be discussed, but not necessarily resolved. Third, parties need to create as much value as possible. This usually involves seeking ways to expand the useable quantity of water or the range of water uses through the introduction of new water management techniques or technologies, consideration of how virtual or embedded water can alter the current situation, and trades of various kinds. Value creation tends to be most successful when a mutual gains approach to negotiation is used and a professional mediator or facilitator organizes the conversation. Fourth, informal problem-solving should be used to ensure that the product of informal deliberations is connected to formal decision-making. The product of water negotiations should take the form of proposals (as agreed to by almost all of the parties) that are forwarded to the appropriate political officials for action once the representatives who have produced them have had a chance to review them with their constituents. Fifth, the parties need to suggest how follow-up efforts ought to be organized to ensure that whatever actions are taken can be modified or enhanced as preliminary results become clear. We call this collaborative adaptive management. Finally, the individuals, groups, and organizations involved in the search for negotiated water agreements should spend some time together reflecting on what they learned, so that further capacity-building is possible.

This chapter analyzes each of these tasks from a prescriptive standpoint, considering "best practices" in water diplomacy. Our analysis is followed by a final set of readings that illustrate how each of these tasks has worked in practice.

Stakeholder Representation

The first task in the management of water networks, representation, was discussed in Chapters 2 and 5. The aim should be to ensure that all network interests are identified and invited to select someone to speak on their behalf. Stakeholders, or nodes in the water management network, should include individuals and groups who are, or expect to be, affected by water allocations, water management, or water policy decisions. Having a full range of stakeholder interests represented is vital to the credibility of every water diplomacy effort. For one thing, excluded individuals or groups, as we noted in earlier chapters, may have important local knowledge that will be left out if they are not involved. Moreover, implementation of whatever agreements the groups reach will be a lot easier if everyone affected has a chance to make their interests known. Groups that are excluded may feel obliged to block implementation of the results of negotiations, arguing that the outcomes are illegitimate because they were not allowed to participate.

To ensure adequate stakeholder representation, a stakeholder assessment, of the sort described in Chapter 5, should be conducted by a professional neutral. Such assessments can help not only clarify who should be involved, but also provide a means of engaging stakeholder groups in the design of whatever informal problem-solving process follows. A stakeholder assessment typically requires a neutral party to undertake confidential interviews with a widening circle of potentially interested parties. The assessor needs to summarize his or her findings in a draft document that goes to everyone interviewed for their review. While they will not see their names mentioned, since all interviews are confidential, they should be satisfied that the concerns they raised and the priority they attached to various issues are incorporated into the design of the problem-solving process proposed by the assessor. Once all the relevant network interests have been identified, hard-to-fill categories—which may require the identification of proxies or surrogates—can then be added (Susskind et al 1999).

After the relevant stakeholder groups have been identified and have had an opportunity to review the proposed agenda, timetable, ground rules, and budget suggested by the assessor, the convening agency must decide whether or not it wants to proceed. If it does, it can follow the process the participants helped to design with confidence that all of the key players have agreed to come to the table. The facilitator may need to caucus groups in a broad stakeholder category, like environmental activists, to let them choose a representative to speak for them in the problem-solving process that follows. It is always best if the parties in a stakeholder category choose their own representative, rather than waiting for the convening agency to handpick the people it prefers.

Some groups may require technical assistance before the negotiations begin. That is, they may need help canvassing their members, getting their representative up-to-speed on technical matters, or thinking through their priority interests and the trade-offs they will be willing to make. Such help can come from the facilitator as long as every group has the option of asking for similar assistance (Susskind and Cruikshank 2006).

Joint Fact-Finding and Scenario Planning

Once the relevant parties have been identified and brought to the table, the negotiators will need to consider the scientific and technical information required to address the water management decisions they face. Such information is usually quite complex, and multiple interpretations are almost always possible. As a result, allowing parties to generate their own forecasts or analyses is likely to generate further disagreement. Instead, computer-assisted modeling and/or other group decision analysis software should be used to make it easier for the group to reach informed agreement (van den Belt 2004).

The most difficult aspects of joint fact-finding are described in Chapter 5. As we noted, it is important to blend local or indigenous knowledge with expert scientific advice. This can be difficult, especially if technical experts do not respect local sources of expertise. The group also needs to confront the fact that its "findings" are likely to be sensitive to a range of non-objective judgments. For example, since there is no "correct" way to value all possible damage to local water resources, the group should consider how alternative methods of monetizing environmental or cultural impacts might lead to very different decisions. Even slight changes in the way visual landscapes or threats to human safety might be affected by water management plans deserve consideration. Also, as we noted in several earlier chapters, the complexity of water networks makes forecasting difficult. We suggest using scenario planning or some other technique that enables the parties to deal with uncertainty and complexity in a contingent fashion. (Wright and Cairns 2011)

Joint fact-finding can be used to blend differing interpretations of policy or management options. Engineers trained to produce integrated water resources management (IWRM) plans will have their own ideas about how to frame the choices that a water network ought to be considering. They tend to think in terms of bounded systems and optimization, as we discussed in Chapter 2. Indeed, funding agencies may insist that the principles of IWRM be included as part of any new water-related investment project (World Bank 2007). Other stakeholders, however, may have different perceptions about the most desirable options and the best way forward.

Once believable forecasts have been generated, the group will have to decide how it wants to handle data gaps and differences in interpretation. Joint fact-finding does not eliminate disagreement, it only makes clearer what the parties are likely to accept as common information and where and why they disagree about interpreting this material. Agency participants often explain that their legal or administrative mandates force them to give priority to certain kinds of analyses and to reject others. Some agency representatives will say that their hands are tied, requiring them to use a particular discount rate or to ignore impacts that might happen far into the future. For example, U.S. Army Corps of Engineers' guidance on hydropower dams limits cost and benefit analysis to "a period not to exceed 50-years except for major multiple purpose reservoir projects," meaning that any costs or benefits beyond 50 years should be set at zero (U.S. Army Corps of Engineers 2004). It is up to the mediator to draw attention to such sources of disagreement and to help the group look for contingent agreements that allow them to proceed even in the face of different assumptions about the future (Susskind et al 1999).

Value Creation

A central task of water diplomacy is value creation. This means searching for more efficient uses of water to meet multiple, often conflicting, interests as best as possible. In other words, value creation involves both understanding each stakeholder's core concerns—or interests—and thinking as creatively as possible about how to expand the available water supply so that non-zero-sum outcomes are possible. This is likely to involve the introduction of new technologies or new patterns of development. As discussed in Chapters 2 and 5, when we think of water as a flexible resource, value can be created by imagining new agricultural or industrial practices that release embedded water for other uses. Once locational trades or trades over time are considered, as discussed in Chapters 3 and 5, the prospect of sharing virtual water becomes possible. Creating value, therefore, implies not just a shift from an adversarial to a collaborative style of deliberation, but also a shift from viewing water as a fixed resource to viewing it as flexible and expandable.

Value creation is most likely to be accomplished if the parties consider ways of meeting their own interests while simultaneously meeting the interests of others. Brainstorming along these lines works if the parties agree that ground rules—including suspending criticism of all proposals until as many as possible have been noted—should be enforced by a mediator (Susskind and Cruikshank 2006). In general, value creation is more successful when efforts are made to build trust among the parties, encourage option generation, and aim for conversations that are respectful of cultural, educational, and political differences.

A mutual gains approach to negotiation helps with these tasks by emphasizing and clarifying the interests of the parties and generating multiple ways of meeting them before committing to particular solutions. It is common to use a professional mediator to support such efforts. This approach allows stakeholders in complex water management networks to generate results that are better than what they would be likely to achieve through solutions imposed by authorities at any level. The goal of the mutual gains approach to negotiation is not an agreement in which everyone "wins" everything they want, but rather an agreement that meets all parties' interests better than if there were no agreement at all.

Value creation is usually the product of trades. If B's support is the key to A getting something it wants very much, and B would gladly give that support as long as A promises something that is equally valuable to B, then neither side is being asked to compromise. Rather, they are creating value through a trade that exploits differences in their priorities, interests or values. It is hard to do this without an extended list of items to trade. That's why value creation is so difficult when parties break their agenda into pieces and deal with issues one at a time. Instead, they need to rely on brainstorming to enrich their agenda. And then, nothing should be decided until everything is decided. This is the way to encourage exploration of possible packages without anyone locking into something they might be hesitant to support. New technologies that permit multiple uses (or reuse) of the same water make mutually beneficial trades possible. This is what we mean when we say that it helps to think about water as a flexible rather than a scarce resource.

Convening

Problem solving in the water domain is most likely to be successful if informal efforts to generate mutually beneficial proposals are clearly linked, from the outset, to formal decision-making by government agencies or officials. Otherwise, parties might not be willing to invest time or energy in ad hoc efforts to search for solutions to cross-boundary disagreements or conflicts. To manage what might be called the "governance connection," informal negotiations ought to be convened by one or more agencies or organizations with formal decision-making authority. It helps if the convening agency is prepared to make an explicit commitment to support proposals that emerge from informal problem-solving as long as representation is handled properly. The agencies or legislative bodies should designate staff to participate in or at least monitor the efforts to produce an informed consensus.

In addition to clarifying the link between ad hoc and formal governance, conveners should play other roles as well. First, they should familiarize themselves with the interests of the key stakeholders. Commissioning and paying attention to the results of a stakeholder assessment will make this relatively easy. Second, conveners should be prepared to contribute or help locate financial and other resources needed to support the involvement of independent experts in joint fact-finding. Third, they should be willing to identify and work with a professional mediator and not try to control all aspects of the process themselves. Joint convening, by agencies at multiple levels, is not uncommon.

Collaborative Adaptive Management

We assume that even the best-intentioned and most carefully structured informal problem-solving efforts will run up against the complexity and uncertainty that challenge almost all water management efforts. Even if the right parties are at the table with an effective mediator, and even if the convener has provided the support the group needs to engage in joint fact-finding, what they do not know about the dynamics of a particular water network will probably exceed what they do know. This is especially true with regard to interactions among the societal and natural variables that comprise NSPD.

By taking a collaborative adaptive management (CAM) approach to each battle over water allocation, water resource development, or water policy, parties in water management networks will have a better chance of implementing the decisions made by the agencies with the formal authority to act. In other words, CAM offers an opportunity for parties to participate in implementing decisions. Accordingly, decision-making agencies should view stakeholders who advise them as allies, not adversaries. They also ought to view the relationships among these parties, which are enhanced when the water diplomacy framework is followed, as a kind of social capital that can support implementation of negotiated agreements (Putnam 2002).

CAM assumes that water network managers will never get everything right on the first try. So, whatever they decide to do (informed by the proposals they receive from the stakeholders) is likely to fall short. If such efforts are viewed as "experiments," however, they can provide information and insights that will allow recalibration of policies, programs, and plans. They might also lead to reconsideration of longer-term goals and objectives. For such adjustments to be successful, network managers need to invest in careful monitoring. Stakeholders can help with this: for example, water users may be in the best position to gather data on whether intended efforts are working. If, as part of negotiated agreements, participants in informal problem-solving specify what they think needs to be measured and how proposed efforts will be assessed, there is a good chance that mid-course corrections will be successful. In sum, collaborative adaptive management offers an approach to implementation that leaves room for ongoing adjustments or reconsideration of earlier decisions (Camacho et al 2010).

Societal Learning

Water diplomacy is not just about resolving specific boundary-crossing conflicts. It also seeks to improve the management of water networks more generally through capacity building and societal learning. This implies enhancing the knowledge and capability of individuals, organizations, and networks over time. Whenever a water management network succeeds in generating a way to resolve a particular conflict (or, even if it fails in its efforts to do that), reflection on that experience should be used to strengthen the underlying capacity of the agencies and actors involved. Even a modest effort along these lines can make it easier to handle similar problems more effectively in the future. Making lessons learned explicit will help inform actors in other water networks and other contexts as well. Water management networks should make an explicit commitment to knowledge transfer and capacity building.

Water management networks should also take advantage of whatever assistance they can get from adjacent agencies and organizations. For example, the Social Learning Group, which includes scholars from nine countries representing multiple disciplines, offers helpful resources for water managers interested in societal learning and capacity building. This group has analyzed theories of social learning and applied them to research on how social learning relates to management of global environmental risks (Social Learning Group 2001). Much of this work reinforces our prescriptions around collaborative adaptive management.

The Social Learning Group is also an example of how a support network can be created to further scholarship, while simultaneously allowing colleagues to strengthen each other's capacities and produce resources that others can easily access. If knowledge transfer and capacity building are explicit network goals, then resources and research, along with lessons learned from practice, should be made as widely available as possible.

Conclusion

These "best practices" of water network management reflect our three core challenges to the conventional wisdom presented in Chapter 1: water systems are not bounded, rather they operate more as open networks; they are constantly changing in unpredictable ways; and water is a flexible, not a fixed, resource. They also reflect our analysis of what others around the world have actually been doing to manage water networks, beyond just what the conventional wisdom would have them do when tough decisions need to be made. On the basis of these propositions and real life inputs, we offer a water diplomacy framework that stresses the need to: account for and engage diverse and multi-sectoral interests; use joint fact-finding and collaborative adaptive management to take decisions in the face of uncertainty and flux, and adjust those decisions as necessary; and take advantage of the potential to create value through a mutual gains approach to multi-party negotiation. We have tried to be analytical and prescriptive, explaining what we think is required to resolve water management disputes and why. We have also tried to be practical, giving guidance on how to proceed in actual network management situations (see Figure 6.1).

The next chapter offers a set of teaching materials to help water managers and stakeholders practice using the water diplomacy framework.

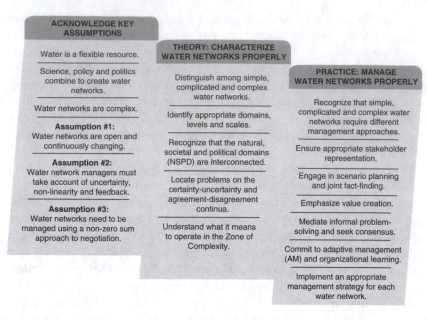

FIGURE 6.1 The Water Diplomacy Framework (WDF)

Selected Readings with Commentaries

B.N. Tapela "Stakeholder Participation in the Transboundary Management of the Pungwe River Basin," (2006)

Introduction

Barara Tapela's article discusses stakeholder engagement in the Pungwe River Basin Joint Integrated Water Resources Management Strategy, an initiative launched by Mozambique and Zimbabwe in 2002, with funding from the Swedish International Development Cooperation Agency, to facilitate joint management, development, and water conservation in the river basin.

The Pungwe River originates in Zimbabwe's Eastern Highlands, flows across the international border with Mozambique, and ultimately empties into the Indian Ocean on Mozambique's coast. Only 5 percent of the river's area is in Zimbabwe, but this region contributes significantly to the river's flow. Water scarcity in areas within and adjacent to the basin, as well as anticipated increases in water demand in both countries, are the primary challenges for water managers. To help address these challenges, and in response to international efforts to promote participatory water management, both countries formed multi–stakeholder committees. These committees included representatives of major water users and local government authorities, as well as regional water authorities and members of civil society groups in Mozambique.

The excerpt describes Zimbabwe's experience with stakeholder participation in the Pungwe initiative, highlighting some of the difficulties the participants faced that might have been better addressed through the approach to stakeholder engagement we have prescribed.

Excerpt (pp. 15–23)

4. Stakeholder Participation: Zimbabwean Experience

Stakeholder participation in the management of the Pungwe watercourse in Zimbabwe formally began in July 1999 when the Pungwe Sub-Catchment Council was established as a constituent of the neighbouring Save Catchment Council. The Pungwe watercourse does not form part of the Save watercourse system, and this arrangement was therefore intended for administrative expediency. Both the Pungwe Sub-Catchment Council and the Save Catchment Council were established with funding from the Swedish International Development Cooperation Agency (Sida), and they became fully operational in January 2001. Both the Pungwe and the Save councils represent various identified stakeholders within Zimbabwe, and are assisted in the more technical issues by the Zimbabwe National Water Authority (ZINWA), through services of a Catchment Manager and a Training Officer. By April 2005, the Pungwe Sub-Catchment Council had registered one hundred (100) water permit holders in Zimbabwe. These constituted 66 percent of all the water users within the sub-catchment, and commanded a share of 60,000 megalitres from the Pungwe. The remaining 33 percent of water users had yet to be issued with valid permits.

At the sub-catchment level, representation in the Pungwe Sub-Catchment Council includes large-scale commercial farmers, various sub-groups of small-scale commercial farmers, communal farmers, farmers in resettlement schemes and local authorities—represented by traditional leaders and councilors elected into the Mutasa Rural District Council (Table 6.1). Representation in the sub-catchment council is either by consensus or by election. Stakeholder groups have the freedom to choose the mode of their representation. Once nominated or elected into council, membership of the Pungwe Sub-Catchment Council is vested in the person of the representative. Membership does not vest with the stakeholder group or institution represented. Effectively therefore, the resignation or exclusion of a member does not automatically entitle the stakeholder group to nominate or elect a replacement. Rather, it is the prerogative of the council to nominate or call for elections of a new member. This strategy is aimed at ensuring the accountability of councilors in their personal and not institutional capacity.

Representation in the Save Catchment Council includes two members of each sub-catchment council in the catchment council's area of jurisdiction. In 2001 (Tapela, 2002) and in 2004 (Kujinga & Manzungu, 2004), farmers were the most

TABLE 6.1 Stakeholder Representation in the Pungwe Sub-Catchment Council, 2001

Stakeholder	Interest group represented
Large-scale commercial farmers	Commercial Farmers Union (CFU)
Small-scale commercial farmers	Banana Growers' Association Coffee Growers' Association Vegetable Growers' Association
Communal farmers	Zimbabwe Farmers' Union (ZFU)
Farmers in resettlement schemes	Various individuals and groups of resettlement farmers
Local authorities	Residents and institutions of Mutasa Rural District Council

represented group, while the local authority represented was the Mutare City Council. However, attendance of meetings by the City Council was erratic, with the council not attending most of the meetings. Furthermore, the City Council appears to have shunted the responsibility of representing the residents of Mutare in the Save Catchment Council to water engineers employed by the council. In view of that the Mutare City Council is a key stakeholder in the management of Pungwe water resources, the lack of its active involvement in catchment council activities has been a source of concern for Save Catchment Council members, who feel that the City Council undermines their authority and efforts in articulating the new water policy. Assuming that elected councilors have greater responsibility to be accountable to their constituencies in Mutare than employees of the City Council, the power play between the Mutare City Council and the Save Catchment Council effectively denies the Mutare constituency an opportunity to voice their concerns with regard to water problems that affect them, such as water pollution, supply and sanitation.

5. Challenges to Effective Stakeholder Participation

Studies of both the Pungwe Sub-Catchment Council (Tapela, 2002) and the Save Catchment Council (Tapela, 2002; Dube & Swatuk, 2002; Kujinga & Manzungu, 2004) show that stakeholder participation is highly complex, and that there is a need to go beyond looking at mere inclusion in decision-making structures and to examine in depth the nuances of stakeholder roles, resources and relationships, in particular power relations.

5.1 Accelerated Process of Forming Stakeholder Institutions

The formation of the Pungwe Sub-Catchment Council and the Save Catchment Council was modeled on institutional design developed in the pilot phase of Zimbabwe's water sector reforms. Studies by Latham (2001), GTZ (2000a) and Sithole (2000) indicate that there was a general lack of effective stakeholder participation in policy formulation in the pilot phase of Zimbabwe's water reforms. Policy discourses effectively remained top-down (Sithole, 2000), and the stakeholder identification process did not involve public participation (Latham, 2001). Despite this, the insights derived from the Mazoe and the Manyame Catchment pilot projects were to be extrapolated to the remaining five catchment areas in the country … Donor agencies were reported to have been particularly keen for the Save CC and its constituent SCCs to adopt the pilot Catchment Council models, as a way of saving costs.

Stakeholders within the Save Catchment Area appear to have been critical of the approach used in the formation of the pilot Catchment Councils. However, certain developments at the national political and macro-economic levels that coincided with the inception of the councils seem to have compelled them to adopt an even less participatory approach. In particular, the initiation of the government's "fast track" land redistribution programme seems to have exerted ripple effects on the water sector and put political pressure on stakeholders to fast track the water redistribution process. At the same time, the insistence by the IMF on the government to cut spending on public service also seems to have accelerated the process of devolution of authority to the river basin institutions.

The result of reducing the scheduled six month inception period for the Save Catchment Council and Pungwe Sub-Catchment Council to a mere six weeks was that the process of council formation was top-down. The fast tracking of the formation of river basin institutions seems to have created a number of difficulties, most of which related to the transaction costs of the reforms. Notwithstanding the acceleration of the devolution process, these transaction costs were also directly related to the persistence of the sectoral approach in water resources management in Zimbabwe.

5.2 Transaction Costs of Water Sector Reforms

The transaction costs included coordination and communication, both within the water governance hierarchy and in terms of related sectors, as well as enforcement of the new water laws. The acceleration of the devolution process resulted in the river basin institutions assuming responsibilities before they had the necessary capacity to implement Integrated Catchment Management (ICM). The Save Catchment Council and the Pungwe Sub-Catchment Council had yet to acquire office premises, communication links and personnel for monitoring and enforcement. These setbacks seemed to have been effectively addressed through funding from SIDA as well as the ingenuity of the councillors. The more difficult challenge, however, related to the coordination of ICM planning.

The study found that there had been lack of effective coordination and consultation in the drafting of the Preliminary Catchment Outline Plan. The problem seems to have emanated mainly from the fast tracking of the inception of the Save Catchment Council and its constituent Sub-Catchment Councils. The first draft of the Outline Plan retained the traditional water management focus on surface water supply, to the exclusion of water demand management and the management of groundwater sources, particularly for primary use in the rural areas. A particular concern was that this focus did not take into account the fact that surface water scarcity severely affects three of the seven Sub-Catchments in the Save Catchment Area, and that rural people in these areas rely mostly on boreholes and wells. The focus on surface water also failed to address the view by the new water policy that all water, whether it occurs as surface water, groundwater or other forms, constitutes part of the same watercourse system, and should be managed as such. The Catchment Planning process that was used therefore went against the ethos of IWRM.

The Department of Natural Resources (DNR), which is the sector mainly responsible for catchment protection in terms of the Natural Resources Act of 1996, viewed the lack of effective coordination and consultation as having resulted in the drafting of a Catchment Outline Plan that was based on inadequate knowledge of the environmental conditions within the Save and Pungwe Catchment Areas. There was possibility therefore that some of the envisaged water development projects might have profound negative impacts on the security of downstream communities and ecosystems during periods of drought.

Local government sector officials considered that the lack of effective coordination and consultation in the catchment planning process had resulted in discrepancies between the needs perceived by councillors in the Catchment Councils and Sub-Catchment Councils and the actual needs perceived by local people. It is perhaps worth noting that in terms of the government's decentralization policy, the local government ministry, through local authorities, has the responsibility for coordinating local level service provision by the various sectors. This role includes the coordination of services related to primary water supply and sanitation. In terms of the new water policy, Catchment Councils and Sub-Catchment Councils are vested with the responsibility for coordinating water resources use, development and management at the catchment level, which transcends the local authority administrative boundaries. The reported lack of coordination of functions between the river basin institutions and the local authorities might therefore have potentially critical implications on social security issues such as basic water requirements, livelihoods, health and sanitation.

The lack of effective coordination was ascribed by some local government officials to the lack of a synergy between the new Water Act and related Acts administered by other sector agencies. Hence, although the legal instruments were not necessarily in conflict, the local level articulation of policies by Sub-Catchment Councils and local authorities tended to dovetail. A closer examination of the mandates of the various water related sectors seemed to indicate that the problem lay also with the institutional actors' failure to develop new protocols of organizational behaviour in line with the recent shifts in the water sector.

Indeed, sentiments were expressed that there seemed to be some resistance by some established local authority actors to the new river basin institutions, who were felt to be usurping the political action space. In some cases, Rural District Council (RDC) personnel were said to have refused to participate in the sub-catchment planning process. Save Catchment Council records also show that a key stakeholder local authority, the Mutare City Council, had failed to attend more than ninety percent of the meetings held up to the time of the research.

The lack of effective coordination was also due to overlaps in the relative alignment of administrative and catchment boundaries. The Save Catchment Council and its constituent Sub-Catchment Councils, including the Pungwe, viewed some of the overlaps as inconvenient to ICM, and considered that certain portions of some Sub-Catchment Councils be managed by adjacent Catchment Councils since the places were more accessible from those Catchment Areas. In the case of the Pungwe Sub-Catchment Council, although the source of the Pungwe River was located in the southernmost portion of the Nyanga Rural District Council, the local authority did not participate in decision-making by the Sub-Catchment Council. This was mainly due to poor accessibility, and the fact that the said portion of the Nyanga RDC was predominantly comprised of a National Park, under the jurisdiction of the Department of National Parks, and the Large Scale Commercial Farming sector. The latter was represented in the Pungwe Sub-Catchment Council. While the Water Act of 1998 identifies local authorities within particular catchments as stakeholders, this situation points to a need for flexibility in the ICM framework in order to balance the legal requirement for stakeholder constituency representation with what is practically feasible at the operational level.

There were overlaps of functions in the institutional arrangement between ZINWA and the Save Catchment Council. The provision for the Catchment Manager to perform water allocation duties on behalf of the Catchment Council during intervals between Catchment Council meetings was made in order to facilitate expedience in the issuing of permits. However, the arrangement was likely to cause problems of coordination. Since the Catchment Manager was accountable to ZINWA, it seemed possible that some decisions might reflect the interests of the water parastatal rather than the stakeholders. Some respondents attributed the absence of conflict, in the case of the Save Catchment Council and the Save Catchment Manager, to the compatibility of personalities between the Chairman of the Catchment Council and the Catchment Manager. It would seem as if although there has been a degree of devolution of authority to lower level institutions, there remains a degree of state control over local stakeholder decision-making processes.

The problem of coordination might be ascribed, to a significant extent, to power relations between institutional actors in the various sectors. Competition for control over the political action space was prominent between the new river basin institutions and established local authorities. The latter have been responsible for coordinating development activities by the various sectors in the local administrative areas since the government's decentralization process started in 1984. By contrast, river basin institutions are more recent structures and they have yet to strengthen their capacity to carry out ICM responsibilities and thereby instill confidence among various interest groups. Since the ability of an institution to perform duties vested upon it contributes to the acceptability of institutional content, procedures and processes, the problem of coordination by the Save CC and the SCCs would therefore appear to be linked also to the issue of institutional legitimacy.

5.3 Institutional Legitimacy

There seemed to be a plausible link between the problem of legitimacy of the river basin institutions and the top-down process of council formation and accession into office by SCC councilors within the Save catchment area by nomination rather than election. However, legitimacy does not only derive from a democratic process of accession into office by stakeholder representatives. Rather, legitimacy in water resources governance derives more strongly from the extent to which the stakeholder representatives are seen to balance the pursuit of the interests of their local constituencies on the one hand with those of the broader watercourse, national, regional and global resource communities on the other.

Sentiment was expressed that most councilors in the Sub-Catchment Council pursued self-interest or the interests of their constituencies at the expense of the interests of the broader local community. While interviews with the councilors could not fully verify this, primary observation showed that representation in the Save Catchment Council and the Pungwe Sub-Catchment Council was heavily skewed towards representation by male members of the commercial farming sector. This seemed to have resulted in the preoccupation by the institutions with issues relating to the commercial use of water, particularly for irrigation purposes. Not much attention was given to issues of primary, industrial and recreational water use. Such omission raised questions on the commitment of the institutional actors to address the interests of stakeholders other than those of the majority constituencies represented.

The dominance of commercial farming interests seemed directly linked to stakeholder power relations and the requirement that water management institutions should finance their operations through levies collected from commercial users. The focus on commercial water use resulted in institutional failure to address some of the major water problems pertaining to Pungwe water resources. These included the wastage of Pungwe water by Mutare City Council and the discharge by Mutare's manufacturing and processing industries of industrial pollutants into streams flowing through Mutare. Problems also included challenges of addressing interests of downstream users, such as Beira City in Mozambique, which relies on the Pungwe River flow to prevent the salt-water intrusions that cause water shortages in the dry season. The catchment management institution seemed more concerned with the failure by Mutare City Council to attend the Save Catchment Council meetings than with the unaccounted for water loss of approximately 50 percent of the water obtained through the Pungwe Mutare Water Supply Project. It seemed as if water pollution, inefficiency in water use and interests of downstream water users in Mozambique, were less important than the levies that the water institutions generated from Mutare's use of the Pungwe water.

Notwithstanding the observed shortcomings of the Save Catchment Council in demonstrating commitment to serve the interests of the broader resource community, the Pungwe Sub-Catchment Council, by contrast, was found to have made a conscious effort to address more broadly the various interests ranging from the local to the international level. Such robustness was evident in the attempt by the Sub-Catchment Council to enhance gender representation and in the expressed objective for the Sub-Catchment Council to be directly involved in the interstate discourses concerning the use, development and management of the Pungwe watercourse. There was still need, however, for the Pungwe Sub-Catchment Council to give the issues of representation and legitimacy a more rigorous treatment. With regard to strengthening institutional legitimacy, primary observation seems to suggest that the discrepancies in the management styles of the Save Catchment Council and the Pungwe Sub-Catchment Council owed more to personalities and stakeholder power relations than to any fundamental differences in organizational culture.

5.4 Power Dynamics

Empirical observations of stakeholders in the Pungwe Sub-Catchment Council (Tapela, 2002; Dube & Swatuk, 2002) and the Save Catchment Council (Tapela, 2002; Dube & Swatuk, 2002; Kujinga & Manzungu, 2004) reveal that the issue of power pervaded relations among stakeholders and between water management institutions and other sector agencies, such as local authorities, other government agencies and non-governmental organizations (NGOs). Among stakeholders, power-distributing cleavages included interests in water resources, political and economic clout, gender, proficiency in the language of discourse and personality.

Language constituted a source of power among stakeholders. Despite that many of the stakeholder representatives spoke Shona as their first language and that most of the other representatives were conversant in Shona, the language used in Catchment Council and Sub-Catchment Council deliberations was English. In the observed Save Catchment Council meeting, this seemed to contribute to the difficulty of expression for some participants. By contrast, although English was also used in the Pungwe Sub-Catchment Council, where 90 percent of the councillors were first language Shona speakers, the debates at the observed meetings were very lively, and councillors showed a remarkable command of the language and confidence in expressing their needs. Indeed, one of the vocal members of the Pungwe Sub-Catchment Council had been reticent at the previous Save Catchment Council meeting. A follow up to this observation revealed that the reticence by some councilors in the Save Catchment Council was due to the dominance of certain personalities, who made it difficult for most other councillors to participate actively in the meetings. This seemed to point to the need for the Save Catchment Council to enhance its capacities and mechanisms in using participatory approaches, so that the representation of stakeholder interests could be more effective.

Competition of interests among stakeholders was manifest in the emergence of alliances among stakeholders belonging to the similar sectors. On the one hand, there had emerged an alliance between small-scale and large-scale commercial farmers, which enabled this group to dominate both the Save Catchment

Council and the Pungwe Sub-Catchment Council. The result was a strong emphasis by river basin institutions on irrigation-related issues. On the other hand, elected and traditional rural local authority representatives were observed to voice the primary water interests of constituencies in Mutasa Rural District Council. However, because perhaps rural local authority representatives are a minority or because their constituencies do not contribute much to levies required for Catchment Council and Sub-Catchment Council operations, their clout in the decision-making processes is visibly lower than that of irrigation farmers. Given that Mutare City Council generates a significant share of revenue raised through levies by the catchment council, it is possible that had the Mutare City Council participated more actively in the Save Catchment Council meetings, there might have been a balance between the interests of farmers and stakeholders such as local authorities.

The power dynamics between the Save Catchment Council and the local government sector also seem to hinge on the issue of political clout. Whereas the latter had either established their authority over the years or through the ballot and networks, the former were only recently nominated. Consequently, the competition for political action space outside catchment council and Sub-Catchment Council meetings tended to be dominated by the local authorities. While the roles of the river basin institutions and the local authorities were indeed complementary, the power relations between the two undermined the integration of water management activities at the local level.

Consequently, water resources management activities at the rural local level were divided into two distinct domains. Groundwater and primary water supply were the domain of the RDCs, who were mandated to coordinate the implementation of the government's Integrated Rural Water Supply and Sanitation Programme (IRWSSP). Despite that the new water policy mandates the Catchment Councils and Sub-Catchment Councils with the integrated management of all components of a watercourse system, these institutions had so far avoided concern with groundwater and focused almost exclusively on surface water sources and commercial use of water. Atmospheric water, particularly as it related to the efficiency of water use in rain-fed and irrigated crop production, was also ignored. By reinforcing the traditional distinction between the various components of the watercourse system, the power politics between the river basin institutions and the RDCs contradicted the philosophy of IWRM.

The lack of an integrated approach in the management of the various components of the watercourse system seems to have been inherited from the pilot phase structure of organizational sector functions [Figure 6.2]. The organizational functions were allocated according to the sources of water traditionally used or managed by the main sectors. Hence, while there was coordination by the Steering Group at the higher levels, there was no integration of functions at the lower levels of the hierarchy.

Despite that women have been identified as playing a central and multifaceted role in the provision, use and safeguarding of water ... their involvement in

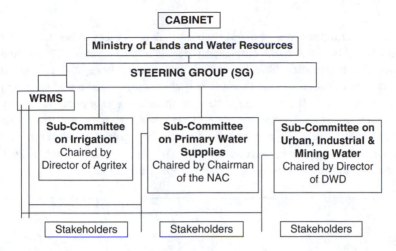

FIGURE 6.2 Zimbabwe water resources management strategy: organizational chart
Source: Zimbabwe, 1995: 10

water-related decision-making structures has been very low. The Save Catchment Council was largely composed of men. Within the entire Save Catchment Area, women councillors constituted 3.5 percent of the total number of Sub-Catchment councilors in 2002 (Tapela, 2002). Of the seven Sub-Catchment Councils, the Pungwe Sub-Catchment Council had made the greatest effort to actively involve women in decision making and planning, with women occupying 20 percent of the Sub-Catchment Council seats out of the council's gender representation target of 60 percent (Ibid.).

TEXT BOX 6.1

Apparently, female members of the Gatsi Irrigation Scheme, near the Mtarazi Falls, staged a strong protest against the Save CC and the Pungwe SCC during a visit by representatives of the donor agency. Many of the women were single heads of households, who eked out livelihoods from micro-scale food production for sale and for household use. Their protest followed a decision by the Save CC to allow 1500 litres per household per day for primary purposes, the excess being levied at commercial rates. The protest also allowed a communication by the Pungwe SCC of the requirement for commercial and semi-commercial water users to pay water permit application fees and water levies. The women responded by expressing a concern to the SCC about their inability to afford to pay the new water prices. When this concern was not given due consideration by the Pungwe SCC, the women insisted that their male leader further engage the SCC about their grievances. When the leader refused to be involved in challenging decisions by Pungwe SCC and the Save CC, the women gave him a vote of no confidence and proceeded to launch a direct challenge themselves. Hence, they staged a protest during the donor's visit. The SCC responded by altering the representation policy to allow two seats for representation of women micro-scale and small-scale farmers. Effectively, women became formally recognized as a stakeholder group within the Pungwe subcatchment. The seats gave women a 20% measure of inclusion in the SCC. (Source: Tapela, B.N. 2002)

In addressing the issue of gender representation, the Pungwe Sub-Catchment Council has been robust enough to adopt a gender-responsive approach, against the prevailing tide of social attitudes that militate against women's involvement in strategic decision-making. However, the inclusion of women in the Sub-Catchment Council has largely been due to women's agency in staking a claim in the decision-making process, with donor support (Text Box 6.1).

Such inclusion of women in decision-making structures, however, does not automatically ensure that women's interests are voiced, as there exist power relations between men and women that result in unequal gender voices. Primary observation of the decision-making process by the Pungwe Sub-Catchment Council pointed to a need for the adopted gender approach to go beyond the issue of gender inclusion, and to enhance institutional capacities and mechanisms that build women's confidence in expressing their views.

Commentary

Many of the challenges to stakeholder participation raised in this excerpt would have been better addressed through a more comprehensive stakeholder engagement process. It does not appear that there was a preliminary systematic assessment of stakeholder groups and their interests. Such an assessment might have generated a more justifiable selection of participants and, thus, been perceived as more legitimate. A comprehensive stakeholder assessment also might have identified issues that were shortchanged, such as groundwater use and the needs of downstream users. Finally, an assessment might have suggested ways to handle overlapping management functions and catchment boundaries, as well as the sharply unequal power distribution among the participants.

The process was also flawed in its approach to stakeholder representation. Council membership was vested in individuals, instead of organized stakeholder groups, making it unclear whether the process was meant to involve individual water users who could speak for various categories of stakeholders or not. It is also not clear what the benefit of "ensuring the accountability of councilors in their personal and not institutional capacity" might mean. In any case, the "erratic" attendance of representatives probably made any effort to hold them accountable for speaking on behalf of a constituency unworkable. Indeed, the author never suggests that the participants should have engaged larger groups before or during their deliberations.

While the article notes that some stakeholder groups, such as small and large commercial farmers, formed alliances in an effort to boost their power, it offers a rather undeveloped analysis of coalition formation as a means of equalizing power and knowledge in joint problem-solving. For instance, Tapela suggests that more active participation by the Mutare City Council could have balanced the interests of commercial farmers with those of other, less powerful stakeholder groups. The author does not, however, point to ways that those "weaker" interests, not to mention Mutare constituents, would have been able to pressure the City Council to increase its participation, or form a coalition, to further their interests.

The author also does not explain who managed the process of stakeholder engagement, or what they thought their goals were. Nor is it clear what responsibilities the stakeholders thought they had, whether they were consulted primarily for purposes of legitimizing decisions that were made without them, or whether joint fact-finding and value creation were given any attention. In other words, it is not clear what the goals of stakeholder engagement were in this case. Relatedly, no explicit measures of success are mentioned. For example, it does not appear that building the capacity of individuals, groups, or institutions to participate in other processes like this was on anyone's mind.

B. Werick "Changing the Rules for Regulating Lake Ontario Levels," (2007)

Introduction

The excerpt is from Bill Werick's report on the proceedings of the 2007 CADRe Workshop on Computer-Aided Dispute Resolution. Werick describes the joint fact-finding process used by the International Joint Commission (IJC), a U.S.–Canadian organization, to inform a study of regulating water levels and flows in Lake Ontario, which straddles the U.S.–Canada border and flows into the Saint Lawrence River. The goal of the process was to make recommendations reflecting the needs of water system users and other stakeholder interests. Topical concerns were segmented into six "impact areas": coastal, navigation, hydropower, municipal and industrial water, recreational boating, and the environment.

A technical working group (TWG) was formed in each impact area. The TWGs were charged with conducting research relevant to their area, which was then used to build a computer model. Stakeholders, experts and decision-makers participated in the TWGs, as well as workshops and modeling sessions. Ultimately, they were able to generate information through joint fact–finding that stakeholders from all of the working groups considered legitimate. This, in turn, helped diffuse persistent conflicts over Lake Ontario water management.

The excerpt describes these joint fact–finding efforts in detail, focusing on the activities of the technical working groups and the use of collaborative modeling to improve water management and reduce stakeholder conflict.

Excerpt (pp. 119–126)

In September 2007, the International Joint Commission (IJC), a joint U.S.-Canadian organization created by the Boundary Waters Treaty of 1909, is expected to promulgate new rules for the regulation of Lake Ontario water levels. Barring extreme difficulties from public review, the rules would most likely go into effect in 2008. This would be the first time (to my knowledge) that the rules for regulating releases on a major North American water system have been changed in the last 30 years, despite the fact that changes to the rules have been under study on almost every major basin. The IJC used shared vision planning to develop and vet these rules, and the Lake Ontario case study now stands as the most technically ambitious and successful shared vision planning application. This paper outlines how the shared vision planning effort unfolded, and highlights the innovations, strengths and weakness in this particular study.

Background

The International Joint Commission (IJC) issued an Order of Approval in 1952 to build the St. Lawrence River Hydropower Project, including a dam across the St. Lawrence River that allows the IJC to regulate Lake Ontario water surface elevations and flows and elevations in the St. Lawrence River. The IJC has since 1963 used a written set of regulation rules called "Plan 1958-D," but about half the weekly regulation decisions are considered "deviations" from the plan. These deviations have been necessary for many reasons, most importantly, because the written plan does not work well when water supplies are much less or much more than the 1860 to 1954 supplies that were used to design and test the plan.

In 1993, the multi-year Levels Reference Study recommended that the "Orders of Approval for the regulation of Lake Ontario be revised to better reflect the current needs of the users and interests of the system." That study did not address environmental impacts, a use of water not identified or explicitly protected in the treaty, nor did it precipitate a consensus on how the current needs could be addressed while protecting traditional uses. In April 1999, the International Joint Commission informed the governments that it was becoming increasingly urgent to review the regulation of Lake Ontario levels and outflows. A plan of study was endorsed in 1999 and the study began late in 2000.

The original plan of study did not define how plans would be formulated, evaluated and ranked, or how researchers would design their work to fit into an overall evaluation scheme. Late in the first year of the study, I made a presentation of how shared vision planning could be used on this study. The presentation included an Excel model based on a STELLA model developed for the five Great Lakes by Phil Chow and Hal Cardwell of the Corps' Institute for Water Resources. Thereafter the Board agreed that all subsequent planning work would be done using shared vision planning. A Plan Formulation and Evaluation Group (PFEG) was formed soon after, and PFEG began to restructure the study with the aim of linking research, public input and decision making. The PFEG reported to the Study Board and the Study Directors. The original membership was made up of those who had pushed the Board to make the formulation and evaluation of alternatives— planning—an identifiable and managed task rather than a natural happenstance of the technical studies. In some cases, PFEG had to realign research that had already begun and assist in the design of studies that were not yet underway. In other cases, work was well along and PFEG used what was done.

The Study

[Figure 6.3], below, shows how the shared vision model fit with the proposed research. Research was conducted by seven technical working groups (TWGs), managing water information (the Hydrology and Hydraulics TWG) and six impact areas (coastal, navigation, hydropower, municipal and industrial water, recreational boating and the environment). The blue boxes collectively were the SVM in design and use, not just the STELLA model. The shape of the research and models mimicked in many ways the relationships between the TWGs and the Study Board.

For example, all TWG work products had to contribute directly or indirectly to the shared vision model. PFEG had no authority over the TWGs but PFEG advised the Board on how TWG research proposals would or would not support the Board's decision making process. The shared vision planning framework connected decision makers to experts and stakeholders:

- **Experts–decision makers.** Our planning process required all TWGs to conduct research that would support a quantitative connection between water levels and economic, environmental or social impacts. For the hydropower and recreational boating TWGs, this was a foregone conclusion and in fact, work along these lines was well along before shared vision planning was in place. But it took a substantial effort to shape environmental studies this way, and considerable tuning to re-shape the navigation and coastal studies.
- **Stakeholders–decision makers.** We asked the Study Board to hold six "practice" decision workshops to iteratively refine the criteria the Board would use to make its decision. Those workshops were conducted with stakeholders and often with commissioners present. These "fire drills" helped make sure

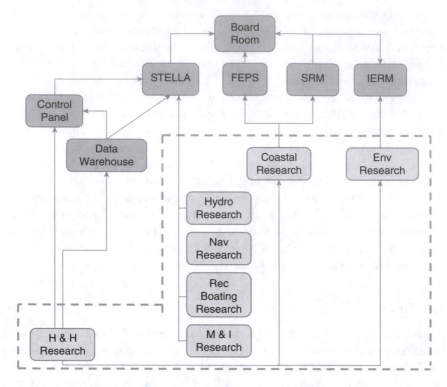

FIGURE 6.3 Integration of the shared vision model with the study research

that the Board understood what stakeholders wanted and helped stakeholders understand why the decisions were made the way they were.

- **Experts–stakeholders.** The study had already allowed stakeholders to participate in technical working groups even before starting the shared vision planning process. In the public TWGs—navigation, hydropower and M&I—stakeholders were represented by paid technical staff; in other impact areas, the stakeholder representatives were not as technically proficient. The shared vision planning process, especially the collaborative model building, had two primary impacts on the expert–stakeholder connection. First, it allowed experts to make sure they understood how stakeholders were impacted. Working with experts and stakeholders, we developed over one hundred hydraulic attributes such as seasonal water level ranges that were used to evaluate plans (especially in the early part of the study, before economic or environmental impact functions were complete). PFEG met with groups of stakeholders around the study area and worked with them to design their own section of the shared vision model that contained the information they told us they would use to rank plans, with tables and graphs they helped design. Second, it gave stakeholders a better understanding of how the impact measurements were linked to water levels, not just in their own areas of interest but also for issues that stakeholders with conflicting interests supported.

PFEG worked with the Environmental Technical Working Group chairs to review over two dozen environmental research scopes and to help establish mathematical relationships between water levels and a biological result. Dr. Joseph Depinto and Mr. Todd Redder of Limno-Tech, Inc. then developed a dynamic model relating water levels to the potential environmental impacts identified in the existing body of research subjects. Although they initially opposed the Integrated Ecological Response Model (IERM), environmental researchers eventually embraced it as their own and in workshop exercises, began to question their intuition when it differed from model results, rather than vice versa.

The Model

There was considerable debate about what software to use to build the shared vision model (SVM). The final structure was a compromise that (in retrospective judgment) worked well, but was bent a little too much in the direction of researchers' preferences. For example, the FEPS model was proprietary and impervious to casual review. PFEG found substantial errors in the FEPS model by close review of its documentation and results, but no one reviewed the code. In the end, many of its processes could have been programmed in STELLA or Excel where review would have been easier. Similarly, the IERM modeled the wetland algorithm in an essentially opaque code. After the study it became apparent that there were small differences between the researcher's coding of the algorithm, the IERM coding, and subsequent attempts to model the algorithms in Excel. While the mathematical differences were not great and the resulting conclusions identical, in retrospect there are three good reasons for modeling the wetlands in easily accessible code. First, it would have allowed us to resolve small differences between modelers' interpretations of the English language version of the algorithm. Second, it would have made it easier to use the model post study in adaptive management because it would have been much easier to modify the code. Finally, the argument that convinced people to use C++ during the study was the looping required to calculate non-wetland environmental performance indicators such as the northern pike model. In the end, the pike model meant it took more than an hour to run the IERM but the pike performance indicator did not help distinguish among plans. The final SVM was a system of models, not just one software or file, but all the results were captured in a sophisticated Excel spreadsheet that became the face of the SVM for most study participants. That spreadsheet came to be known as the Board Room. The PFEG led the development of the model, with STELLA and Excel coding being done primarily by Bill Werick and Mark Lorie of IWR, and David Fay and Yin Fan of Environment Canada. A few other agency experts added elements to the STELLA and Excel models. Lay stakeholders sometimes were engaged in modeling workshops, but by their choice, none did any coding. Stakeholders such as David Klein of the Nature Conservancy trusted the models because they were very familiar with the modeling effort, not because they performed it, and because they knew there was no censoring or significant time delay in reporting modeling news. When we found a big mistake, everyone knew about it the next day because the modeling process was very public and the model results were used directly in activities that stakeholders and decision makers took part in.

The planning process percolated through various models in this fashion:

- Researchers developed algorithms connecting impacts to water levels or flows using field data and their own analytic procedures. For instance, stage-damage relationships in the lower St. Lawrence River were developed using GIS that estimated the level of flooding on individual homes at a range of water levels. Information from these models was then used to develop damage functions in the shared vision modeling system.
- Board members, stakeholders, experts in various fields other than regulation plans and paid plan formulators would propose new regulation plans in con-ceptual terms and then the plan formulation team members would code the concepts. There were four formulation teams that experimented in four catego-ries of plans: modifying the existing rules; optimization schemes; "natural" regulation, and coding of plan concepts offered by others. Each team would use whatever software they wanted to code the rules. The four teams would meet every few months to share successes and challenges; they were competi-tive but were part of PFEG and ultimately wanted to see a great alternative produced more than they wanted the alterative from their team to be the best of a mediocre lot. Each team's model output, a 4,848 quarter-month time series of releases, was then pasted into an Excel model, a part of the shared vision model called the "Control Panel." That release set defined a unique alternative plan.
- Plan formulation was also used to explore the potential to solve problems, even with plans that would be impossible to implement. "Fence post" plans were also developed, with each fencepost defining a plan that was designed to serve one interest no matter the effect on other interests. These fence posts defined the decision space, and showed the limits of our ability to control water level related impacts. Most importantly, we showed that we could not reduce damages to Lake Ontario shoreline properties much more than we already had. In a similar fashion, we formulated "perfect forecast" versions of alternative plans so that we could quantify the potential benefits of better forecasts.
- Water levels and most impacts would then be calculated in a STELLA model dynamically linked to two spreadsheet input models, the Control Panel and Data Warehouse. After the STELLA model was run, tables from that model would then be copied and pasted into a third Excel model called the "Post Processor." The post processor included macros and tables that could be used

to call external models that did the rest of the impact evaluations including Lake Ontario coastal impacts (FEPS), St. Lawrence River shore protection damages (SRM) and the environmental impacts (IERM). Those three models are described very briefly below.

- FEPS (Flood and Erosion Prediction System) is a proprietary C++ model developed by Baird Engineering during previous investigations into Great Lakes erosion and flooding research. FEPS uses water level erosion relationships developed using a very data intensive erosion model called COSMOS at several representative cross-sections around the lake and then applies the results over and over using reach specific parameters around the entire Lake Ontario coastline. Flooding damages are based on water levels and wave heights, capturing both inundation and wave impact damages, and shore protection structure damages are assessed using erosion and flooding models. Erosion at any moment in time is serially dependent on the water levels experienced in the years preceding that moment. Hence, a shore protection structure becomes more vulnerable to damage as erosion eliminates protective beachfront, and it may fail in the eighteenth year of simulation under one plan and in the twenty-fifth year under a different plan. Run time for the FEPS model was about three minutes.
- SRM (Shoreline Response Model) was a proprietary model developed by Pacific International Engineering to assess the effects of different releases on shore protection built along the banks of the St. Lawrence River. Our evaluations showed that all regulation plans being seriously considered had about the same amount of river shoreline damage. Once this was established, this model had little additional relevance in the evaluation process. Runtime was about a minute.
- IERM (Integrated Environmental Response Model) a Visual Basic model, was itself a collection of sub-models. When called from the post processor, the IERM would present a window announcing which sub-model was running. Run time was about 80 minutes on a modern laptop.

While there were four primary formulators, several more PFEG members had the model suite on their (personal?) computers and used it to evaluate models and to check the evaluations other people had done. This work was done methodically and on an ad hoc basis. As an example of the former, a non-formulator might question the results of a formulator and re-run the evaluation checking that all the agreed conditions (for example, was the FEPS model set to use the agreed application of the wave data, did the formulator use the recent revision to Plan 1958DD to define the baseline) were being honored. All of the modeling described above was used to evaluate plans using 101 years of quarter-monthly data. All these evaluations were designed around the 101 year, 4,848 quarter-month structure. When we first tested the alternatives with climate change and stochastic information, we had to manipulate the hydrologic input datasets to this structure. Twenty-nine year climate change datasets had been developed using the 29 years of historic data for which we had enough collateral information, such as precipitation and evaporation, to downscale and interpret global circulation model outputs. We simply repeated these 29-year datasets until we had 101 years. The study developed a 50,000-year stochastic hydrology, and at first we snipped particular 101 year "centuries" from this large data file to form four 101-year quarter-monthly datasets that represented extremes in the stochastic data, and put these snippets in the Data Warehouse spreadsheet so they could be used in this same way to evaluate plans with alternative hydrologic assumptions. Later, a full stochastic analysis using 495 101-year sequences was also done using FORTRAN code translated from STELLA equations and a variation of the FEPS code.

The four plan formulation teams compared results and benchmarked each others' plans, both over the internet and in face to face workshops. This developed a rich understanding of how the system worked, and allowed us to share break-throughs wherever they were made. Stakeholders had complete access to these sessions, and while few took part in them, stakeholders who did take part helped spread news of plan formulation, and this helped people trust the process. Hundreds of alternatives were tested with the historic evaluations, which could take from two minutes (STELLA only) to ninety-minutes (STELLA, FEPS, SRM and IERM). The evaluations produced economic benefits as traditionally calculated for navigation (shipping cost changes); coastal (changes in expected damages); recreational boating (changes in the value of recreation-day values); hydropower (changes in the value of energy at marginal market based rates) and municipal and industrial water (changes in operating costs). The environmental impacts of each plan were calculated as the ratio of the score achieved by an alternative for a particular parameter to the score achieved under the current regulation plan (in Corps parlance, the "without project" condition). For example, the wetland model produced the acres of meadow marsh present each year after a specifically defined low water supply condition. Those acreages were averaged over the entire 101 year run for each alternative and then divided by the number of acres produced by Plan 1958DD, the baseline plan. In addition to the performance indicators, statistics on over one hundred stakeholder designed "hydrologic attributes" was calculated for each plan and displayed automatically in the Board Room in both central locations with each or all attributes, and in "Interest" corners designed based on focus group like meetings with several stakeholder groups, For example, the navigation industry had a place in the Board Room with graphic comparisons they helped design of the hydrologic attributes they said they would use in deciding which plans were their favorites.

The full stochastic analyses took over a day of computing time to run and these runs were done only for plans that were of particular interest. But the final economic benefit analyses were based on discounted values using the full stochastic evaluations. The discounting captured the reality that erosion happens no matter the regulation plan, so the only difference was how fast it happened (plans that slowed erosion down had positive economic benefits). Had we simply discounted damages using the 20th century "historic" hydrology, the differences between plans would have been muted and distorted, since the wettest and most damaging period came in the last three decades of the century. Instead, the stochastic version of the model recorded damages for each quarter-month of the 4,848 quarter-months in each of 495 101-year "centuries" and so was able to produce an average expected damage for each quarter-month into the future. These average damages were then discounted. A sensitivity analysis allowed various planning horizons and interest rates, but the final report was based on 4 percent discount rate and 30-year evaluation period. Figure [6.4], below, shows that Lake Ontario water levels could be nearly three feet higher and lower than recorded levels even under the current regulation plan, which seeks to compress lake level variation.

The Essential Conflicts

The IJC receives a fairly reliable stream of complaints from some stakeholders because they live or keep their boat in a place which cannot be made satisfactory through regulation. As is true in most places, people have built along the coast based on recent water levels, not on the inevitably higher and lower levels that will come after building. There are a few hundred homes along the Ontario and St. Lawrence coast that will receive at least nuisance flooding no matter how Lake Ontario is regulated. Similarly, there are a few hundred boat slips that will not offer enough draft when water levels are merely normal. This was probably exacerbated by the generally high levels in the last few decades, which coincided with the increase in boating ownership and use. On the other hand, drought management plans that held water on Lake Ontario as long as possible worked for people around the lake and along the river; large short term releases to create normal depths

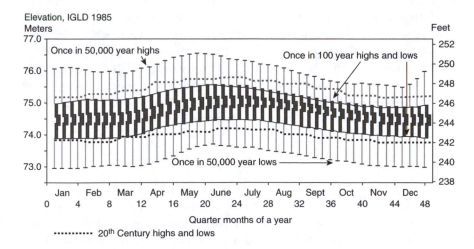

FIGURE 6.4 Range of possible Lake Ontario levels through the year under the current
regulation rules

in the river often hurt people along the river because those releases drained Lake Ontario so much that severe release restrictions were needed when natural flows were even lower.

The main conflict that could be affected by regulation was between shoreline property damage and wetland plant diversity along Lake Ontario. Compressing lake level variations helped property owners but created a narrow band of transition between submerged and upland species. There is also a conflict between coastal damage above and below the dam. The damage risk on the river is by far the greatest when winter ice and snow in Quebec melts. If Lake Ontario is high at the same time, the release decision must balance the near certain river damage from higher releases against much larger potential damages along the lake if wind storms occur while lake levels are high.

The Results

The IJC asked the Study Board to provide options, not one recommendation for a new regulation plan. In their final report, the Board gave the IJC three regulation plans labeled A+, B+ and D+. Of all the plans that met the Study Board requirements, A+ maximized economic benefits, B+ maximized environmental benefits, and D+ minimized sectoral losses. All three plans created millions of dollars per year in net economic benefits, but Plan B+ created more positive and more negative benefits. The implementation costs for any of the plans will be relatively small, with all plans about the same, so no benefit–cost ratio was calculated, No plan was found that improved on the current plan in every sector; tradeoffs, sometimes fairly small, seemed unavoidable. We tried but could not reduce coastal damages from the B+ plan; it would cause an average of about $2.5 million per year in damages, an average created by no damage in most years but tens or hundreds of millions of dollars of damage every 20–30 years. We showed that we could eliminate these damages with perfect forecasting in the fall of local spring runoff into Lake Ontario (that is, not the flow from the Upper Lakes, which is fairly predictable). That creates hope that better forecasting, even if not perfect, would allow us to develop a risk management strategy for fall levels that would keep most of the environmental benefits and not cause more coastal damage than we would expect to experience under the current plan.

Commentary

This article illustrates the importance of joint fact-finding and transparency. As Werick suggests in referencing the Nature Conservancy representative's perception of the model, the legitimacy of the data gathered depended more on the stakeholders' involvement than on their ability to understand the modeling efforts in a detailed or technical way. The transparency of the effort also helped build the credibility of the information that was gathered, as did the use of multiple models to assess several plans.

Another useful aspect of the process was stakeholder interaction with experts of various kinds. The participants learned more about the issues under consideration, and these interactions enhanced their understanding of each other's concerns. The joint fact-finding process could have been a way to merge local knowledge and the technical insights of experts, but that does not appear to have been an objective.

The interaction between stakeholders and decision-makers appears to have been helpful. There is little doubt that the effort enabled stakeholders and decision-makers to increase their understanding of each other's priorities. While the participants used modeling to help frame policy options, they did not use it to generate a comprehensive proposal or package, or to create value through "trades" across issues they valued differently. Thus, the modeling effort did not contribute to generating a proposal that those in authority would have had to accept or reject.

B. Fuller "Trading Zones: Cooperating for Water Resource and Ecosystem Management when Stakeholders have apparently Irreconcilable Differences," (2006)

Introduction

This excerpt is from Boyd Fuller's doctoral dissertation at MIT, in which he looks at two collaborative water management cases in the United States: the CALFED processes, in the state of California, and efforts to address water management in the Everglades, in the state of Florida. The excerpt focuses on a joint fact-finding process in Florida.

The Everglades is a vast area of wetlands in the central and southern regions of Florida. The water system has deteriorated badly, primarily because of over-development, detrimental water uses, and poorly-managed flood control measures. Although the Everglades do not cross state or international boundaries, management of this water resource is clearly multi-jurisdictional, falling under the purview of both federal regulators and state water-management authorities. Fuller describes a range of contentious efforts to manage and protect the Everglades. These eventually ended up in a protracted legal dispute between state and federal agencies after multiple efforts to reach a negotiated settlement, with the help of a mediator, failed. Disputes over scientific considerations were a key barrier to reaching agreement. We learn about an initial mediation process that included environmental and sugar industry representatives. A technical group was formed to help resolve some of the scientific disagreements.

This excerpt discusses the formation of this technical working group and describes how a long-standing disagreement over the impact of phosphorous on downstream ecosystems was resolved with the help of a mediator.

Excerpt (pp. 235–239)

Mediated Technical Plan[1]

Beyond expanding the scope, the parties agreed in the initial meetings to the general procedures of the mediation at the beginning. All parties would be included and accorded a seat at the table. The scientific disputes would be handled by a group of experts from all sides. This technical group was designed with several considerations in mind. First, it was decided that the scientific group would seek consensus. In part, this was due to the recognition that there was a great deal of scientific controversy and uncertainty surrounding the science of the Settlement Agreement and the whole phosphorous issue. Second, there was a real recognition that this group would have to do more than develop a technical plan for saving the Everglades. They would also have to sway constituencies divided by animosity to support the eventual plan. To increase the likelihood that the technical group would have influence on decision-makers, all the stakeholder groups participating in the mediation were invited to send their experts to join the group. The selection of the technical group's members was vetted through the policy-making group using criteria of expertise and ability to sway others in and outside their communities. The technical group was given the task to not only to resolve the disagreements about standards and solutions for phosphorous, but also to propose some changes to the water management that might improve the hydroperiod of the Everglades. Furthermore, they had a broader geographical area to address, both in terms of problem sources and solutions.

With help from the mediator, the technical group reached agreement on both the range of acceptable phosphorous levels and a "Mediated Technical Plan." This plan included the group's agreement on (a) possible rates of decrease in pollutants that the group felt could be achieved, (b) suggested sites for the treatment areas that would remove phosphorous from the EAA drainage flows; and (c) some measures to improve the hydroperiod—for example by including Stormwater Treatment Area (STA 1-East) into the program.[2]

The Mediated Technical Plan also outlined possible secondary benefits that the Mediated Technical Plan would have, including the reduction of salinity in downstream estuaries and potential improvements for Lake Okeechobee.

Most of the technical participants in the technical committee supported the Mediated Technical Plan; however some parts of the plan were not supported unanimously. For example, the Mediated Technical Plan set a standard of 50 ppb for phosphorous to be reached over the next decade. This deadline represented a delay in phosphorous reductions as compared to the original Settlement Agreement; furthermore, the Mediated Technical Plan had no provisions stating when and if phosphorous would be reduced to the lower levels (~10 parts per billion or 10ppb) that many scientists had already said would be required to end phosphorous' impact on the Everglades. Thus, environmental support for the Mediated Technical Plan was partially contingent upon the parties' reaching agreement on other policy-level issues such as cost-sharing, land acquisition, and meeting appropriate standards for phosphorous removal. Similarly, Sugar stakeholders also wanted to see how much they would be expected to pay before they gave their full support for the proposed numerical standards. Many Sugar stakeholders were dubious about the science behind the proposed numerical standards and so they were waiting to see whether or not the other elements of the deal were worth putting those concerns aside.

Procedures for Science Contested

The Settlement Negotiations' technical group had put together the rudiments of a technical solution, including a numerical standard (~10 parts per billion or 10ppb)[3] and the use of STAs and best management practices (BMPs). In the mediation's technical group, many of the basic concepts and components found in the Settlement Agreement remained the same but the details were modified and expanded. For example, the Mediated Technical Plan agreed on an interim target of 50ppb for phosphorous, but did not include any discussion of how they might lower the phosphorous concentrations to 10ppb, which most scientists from the Settlement Negotiations agreed was appropriate for waters entering the Everglades. That they did not do so was one contention that environmentalists had with the plan.

Similarly, the idea of using STAs to clean water of phosphorous remained largely the same but the mediation's technical committee hired a consultant to help them create a more detailed plan that included engineering drawings.[4] In fact, throughout the mediation, the assumption by most parties was that STAs would be used to treat phosphorous. The main remaining questions were (a) how many acres would be set aside for STAs; and (b) who would pay for them? In the original Settlement Negotiations' technical plan was a repository of knowledge and ideas that influenced and constrained the options of the parties in the Settlement Negotiations and now in the mediation. For example, consider this next stakeholders' description of the issues around numerical phosphorous standards.

> The sugar companies were obviously very concerned about setting numerical standards. Number one, because it would be easily enforced and they felt that once the number was set, you can never go back. It only gets stricter over time. I've never heard of a water quality standard being loosened [laughs]. [Second,] without the science, they felt it was unsupportable to go ahead with a number. On the other hand, I think the agencies and a lot of other group felt very strongly that there should be a numerical standard, not a narrative standard. Otherwise, they would be perceived as having being sold a bill of goods.[5]

Despite Sugar's challenges of the science behind the setting of numerical phosphorous standards, many of the agencies were unwilling to move away from the idea. They had made public commitments to numerical standards in previous products and it was something they could not change their minds about without fueling fears of cooptation. Furthermore, state and federal agencies had put signifi-cant efforts into creating and agreeing upon those standards, so they had much invested in them. In the end, Sugar agreed to the numerical standards when the other parties agreed that (a) phosphorous levels would be measured more broadly throughout the Everglades (rather than just at convenient locations), and (b) the standard would be interim, contingent on the outcomes of further research into appropriate levels.

Many in the Sugar industry remained unconvinced by the science premises under-lying the importance of phosphorous as compared to hydroperiod modifications, but they saw the plan as a way to move forward to an agreement that would promise some stability to the industry. In other words, their willingness to go along with the Mediated Technical Plan rested more on reducing costs and public image problems than a strong agreement about the validity of the plan's underpinnings. That is another reason why they were willing to accept the numerical standards.

Notes

1 Sometimes stakeholders refer to this as the Technical Plan.
2 This additional area was the site of an approved but delayed project to improve flow to the Everglades.
3 The number varied according to the region under consideration.
4 Several stakeholders were quite impressed with the fact that the Mediated Technical Plan had included these detailed drawings.
5 Interview with state stakeholder, Fall 2004.

Commentary

A key strength of this joint fact-finding effort was its ground rules: the process included all stakeholders, stipulated that decisions would be made by consensus, and allowed stakeholders to select their own experts to participate in the commit-tee's work. The ground rules boosted the legitimacy of the information that was gathered, and ensured that linkages were made among the information gener-ated, key stakeholder interests, and the political constraints the parties faced.

The excerpt also illustrates that joint fact-finding does not eliminate disagree-ment over science and data. Rather, as in this case, it can clarify sources of dis-agreement and help parties identify contingent agreements or other negotiated ways of handling them.

B. Fuller "Trading Zones: Cooperating for Water Resource and Ecosystem Management when Stakeholders have apparently Irreconcilable Differences," (2006)

Introduction

This excerpt is also from Boyd Fuller's dissertation. It again focuses on the Everglades, but this time Fuller focuses on how a mediator helped the group create value by reframing the problem and by broadening of the scope of issues and the geographic area under consideration. Even though the mediation ultimately collapsed after the group's proposal was rejected by national-level policy makers, the mediator's focus on value creation helped the stakeholders reach agreement among themselves.

Excerpt (pp. 229–234)

Everglades Mediation

The mediator started the process by meeting all the stakeholders—including not only the state and federal agencies but also environmental, tribal, and sugar and other agricultural interests. From those interviews, he identified several important areas of agreement and disagreement among stakeholders about the issues and determined that parties were generally positive, although cautious, about participating in a more inclusive process.

The various stakeholders seem to agree that the settlement was not sufficient to save the Everglades; something needed to be done about water flows as well as water quality. Most parties also agreed that they would rather not continue their battles in court. Even a "victory," many realized, might be detrimental because of the time and effort it would take away from dealing with the problems of the whole system. For example, would it benefit sugar to win a victory that reduced their financial obligations with regards to phosphorous if at the same time they lost access to water supply as disgruntled urban water users gained political power?[1] In addition, there was considerable uncertainty about who might win the litigation and how long it would take. From these interviews, the mediator convinced the settlement parties to open the mediation to all interested parties.[2]

An Initial Vision—Saving the Everglades

The mediation started with meetings among the principals to define a general scope for the issues that the mediation would tackle. The parties agreed here to include a broader array of issues that were appropriate to a discussion about the preservation and restoration of the Everglades' health. With regards to expanding the set of substantive issues, all the parties seemed to agree that removing phosphorous from the water entering the Everglades National Park would not be sufficient to save the Park nor the Everglades system as a whole. Problems in these areas were also being caused by changes to hydroperiod—the timing, quantity, and "shape"[3] of the water flow through the Everglades—imposed by water control projects such as the Central and South Florida Project.

The parties also agreed that the geographical area covered by the Settlement Agreement did not include all of the relevant areas—neither all the areas contributing to the problem nor all the areas being impacted—that should be addressed if the health of the Everglades was going to be addressed effectively. So, for example, the parties recognized that the Everglades National Park (Park) and the Loxahatchee National Wildlife Refuge (Refuge) were only a small part of the important ecosystems under threat. There was also significant land in between the Park and Refuge areas that also was part of the overall ecosystem that supported much of the biological function of the Everglades and which was being impacted by manmade water management. In addition, there were multiple areas upstream that were not only introducing phosphorous to the watershed but also altering the hydroperiod. For example, there is a significant dairy farm industry north of Lake Okeechobee that contributes phosphorous to the watershed. Similarly, by damming Lake Okeechobee, the Corps through the C&SF Project had fundamentally altered the hydroperiod of the water flowing through the Everglades to the south.

So the mediation took two important initial steps. First, they moved the question they were addressing from the legal one, namely about phosphorous and the health of the Everglades National Park and Loxahatchee National Wildlife Refuge, to a broader policy question about how the Everglades might be best "restored." This allowed the parties to consider questions and areas outside the scope of the litigation to see if there might be solutions that made sense to all parties that addressed the overall health of the Everglades and the bays into which its water drained. It also allowed for the inclusion of at least some consideration of hydroperiod concerns in addition to the current questions over phosphorous. So in this way, it provided a framework for future discussions and negotiations with which all parties felt at least somewhat comfortable.

Second, parties in these initial discussions spelled out what procedure would follow next as they worked towards a solution, the convening of a technical group that would hammer out some kind of agreement, if possible a consensus, about the scientific and technical elements of a plan to "save the Everglades." After that, the principals would reconvene to discuss how that plan would be implemented, including the financial arrangements and a schedule.

Choosing to adopt this new framing of the problem as one of "saving the Everglades" had the potential to open up the dialogue significantly. Each side had already acknowledged that the hydroperiod changes caused by the C&SF Project was a problem and probably causing significant damage to the Everglades. Framing the mediation this way took the focus away from the litigation and issues of who would pay and how strict would the standards be—both issues being largely zero-sum. Saving the Everglades itself was a goal that all parties could agree upon. Even if stakeholders varied in their reasons for supporting the goal varied, there was still power to the rhetoric, especially when much of Florida public now favoured the restoration of the Everglades.

Talking about saving the Everglades also opened up new potential avenues for cooperation that raised some hopes that agreement could be found. Sugar and other parties saw the potential to spread the "blame" and reduce the zero-sum nature of phosphorous. By broadening the scope of issues, Sugar hoped that the parties could consider additional opportunities and benefits could be added to each party's calculations, making the negotiations easier (more benefits for everyone) and lessening the link between the deterioration of the Everglades and Sugar's farming practices (additional sources of harm). Similarly, the expansion of the area could bring in areas and issues that were having an impact on the Everglades that had not been considered in the litigation. In fact, Sugar was purportedly one of the most proactive stakeholders in trying to introduce both possible hydroperiod improvements and additional areas of concern, as this stakeholder explains.

> The Sugar industry actually recommended some of the best enhancements to the plan. They wanted to make it more inclusive. So the solution got much bigger—and getting much bigger it brought in more people than just Sugar and so there was a need to address the funding. They wanted to address the water quantity issues and tie this plan with getting more water to the Everglades. …They pushed hard to bring in what became STA 1-East, which was a component of a federal project that had been authorized and designed, but had never been built. This was a way to get that built and get federal civil works monies to do it. …If you go on the other side, there was the C-139 basin which is west of the Hendry County which is west of the main Sugar area. It also flows to the Everglades, it flows right by a couple of the Indian reservations so it was an area that was very important to get solved. It was not in the lawsuit at all. So they basically expanded the watershed east and west to bring in more water and more property and treat more phosphorous.[4]

As we see by this stakeholder's comments, Sugar took proactive steps to bring in other areas because they saw different opportunities. First, Sugar believed that hydroperiod modifications were having more significant impacts on the environment than phosphorous was. By bringing in that issue, they could see if those problems could be addressed by methods that other parties could pay for—including STA 1-East. If so, then they could reduce the risk that other stakeholders would come after them for more money at some later date. Furthermore, the issue of who pays for what became more uncertain as the number of problem sources and their links to impacts was expanded. So the expansion not only included hydroperiod modifications as a problem source, but also other geographical areas as phosphorous sources.

These modifications were generally in the interests of other parties as well. Congress had already approved the proposed federal component (STA 1-East) and the Corps had completed the necessary design work. Environmentalists wanted to see the Everglades saved. State stakeholders wanted to find some way to meet their obligations under the Settlement Agreement and had been concerned about the problems of water quantity management for some time. The Miccosukee tribe was also concerned with phosphorous levels and water quantity management in their lands as well. In these ways, the broadening of the scope to "saving the Everglades" offered real possibilities for cooperation.

Using the principled negotiation perspective, we can understand each of these moves represented opportunities to "expand the pie," a classic negotiation concept that refers to moves in which stakeholders can imagine and talk about tradeoffs and solutions that yield more for each stakeholder individually and the group as a whole.[5] The importance of these moves to make the upcoming negotiations more palatable speaks directly to findings we saw in the previous chapter, namely that trading zone theory does not speak directly to the cooperating parties' willingness to cooperate.

Notes

1 In fact, this concern was raised by some of the Sugar stakeholders I talked to in interviews. This fear was also a major reason why Sugar participated in the subsequent Governor's Commission for a Sustainable South Florida.
2 Other small agricultural producers were also invited to the mediation, but chose not to participate.
3 The shape of the flow relates to how much of the flow is "sheet" flow—flowing overland as a flood—versus channeled flow—concentrated flows through rivers, canals, and pipes. In other words, not only is it important to achieve certain quantities of flows at different points of the watershed at certain points in time, but it is also simultaneously important how that flow is distributed across the landscape of the Everglades and how deep it is.
4 Interview with district stakeholder, Fall 2004.
5 See for example, Fisher and Ury (1991) and Susskind and Cruikshank (1987).

Commentary

In this excerpt, Fuller clearly illustrates three successful value-creating moves—bringing in additional stakeholders, broadening the scope of the discussion, and reframing the problem—that a neutral party is in the best position to carry out.

First, the mediator began by meeting with all the stakeholders, not just the parties to the litigation, to learn about their interests and gauge the possibility that a collaborative problem-solving process might be successful. This initial work helped to lay the foundation for a more inclusive process, and built trust in the mediator as a process manager.

Second, the mediator helped the group widen the range of issues on its agenda and the geographic areas under consideration. As Fuller correctly highlights, these value-creating steps helped the parties move beyond the narrow set of issues that had been the focus of protracted litigation, and allowed them to consider fresh trades that helped them reach agreement.

Third, the mediator was able to reframe the problem in a way that emphasized common interests. As Fuller describes, shifting the focus of the conversation to "saving the Everglades" enabled the parties to establish a shared vision for the mediation that they could all support.

G. Radosevich "Mekong River Basin, Agreement and Commission," (2011)

Introduction

This excerpt by George Radosevich describes the facilitated negotiation that produced the 1995 Mekong Agreement, a framework for cooperation in the Mekong River Basin.

The Mekong River originates in the Tibetan Plateau in China, forms the boundary between Laos and Myanmar, forms part of the boundary between Laos and Thailand, passes through Cambodia, and finally flows through Viet Nam and empties into the South China Sea. It is a principal water source for several countries and communities, and is heavily utilized for hydropower and agriculture. The four lower basin countries—Cambodia, Laos, Thailand and Viet Nam— negotiated an agreement with help from Radosevich, who facilitated the process on behalf of the United Nations Development Programme (UNDP). (China and Myanmar did not participate, but were designated as "dialogue partners" and have subsequently been involved in some joint management efforts.)

This excerpt describes the negotiation process, highlighting some of the facilitation techniques that Radosevich used to help the parties create value, engage in perspective taking, and utilize "objective criteria" for decision-making. The author makes important observations on how to select a facilitator, emphasizing UNDP's comparative advantage in offering good offices for communication and the importance of relying on someone with substantive knowledge of the issues under discussion.

Excerpt, pp. 4–7:

2. International Water Law and the 1995 Mekong Agreement Negotiations

In December 1992, representatives of the four lower riparian countries gathered in Kuala Lumpur, Malaysia to attend a meeting facilitated by the United Nations Development Programme (UNDP) to discuss the future framework of cooperation in the Mekong River Basin.[1] The countries had before them three options: i) amend the existing two basic documents (1957 Statute and 1975 Declaration), ii) negotiate a new framework of cooperation, or iii) suspend cooperation in a formalized manner but still adhere to the principles of customary international water law. Following intensive discussions which included high-level foreign policy representatives from the four states, the governments agreed to work together to negotiate a new frame-work of cooperation while continuing the Mekong Committee Secretariat's ongoing activities. Furthermore, it was decided that once a new framework was established, China and Myanmar would be invited to participate. A Joint Communiqué was issued by government representatives from Cambodia, Laos, Thailand and Viet Nam at Kuala Lumpur which declared:

> We reaffirm the resolve of our respective countries to continue to cooperate in a constructive and mutually beneficial manner for the sustainable utilization of the Mekong river water resources. ...We recognize that various developments since the original establishment of the Mekong cooperation mechanism necessitate further efforts to define the future framework of cooperation. We have agreed to continue our dialogue and work towards such an appropriate framework through the process of a Working Group under the auspices of UNDP.

The Kuala Lumpur Joint Communiqué established the political commitment of the countries to negotiate in good faith with the aim of reaching a new, mutually acceptable framework of cooperation. However, the complex task of actually formulating an acceptable agreement still confronted the countries. UNDP offered to facilitate the negotiations by providing a Senior Advisor (Dr. George Radosevich) and financial support, as well as using its good offices as a channel of communication. The four countries established a Mekong Working Group (MWG) of five representatives from each country including at least one from Ministry of Foreign Affairs, which held five "formal" meetings and two "informal" technical drafting group (TDG) meetings between February 1993 and November 1994, culminating twenty-six months later in the signing of the Mekong Agreement for Cooperation for the Sustainable Development of the Mekong River Basin in April 1995.[2]

The MWG drew upon the basic principles of customary international water law in formulating the Agreement. The MWG participants did not question "if" they should follow customary international water law, but rather "how" to apply the general principles to the specific circumstances of the Mekong basin. Hence, customary international water law served its function of providing a framework for negotiations as well as fall-back position in the event an agreement could not be reached. The Kuala Lumpur Joint Communiqué did not signal an end to differences in positions among the countries, but rather the start of a negotiation process which was bounded by the rules of customary international water law. Radosevich in a speech delivered to the International Water Resources Asia Forum on the Mekong negotiations process, stated in 1995 that:

> The following remarks are prefaced by the appreciation and respect of the negotiating process involving four sovereign nations who have the right, as such, to take varying positions reflecting their interests on issues regarding the development and use of the transnational waters of the Mekong River; and whose positions, in absence of mutual agreement, are subject to interpretation under the prevailing rules of international law. The unique positions of each country exist partly due to their relative locations in the Mekong River Basin,

and partly as a result of 37 years of "Mekong Cooperation" under the 1957 Statute.

Four factors assisted the MWG in its understanding and application of international water law to their specific case. First, through the long existence of the Mekong Committee and its data gathering and planning activities, the negotiators had an excellent common understanding of the physical and socio-economic characteristics of the lower Mekong basin. Second, some members of the MWG had already received training on international legal issues through participation in a Legal Studies Group in the Mekong Committee Secretariat which was funded by the Asian Development Bank (ADB) and the European Union (EU) from 1990–1992. Third, the UNDP-provided Senior Advisor to the MWG was an international water law expert. During the course of the MWG, he reviewed and explained the various tenets of customary international water law, and as well numerous existing international and federal water agreements.[3] Finally, some key members of the MWG were active members of the International Law Commission's working group for the formulation of The Law of Non-Navigational Uses of International Watercourses.

Now some serious thinking and discussions had to take place to reach issues of mutual consensus. To acknowledge each riparian's particular rights and interests in the Mekong River system, monthly flows in the mainstream at the nine established gauging stations in the LMRB were discussed by examining a series of time and location hydrographs. The various uses and needs of water as well as potentials and expectations for development throughout the basin mainstream and tributaries were discussed. In an attempt to disengage each riparian's beliefs and perceptions based partly on their location in the basin (upstream/downstream, right bank/left bank), further discussions were undertaken applying a "one-nation basin" scenario in which it was assumed there was only one nation in the LMRB, and in turn each would play the role of other national riparian representatives to work toward an "optimum" utilization and protection of the water and related resources and the basin's environment. Later, the national boundaries would be overlain to adjust the conditions in conformity to the rights and interests of each riparian. This approach helped upstream riparians understand the concern of downstream riparians and vice-versa.

Another negotiating approach applied was to discuss and evaluate the options first as to the kind of agreement and basin organization would be desirable and suitable, the denominator of the future agreement fraction equation; and second the options on the range of objectives, principles, specific issues, processes, etc. that would be included in the agreement and implemented by the organization to be set up, the numerator to the fraction equation. For each of the denominators and numerators, the objective was to seek the highest common mutually acceptable option—the HCD and HCN. As described below, for example, regarding the options for the denominator of scope of agreement ranged from the mainstream or watercourse agreement to the waters within the MRB/LMRB to going beyond water and including the water uses within and beyond the MRB and related resources and the environment of the basin. That was the HCD adopted by the MWG that went beyond traditional customary international water law and the draft 1974 UN Convention. With that basic decision and concurrences on nature of the agreement set out in paragraph one of this case study, the HCN was negotiated for content of Chapter 1 Preamble, definitions, types of uses (that includes in-basin uses on-stream, in-stream and off-stream and inter-basin diversions) that combined navigation uses in-stream and minimum flows to protect uses and the environment, the range of objectives and principles and their content, the structure and functions of the implementing organization, and other provisions to include extending membership to the other two riparians and bilateral agreements amongst riparians. During the negotiations and drafting, there were many times when one or more country would have a higher level numerator, but during the discussions, the MWG would reach the highest level that all were mutually and unanimously willing to accept. This brings in the "agreement to agree" and in reading the Mekong Agreement, it should be clear that it only reflects what the MWG representatives and subsequently their respective governments could mutually accept—it does not reflect the full range of issues and options considered.

2.1 1995 Mekong Agreement

The 1995 Mekong Agreement took 21 months to negotiate and draft (a record for a complex international agreement) and only three months for ratification by the participating four countries (an accomplishment that even exceeded the expectations of the MWG members). It is a relatively short document, consisting of six chapters with 42 articles, and complimented with the Protocol to The Agreement on the Cooperation for the Sustainable Development of the Mekong River Basin for the Establishment and Commencement of the Mekong River Commission. The Mekong Agreement represents a "constitution" for a framework of cooperation. It focuses on fundamental issues such as general principles, areas of functional responsibility, decision-making procedures, eligibility, and organizational structure. The purpose of the Mekong Agreement is to establish a basis of cooperation under an institutional framework which will be robust enough to make operational decisions under a variety of future conditions. In addition to setting out substantive principles and objectives, the Agreement provided for a new international organization to implement the terms of the Agreement—The Mekong River Commission— substantially different in structure and function to its predecessors.

In many areas, the countries went well beyond the requirements of customary international law and reaffirmed the "Mekong Spirit of Cooperation" which had been in existence for almost forty years.

Notes

1   In early 1992, discussions among the four countries over an acceptable course of action to take to address member concerns about the existing basic documents

and reactivation of the Mekong Committee came to a standstill - an impasse. UNDP offered to provide neutral assistance in facilitating a solution by proposing an informal consultation, which took place in Hong Kong on 6 October 1992. The success of that meeting led to the historic meeting in Kuala Lumpur in mid-December, during which time, the four parties drafted key points that form the basis and commitment of each to work out a future framework of Mekong cooperation in a Communiqué and Guidelines, officially approved in Hanoi on 5 February 1993 at Mekong Working Group (MWG)-I. This "mandate" of the MWG served to guide the preparation of the draft agreement, along with the various subsequent papers and discussions. In the Communiqué of 17 December 1992, the commitment of each country was reaffirmed "to continue to cooperate in a constructive and mutually beneficial manner for the sustainable utilization of the Mekong river water resources," and in recognizing changes have taken place since the original mechanism was adopted, agreed to continue the dialogue to create an acceptable "future framework of cooperation." The Guidelines drafted in Kuala Lumpur, contain many important provisions of common interest and mutual acceptance to the parties. Acknowledging the "great political, economic and social changes" that have taken place in the sub-region, the countries are "part of the most economically dynamic region of the world," but also "faced with major challenges of natural resources management and environmental protection." Recognizing that "certain elements of cooperation already exist" that may need redefining, six elements for the future framework of coordination" were set out:

- A set of principles for the sustainable utilization of water resources of the Mekong river system;
- An institutional structure and mechanism for coordination;
- A definition of the functions and responsibilities of the structure and mechanism;
- The legal basis for the governance and financial operation of the structure and mechanism;
- Future memberships of the structure; and
- Management of the structure.

2 An important technique in guiding negotiations of water treaties is to insure the parties can have an opportunity for open and frank discussions to better understand each other's positions and to explore mutually acceptable options. The MWG structure was formal with flags, neckties, seating arrangements and minutes kept. To provide an informal environment (no flags, ties and minutes) we convened the Technical Drafting Group (TDG) meetings – these became "life savers" to the outcome.
3 Specific Agreements reviewed by MWG in detail included: Boundary Water Treaty between Great Britain and USA (1909), Columbia River Treaty between Canada and USA (1964), Great Lakes Water Agreement between U.S. and Canada (1978), Colorado, Tijuana, and Rio Grande Treaty between U.S. and Mexico (1944), Indus Water Treaty between India and Pakistan (1960), and the Senegal River Convention between Mali, Mauritania and Senegal (1972 and 1978).

Commentary

The author points to several things that accounted for the mediator's success. For one thing, he represented a neutral organization (the UNDP) that had successfully mediated a previous agreement in the same context. The UNDP was credible because of its role in the international community and the fact that it makes financial contributions to help implement whatever agreements it assists in formulating. So, the UNDP was well placed to facilitate communication among the parties. Moreover, the mediator himself had expertise in the content and context of the negotiation. This knowledge enabled him to facilitate the conversation among the stakeholders on technical, legal, and other specific details.

Radosevich's knowledge of customary international law also appears to have been particularly useful, since it was used as a basis for framing the discussions and for resolving disagreements. Indeed, it served as a source of what Fisher and Ury call objective criteria, a key means of generating agreement when the interests of the parties are in conflict (Fisher and Ury 1991). He also used customary international law to frame a common goal: an agreement in line with tenets of customary international law.

Another important value-creating move was the mediator's encouragement of perspective-taking through the "one-nation basin" scenario Radosevich describes. This is an example of using a fictionalized scenario or simulation to help parties understand one another's interests. We strongly endorse the use of fictionalized but tailored role-play simulations as a means to enhance the skills of the parties involved in joint problem-solving (Plumb, et. al. 2011).

Finally, using the denominator–numerator approach—reaching agreement on a broad scope first (the "denominator") and then more specific provisions (the "numerator")—appears to have been quite helpful. It is not clear, though, to what extent the mediator encouraged the group to brainstorm a wide variety of options before committing to a particular package.

S. Pollard, D. du Toit, and H. Biggs "River Management under Transformation: The Emergence of Strategic Adaptive Management of River Systems in the Kruger National Park," (2011)

Introduction

This excerpt by Sharon Pollard, Derick du Toit, and Harry Biggs describes the shift to strategic adaptive management of the river system in South Africa's Kruger National Park (KNP).

Within South Africa, the KNP is a "downstream user" of six major rivers: the Shingwedzi; Letaba; Luvuvhu; Olifants; Crocodile; and Sabie. Internationally, most of these river basins are shared by some combination of South Africa, Mozambique, Swaziland, and Zimbabwe. Increasing water use outside of the KNP, especially for agriculture, is considered a main cause of the continued degradation of water quality and quantity observed in the park. This is largely what prompted the park to shift to an adaptive management approach. The excerpt focuses on the KNP Rivers Research Programme (KNPRRP), the Shared Rivers Initiative (SRI), and use of multi-stakeholder catchment management agencies and water user forums. The authors highlight important elements of each of these programs that contributed to successful adaptive management, including: joint fact finding by park researchers, managers, and field staff; inclusion of water users and stakeholders in corrective actions; and the KNP's role in facilitating a "feedback loop" to link research findings and management.

Excerpt (pp. 6–11)

Institutional Arrangements

Against this precarious situation, we explore the response of the KNP, which, as for the examples of fire and borehole provision, has transformed over time. What is distinctive about rivers is that owing to their fugitive nature, KNP management was forced to look beyond the park borders for potential solutions (Venter et al. 2008) and to devise monitoring and management responses that were embedded in a wider socio-political landscape. The KNP has influenced institutional arrangements (Biggs et al. 2008) and has undertaken, or at times initiated, engagement in wider water management actions such as catchment strategy development, international agreement revision, water quality monitoring and even legal action. Mitigatory actions taken with respect to, for instance, maintaining the flow of the Sabie during the 1992 drought (Venter & Deacon 1995) and in the Letaba River (Pollard & Du Toit 2008) would otherwise not have succeeded. Indeed, the role of the KNP as "watchdog" (the first agency to alert water managers and society in general of a river problem) has been highlighted by Pollard and Du Toit (2008) as essential to functional feedback loops.

In the political climate that has prevailed since 1994, stakeholder involvement, transparency and accountability are regarded as key tools for achieving equity and sustainability. This means that the KNP can no longer operate as a conservation island, because such policies compel it both to be involved and to partake in wider stakeholder discussions within its expanded, albeit "informal", borders for water resources negotiations. The National Water Act (1998) outlines institutional arrangements for the management of water through the Catchment Management Agencies operative for 19 Water Management Areas (WMAs). The KNP straddles three WMAs:

- the Inkomati (incorporating the Sabie and Crocodile rivers within the Inkomati River basin, an international drainage basin shared by South Africa, Swaziland and Mozambique)
- the Olifants
- the Luvuvhu–Letaba.

Both the Olifants and Luvuvhu–Letaba systems form part of the Limpopo Basin in Mozambique. Although only the Inkomati WMA has been gazetted thus far, it has set a precedent in that a seat for conservation is reserved in the composition of the board. This places the onus for participation on the conservation sector (and especially the KNP) and affords a much stronger voice than in the past.

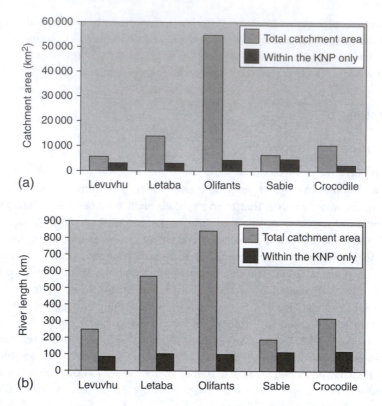

FIGURE 6.5 A comparison of (a) total catchment areas and (b) river lengths of the five Perennial river systems found within the KNP

Source: (from Pollard and Du Toit, 2007)

KNP, Kruger National Park; km², 1 square kilometer.

The Formal Adaptive Response: Developing a Strategy to Respond to the Challenges Posed by Declining River System Integrity

The KNP Rivers Research Programme and Adaptive Management

Until the late 1980s, river management per se was not explicit as a park objective other than park authorities asking the (then) Department of Water Affairs and Forestry (DWAF) for special releases from upstream dams. Nonetheless, by the late 1980s the development of water quality guidelines seemed to signal a resurgence of belief that the KNP could exert constructive pressure on external agencies above ad hoc requests for releases, for instance, from the Tzaneen Dam. Additionally, DWAF announced its intentions to allocate water for environmental flows in rivers as it became obvious that if demands were unchecked, the integrity of the rivers would be threatened ... Despite these intentions, estimates were limited to preliminary calculations and based on absolute amounts of water. Indeed, the first formal recognition of water for instream flow needs for South African rivers was by Roberts (1983), who used an allocation for "conservation" of 11 percent of the country's mean annual runoff (later modified to 8 percent of the exploitable water resources by Jezewski and Roberts [1986]). Roberts acknowledged that this figure was simplistic in that it was based on coarse, countrywide estimates of water for estuaries, lakes and nature reserves. As such, it could not be used for individual rivers (see also Breen et al. [1994]), but nonetheless provided the catalyst for future work. Researchers contested this figure and pointed to the paucity of understanding regarding Lowveld river systems as a major challenge to management for sustainability of the rivers. Ultimately a much more sophisticated and ground-breaking approach for calculating riverine water requirements was developed by a South African team (King et al. 2000) and prototyped through the KNPRRP and other research initiatives in South Africa (see later).

By 1988 the KNPRRP was conceptualised and initiated as a co-operative undertaking by managers or resource users, funding agencies and researchers (Breen, Dent et al. 2000). The KNPRRP consisted of three phases ... Phase I (1989–1993), which ran for four years, was largely limited to scientific research. The focus was on a range of research topics relating to environmental water requirements but was unstructured in detail (O'Keeffe & Coetzee 1996). Moreover, managers within the KNP were not convinced that, in practical terms, the research outputs supported the continuing management crises that they experienced (H. Biggs, pers. comm.). In 1991, the newly constructed Zoeknog Dam in the upper Sand catchment collapsed, delivering sediment for weeks into the river, a phenomenon that persisted through the KNP and into Mozambique (Weeks, Pollard et al. 1992). Such a patent demonstration of undesirable consequences of poor design or construction compelled the Park to respond through radio interviews and public platforms. It had not been customary before this for the Park or their associated researchers to respond, signalling the start of a more public voice for the KNP. It again highlighted the need for directed research that could support managers in their response to short-term crises. Other research in the programme examined the potential fragility of the system, such as the effects on fish of being confined to shrinking pools (Pollard, Weeks et al. 1994).

A comprehensive review of Phase I recommended a second phase (1994–1996), with greater emphasis on predictive capabilities and management action, which was to be more intimately linked to a decision making system. It was during this phase that collaboration between managers and researchers improved with some co-learning. Researchers began responding more explicitly to short-term crises experienced by managers, and managers benefited from the longer-term view provided by researchers. The political transformations that accompanied democratic transitions in 1994 were also major drivers for change, creating opportunity for far more effective international engagement. In 1995 the KNPRRP hosted an international conference on Integrated Catchment Management in Skukuza, a concept which was receiving increasing attention within the DWAF itself. This served to focus interests on holistic water resources management and, interestingly, raised the profile of international issues associated with water sharing across country borders.

At about the same time, research interest in complexity theory and adaptive management within natural resource management started to grow. These ideas arose as a critique of approaches based on averages and the propensity to view nature as balanced, linear and predictable. Variability, in fact, was highlighted as the key characteristic of semi-arid systems (Davies, O'Keeffe et al. 1995). Even where ranges were recognised (e.g. introducing a variation of between 7000 and 9000 in KNP elephant numbers), it was still not appreciated that savanna ecosystems need more extremes than these slight fluctuations to build resilience. This paradigm suggests that the recognition of variation and extreme events are fundamental for biodiversity management. This idea was central to the determination of environmental flows where variation in flow regimen was seen as a key driver of the system. The building-block approach (King et al. 2000) introduced the concept of incorporating freshes (small and intermediate floods) into a flow regimen, which were seen as essential linkages to certain key biotic or abiotic events such as spawning or sediment flushing.

Other concerns at the time centered on the entrenched and "command-and-control" nature of management within the KNP (Biggs & Rogers 2003; Du Toit, Rogers et al. 2003). The imperative of political transformation necessitated change from one of an insular approach of managing the park, separate from its neighbours, to one which attempts to embed the KNP within the socio-economic landscape and encourages wider participation, transparency and public ownership. Moreover, the entrenched science–management activities such as monitoring were aligned with a facilitated, "learning-by-doing" approach. The conservation of protected savannas in Africa has been dominated by a focus on charismatic species and, as mentioned, influenced by stable-state concepts such as carrying capacity, with less emphatic regard for scale or the inherent dynamics of ecosystems. The previous approach has been challenged for its failure to embrace spatial heterogeneity and flux in ecosystems and for not always recognising a fuller array of compositional, structural and functional elements of biodiversity and ecosystems (Noss 1990). Indeed, a recent publication centred on the theme of heterogeneity in the KNP (Du Toit et al. 2003) bears testimony to this fundamental shift in thinking. This raised a number of questions and challenges for the research and management community. Firstly, what research was needed to elucidate the important characteristics of heterogeneity? Secondly, how was management to embrace such variability and flux as the norm and when would the "variability norm" be unacceptably exceeded? As noted by Rogers (2005), strategic adaptive management (SAM), and its associated objectives hierarchy, is one of the few recognised models for managing uncertainty in interactive social and ecological systems, whilst still aiming purposefully at a carefully articulated (but assumed to be shifting) desired state.

As explained, river management was in crisis during the early 1990s and despite a vigorous initiative on the part of the KNPRRP and a few SANParks associates, most managers in the KNP had not internalised that river management was an explicit part of their brief.

Interestingly, another important co-driver of the change in KNP management was the "impasse" on elephant culling, which came to a head at about the same time as the KNPRRP underwent a major reorientation (Freitag et al. in review). Heated public debate and scrutiny called for reforms to the culling programme and a moratorium on culling was introduced (Van Aarde, Whyte et al. 1999). This essentially set the scene for other programmes to be influenced by the thinking that had developed in the KNPRRP. Notably, a conference to discuss elephant management was held in Skukuza and this provided an opportunity for participants to examine progress that had been made within the KNP management framework. The key conclusions were that (1) the vision and objectives cascading from this vision needed to be improved and (2) elephants need to be managed as part of an ecosystem (Braack 1997a).

This led to the revision of the entire KNP management plan (Braack 1997a, b), starting with a visioning exercise of which the learnings and elements were already available through the KNPRRP. Under the theme of accountability, the KNP had to go public with its objectives. At the time, the KNPRRP was in the process of exploring and prototyping the concept of defining and operationalising the desired future state (DFS) of rivers (Rogers & Bestbier 1997). Through collaborative efforts between the KNP management and the KNPRRP, the application of this concept was explored for use beyond river management alone. Much of the philosophy behind the DFS is that, as a public participation process which arrives at a joint agreement, much of the potential conflict can be reduced. After clear objectives were set as an objectives hierarchy, questions arose as to what needed to be monitored to achieve these objectives. A large collaborative meeting between managers and researchers in the KNP heralded the start of measurable endpoints, known as TPCs (see, for example, Braack et al. [1997a], McLoughlin WRC 2011 and Pollard and Du Toit [2007]). As described elsewhere, these TPCs are intimately embedded in an adaptive management framework. Critically, they are set against the background of complex systems, representing spatio-temporal flux, often with lower and upper limits (see McLoughlin et al. [2011] for a comprehensive review of river-related TPCs).

The third phase of the KNPRRP (1998–2000) was designed to enable the completion of first-generation procedures and technologies to support SAM of rivers and to promote corrective action through the participation of stakeholders, especially those who had previously been marginalised. The need for a more holistic approach also prompted creative thinking around the issue of integrated catchment management and the role of SAM in this regard. It could be argued that, given the leadership and close relationship between key individuals in the KNPRRP, DWAF and the Water Research Commission (WRC), who funds water research in South Africa, a strong learning alliance was formed, albeit informal. Many of the ideas emerging from the KNPRRP and the process of water law reform were echoed in WRC research reports and were mutually reinforcing. For example, today notions of SAM are embedded within certain strategic documents and guidelines of DWAF. Moreover, the approach is now firmly embedded in Scientific Services within SANParks on an ongoing basis. Although not named as such, a new phase of work was initiated in 2007 (after being discussed for several years) to build on that of the KNPRRP. It was driven by scientists' and managers' questions as to the apparent lack of improvement in the status of the rivers despite the advent of the National Water Act. The intention was to deepen the understanding surrounding the causes for the lack of improvement. About a year later the KNP also recognised the need to strengthen its own adaptive management of the rivers and also initiated an associated project, both of which will now be discussed.

A New Phase of Research: Linking Outputs to Management

The KNPRRP was followed by a hiatus in programmatic river research until the conceptualisation of the two initiatives, both quite distinct from the earlier rivers programme, and both strongly focused on action research and adaptive management processes. One of these, introduced earlier as the SRI (Pollard & Du Toit 2008), focused on understanding the factors that enable or constrain meeting the commitment to the Ecological Reserve in six river systems flowing through the KNP (Luvuvhu, Letaba, Olifants, Sabie–Sand, Crocodile and Komati). The intention was to build supportive programmes in Phase II due to commence in 2010. The other closely linked project aimed to consolidate the SAM process for freshwater management in the park, mainly by operationalising the TPCs through effective science–management links (Biggs et al. 2003; 2008; 2010). Primarily as a result of findings from the SRI, which identified case situations that were amenable to effective study, the current focus is on the Letaba and Crocodile rivers. In both cases there is strong evidence of feedback loops between key role players. Feedback loops and self-organisation are considered to be essential components of resilient systems and adaptive management (Biggs & Rogers 2003; Holling 2001; Holling & Gunderson 2002). Feedback is the basis for learning in a reflexive system. Where systems often fail is where one or more of these steps fail such as in cases where the learning is not passed on or is passed to an inappropriate body. When functional, these loops set up a self-organising system that is responsive to change.

As recognition for this approach grows, so does the interest in what makes the feedbacks work (see Pollard et al. 2008). In the Letaba catchment, for example, a number of key feedback loops of self-organisation and self-regulation are evident (Figure [6.6]). The KNP monitors flows against the Reserve requirements (which have until recently been static; that is, not actively dynamic in line with current exact rainfall) and, on detecting problems, the Water Affairs manager (who manages the Tzaneen Dam), in turn, alerts the Groot Letaba Water Users Association to curtail use. They, in turn, inform users of curtailment rules and monitor adherence.

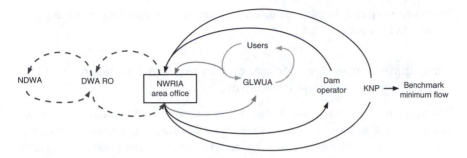

FIGURE 6.6 Functional feedback loops in the Letaba catchment

NDWA, National Department of Water Affairs; DWA RO, Department of Water Affairs Regional Office; GLWUA, Groot Letaba Water User Association; KNP, Kruger National Park; NWRIA, National Water Resources Infrastructure Agency.

Although not always popular, the regulatory system is respected and adhered to by the members.

There are a number of causal factors behind the success of these two loops, including the requirements of the law (the Reserve), the availability of benchmarks against which to monitor (the Reserve), the presence of a "watchdog" (the KNP in this case), the responsiveness of the manager and users, and the ability to self-organise. Whilst an in-depth analysis of these is beyond the scope of this paper and is examined by Pollard and Du Toit [2011], a key point is that the "watchdog" role, which is as important as any of the other roles, is often overlooked and hence needs to be recognised as critical. The SAM project is now refining, together with users, an adaptive monitoring management system. The essence of this system is that there are different levels of concern related to the status of a resource in question (e.g. river flow) and hence different management actions linked to each. The severity of the "worry level" is given via an indicator or TPC, which is collaboratively determined (McLoughlin [et al] 2011). The important principle, therefore, is that there is an envelope of levels of concern—supported by a clear rationale—and each is linked to different management actions.

The SRI (Phase I) has demonstrated that the requirements of the Ecological Reserve are not being met with regard to quantity in all six rivers, despite an improved policy environment and the initiation of integrated water resources management (Pollard et al. 2010). This can, in part, be attributed to lags that are an inherent part of the process of reform in a complex environment; setting the Reserve today does not mean that it will be met tomorrow. However, it is important to consider what makes certain delays unacceptable. In many cases, especially in the northern WMAs, issues such as tardiness in authorisation, unlawful use, the lack of integration of water resources management and supply, weak monitoring and enforcement, and the dearth of skills and capacity all need to be addressed as a matter of urgency. In others, such as in the Crocodile and Komati rivers, recent advances in water resources management provide real possibilities for improvement. The KNP has again been an important roleplayer in this regard, acting both as a catalyst for change and as a constructive stakeholder.

Participation in Wider Catchment Forums: The Development of the Catchment Management Strategy

The commitment to manage water holistically is captured in the National Water Act, which requires that water resources are managed from a catchment perspective. Ultimately, CMAs will take over the management of the water resources, especially with regard to water allocation and protection of the resource. Representation is secured through various structures such as catchment management committees or forums (CMCs or CMFs) and water user associations (WUAs). In many cases, the KNP has initiated forums that could be considered CMF precursors or prototypes. In the case of the Crocodile River catchment, the KNP has spearheaded the establishment of the Crocodile River Forum (CRF) and for the Sabie River, the Sabie River Working Group, which started in 1991. Today, the KNP participates in fora that cover all major rivers entering the park and plays an important role in tabling its position and interests in water resource decisions.

The relationship of the KNP with and its influence on the agriculture or forestry sectors are also worth mentioning. An example is the Sabie River Working Group, which managed to save the Sabie River in the KNP from a flow stoppage during the 1992 drought (Biggs et al. 2010). The KNP initiated this forum and was an active member but it was chaired by an irrigation farmer from the Hazyview area for many years. Another example is the Marula Weir, which was to be constructed in the Crocodile River for irrigation purposes. The KNP managed to stop the building of the weir even though the foundation had been started (Venter et al. 1995). Although the relationship between the irrigation farmers and the KNP was strained as a result, the two parties have managed to build a good relationship subsequently as mutual understanding improved. More recently, the KNP supported the efforts of a local initiative, namely the "Save the Sand" programme, in advocating the withdrawal of poorly managed afforestation in the upper Sand River catchment. The plantations, conceived as labour-creation schemes under the Bantustan regime, covered excessively steep slopes, wetlands and riparian zones, which caused sediment problems and reduced base flows ... The support from the KNP was not so much about actual effects (which were most heavily felt before the rivers enter the KNP boundaries), but rather about the principle of wise use and management of natural resources. Many more such examples are available in both peer-reviewed (e.g. Kingsford et al. 2011; McLoughlin [et al] 2011) and grey literature (reports submitted to the WRC under project K5/1797).

Lessons that Emerged from Learning in Action

This section deals with the emerging outcomes of the transformation described in this paper, which can now be reflected upon and used as guidance for a way forward.

Many important learnings in this field are already available, for example, descriptions and discussion of the:

*   transformation from a more closed to a more open style of management in the KNP (Venter et al. 1995)
*   general steps involved in SAM and the related progress (Kingsford et al. 2011)
*   spread of SAM from a rivers application to a wide range of domains, not only in the KNP but also across general conservation applications in South Africa (Freitag et al. in review)
*   mechanics of actual operationalisation of feedbacks built around thresholds (McLoughlin [et al] 2011).

This paper has taken a more direct look at the philosophical and paradigmatic changes, and styles of management and research, that have characterised the transformation we describe. The overview considered the fuller range of initiatives relevant to water resource management in the park and its surrounds, both before and after the inception of explicit or formal adaptive management. The particular lessons elucidated by this review thus overlap with several lessons from each of the studies described, but also reinforce or complement those lessons with additional value.

The observation that the SAM approach has been widely accepted within the KNP and that no fundamental alternatives for river management have appeared to date, may mean that the KNP and other active collaborators are beginning to understand the complexity required to broker decisions effectively on a continuing (dynamic) basis. The fact that there have been ongoing structured research programmes on rivers seems to imply that, at least in the context of the KNP, active levels of research involvement may be a prerequisite for coping in a fast-changing world with difficult resource management issues.

The process of adopting SAM was, as described earlier, a process of recursive action over time. Although the SAM approach began mainly with biophysical aspects within the park, over time, the recognition of wider socio-ecological systems (initially catchments) became central, with more active systemic connections being realised. This meant that, initially, management procedures became more inclusive of issues as they emerged and ultimately more complicated. There came a point where managers were overwhelmed, which resulted in a retraction to requisite simplicities (Holling & Gunderson 2002), so that the management process did not become untenable.

What is the overall meaning of findings of this particular overview for the KNP and for the wider community who are engaged in SAM of rivers? The main achievement for the KNP has been the development of a new way for approaching its conservation mandate based on complexity principles. This led to the emergence of a management framework over nearly 15 years (detailed in Pollard and Du Toit [2007] and described earlier). The SAM framework, although developed through a focus on rivers, can be embraced for the management of ecosystems as wholes. In summary, the framework requires that management be directed towards achieving a desired state (Biggs & Rogers 2003; Breen at al. 2000). Indeed, this has fundamentally re-orientated the management of the KNP, staff and resource allocations. As explained earlier, once this higher-order statement had been debated and captured in a vision, it provided the basis for the development of objectives and endpoints that could be readily traced back to the vision. This process has allowed a much closer partnership to develop between researchers, managers and field staff, with a strong sense of buy-in and collective learning made possible (Pollard & Du Toit 2005).

An important shift in the management principle governing semi-arid savannas is that the desired future state is not a stable state but one that is based on a fundamental recognition of variability as an overarching characteristic to confer resilience. Thus, judicious management is predicated on understanding the underlying ecosystem drivers and characteristics of the system in question. Moreover, since river systems are dynamic and in a continual state of flux, it is necessary to monitor conditions and to revisit management objectives. System dynamics need to be understood in the broader context of events both inside and outside of the protected area.

Adopting the SAM framework, with its key features being a clear vision informed by stakeholder involvement, an objectives hierarchy, a consideration of management options, an apparatus of TPCs and a reflective evaluation process consisting of feedback loops, has been challenging but important for KNP staff. Pollard and Du Toit (2007) noted that the collaborative role of researchers and management in developing TPCs and ensuring they are met has been cited as a powerful motivation for monitoring staff, such as rangers and wardens, who then become a key link in the iterative SAM cycle. The value of involving field staff in setting management objectives cannot be underestimated with regard to developing commitment and buy-in. The TPCs are hypotheses and hence the TPCs and the associated "desired state", should be audited and refined in a reflexive manner (McLoughlin [et al] 2011; Pollard & Du Toit 2006).

Knowledge management is a challenge that needs to be addressed. Biggs and Rogers (2003) point out that after a TPC has been tabled, several unpredictable threads of information tend to emerge as implementation proceeds. These threads may or may not be documented at the appropriate level of quality (i.e. everything is taken to be equally relevant). These authors recommend a continual "roping together" of the information so that the organisation benefits as a whole, thus averting disparate and isolated approaches. The SAM approach is likely to generate a wealth of field data that need to be recorded, captured and made accessible (McLoughlin [et al] 2011; Pollard & Du Toit 2006). This is seen as one of the challenges for the KNP. At present, the Park is developing a knowledge management system based on a geographic information system as well as non-spatial databases. The intention is to draw science and management together by putting data to productive use rather than archiving for historical purposes only. Once the challenging aspects of knowledge management have been negotiated they can lead to the need for shared learning. Here the KNP has experimented with the formation of "Communities of Practice" (Lave & Wenger 1991) from, initially, a core of enthusiasts whose task it is to continually rework and improve the SAM system and make it more accessible for use by others. Experience shows that there is a need for programmes run by the KNP to be integrated so that, by drawing on a wider variety of specialists and practitioner experiences, more realistic TPCs can be set in the future (Pollard & Du Toit 2006). Nonetheless, lessons for integrating new concepts, such as ecosystems services and social ecology, with more traditional approaches are yet to be learned.

Pollard and Du Toit (2008) argue that the legislative environment for water resources management and the approach of integrated water resources manage-ment afford a particularly strong basis and coherent currency for the adaptive management of river systems. The approach in the KNP thus complements—and puts into practice—the spirit and intent of the National Water Act. In the case of river management, an additional challenge has been to broaden horizons and deal with the realities of conflicting drivers and objectives. River systems are common-property resources (Pollard & Cousins 2008). In South Africa, there is no private ownership of water and flow through a portion of land does not confer inalienable rights on that land owner. Moreover, because demand is viewed from a catchment-scale perspective of the total water resources, there will inevitably be tradeoffs and compromises in working towards more equitable and sustainable con-figurations for catchments (Pollard & Du Toit 2008). These two factors necessitate that stakeholders, including protected area staff, participate in water resources management where different interests and demands on the water resources are used to negotiate water sharing. Fortunately for the KNP, the new legislative environ-ment has provided strong support for the concept of sustainability through the provision of the Reserve, which not only provides a benchmark for monitoring, but also carries legislative "clout", strengthening the KNP's position as "watchdog" (Pollard & Du Toit 2008). This is critical given that infringements of the Ecological Reserve are evident in all rivers flowing through the park (Pollard et al. 2010).

Importantly, the KNP staff do not only monitor the rivers, but link outputs clearly to different actions according to the severity of the infringement (McLoughlin [et al] 2011), the transparency of which is important for monitoring staff (Pollard & Du Toit 2006). Although the systems are still being strengthened and successful responsive action nonetheless varies, the basis for building feedback loops is in place. Indeed, as mentioned earlier, these feedbacks are essential for adaptive management, for without these, learning cannot happen (McLoughlin [et al] 2011; Pollard & Du Toit 2006; Pollard & Du Toit [2011]).

Conclusion

In conclusion, it is important to remember that adaptive management is not an end in itself, but a process that evolves as new learnings are brought to bear. As a result of the challenges confronted in addressing changes in rivers, the KNP has charted new ground in management, research and outreach. The approach embraces the challenge of managing a sensitive, complex system in a context where uncertainty is always an underlying factor. It encourages the "first bold step forward" where "imple-mentation paralysis" can often hamper decision making. By using the best available information to set TPCs, SAM monitors trends and then demands reflection on collaboratively defined goals before jointly agreed action is initiated. The collaborative nature of implementing the SAM system forges a partnership between science and management—an approach that is seen as a way forward for parks, conservation and science (Folke, Carpenter et al. 2002; Van Wilgen & Biggs 2010). Equally important is that such thinking has an institutional home for its eventual mainstreaming as management discourse, which the KNP may well provide.

Commentary

The adaptive management approach the authors describe was developed through internal collaboration and joint fact-finding among park researchers, field staff, and managers (i.e., internal park stakeholders). They rightly note the contribution this collaboration and co-learning made to setting up a management system that connects scientific findings and decision-making.

The authors do not, however, comment on the failure to engage any outside stakeholders. Although they do note the role of water users in spreading the word about required adaptive measures, it is not clear how the water user associations and catchment forums, agricultural and forestry sectors, and other outside stakeholders participated or will participate in the future in adaptive management of the rivers that run through the KNP.

The authors emphasize that setting performance measures, and creating feedback loops to respond to successes and failures in meeting targets, are essential for successful adaptive management. In this case, KNP stakeholders set measurable endpoints and engaged internal and external stakeholders in responding to their findings. This appears to have made it easier to implement adaptive management. Moreover, indicators and principles of adaptive management were written into a number of framework documents and strategic plans. As we would predict, such explicit documentation made implementation easier.

Finally, the authors acknowledge the persistent challenge of making data accessible to a broad range of users and enhancing long-term knowledge management capabilities. The idea of capturing various things that were learned so they could be added to a learning management system, and of making that information more accessible to outsiders, seem to us to be steps in the right direction.

N. Odeh "Towards Improved Partnerships in the Water Sector in the Middle East: A Case Study of Partnerships in Jordan's Water Sector," (2009)

Introduction

The excerpts by Nancy Odeh are taken from her doctoral dissertation at MIT, in which she looks at four partnerships that were created to improve water management efforts in Jordan. These excerpts examine one of those partnerships, the operation of water user cooperatives in the Jordan Rift Valley.

Water demand in Jordan far outpaces the country's renewable supply of freshwater and groundwater, and population growth and agricultural expansion are contributing to an increasing water deficit. This water deficit is a major concern in the Jordan Rift Valley, which accounts for approximately 40 percent of Jordan's irrigable land. To improve irrigation efficiency in the Valley, the national government implemented a modern, pressurized irrigation system. Failure to consult with and provide technical assistance to farmers, however, detracted from the efficiency of the system, and adversely affected the relationship between farmers and the Jordan Valley Water Authority (JVA).

The excerpts describe an initiative, proposed by the German Technical Cooperation agency (GTZ), to establish water user cooperatives in the Valley to increase irrigation efficiency by including farmers in the management of the new system. Eighteen cooperatives were established in four parts of the valley. They were organized around pumping stations, crossing boundaries between individual farm units. The effort also included significant engagement of both the JVA and GTZ. These excerpts identify the drivers behind the formation of the water user cooperatives, their contribution to enhancing the sustainability of Jordan's water supply, and how partnerships contributed to both capacity building and relationship building.

Excerpt (pp. 197–202)

[T]here were two principal forces driving the establishment of the water user cooperatives spearheaded by GTZ in 2001. The first, was the growing awareness among water users, government, and donors of the imminent water crisis facing the country in the early 1990s, as a result of the events that had unfolded over preceding decades (Interviews 5; 13; 73; GTZ, 2002; 2006b). There were a number of indicators of this imminent water crisis. For example, Syria has been exceeding its share of water from the Yarmouk, as agreed to in the 1955 Jordan Valley Plan. Evidence of this is that prior to 1950, the total available surface water resources in the Lower Jordan River basin (which includes the Jordan Valley and Amman) was on average 550 MCM. The Yarmouk River's flow—the main source of water into the Jordan Valley, once at 470 MCM, dipped to 360 MCM in the mid-1990s, and has been about 150 MCM for the past five years (Van Aken et al., 2007; Interview 49). In addition, the transfer of water from the King Abdullah Canal to Amman in order to meet the City's increasing demand, has grown steadily since the initial decision was made to transfer this water to the capital in 1985. It reached over 41 MCM in 1999, and was approximately 60 MCM/year in 2007. This accounts for 40 percent of the water that enters the King Abdullah Canal, and 46 percent of the total amount of water supplied to Amman each year (Interview 49; LEMA OPS, 2007). Another manifestation of Jordan's water crisis was the severe drought in 1997, which lasted until 2001. It was another reminder to farmers that they are among the worst affected in Jordan since droughts mean lower yields and virtually no income from summer crops (GTZ, 2002). A further stress on water resources in the Jordan Valley is that since the mid-1980s, because of technical improvements in agriculture, through the likes of greenhouses, drip irrigation, fertilizers, an influx of Egyptian workers, and more market opportunities, farming in the Jordan Valley has become much more intensive and production has soared. This has happened at the expense of sustainable water use thanks to the increase in cultivation of very water-intensive agricultural undertakings such as small-scale olive and citrus groves, and banana plantations (Venot, 2004; Courcier et al., 2005).

The second driving force behind the GTZ project, which helped create these water user cooperatives, was the inability and unwillingness of both farmer and the JVA to operate the pressurized network in an optimal way. In 2000, approximately 40 percent of the farmland was still being irrigated using surface application methods that required higher water flow rates. Consequently, the system's delivery rate was raised to 9l/s, at the cost of overall system stability. The resultant line pressure losses made it impossible for some of the farmers to get the volume of water within the allotted time that they were guaranteed through their respective FTAs (GTZ, 2000a and 2000b). As both senior officials in the JVA and members of the water user cooperatives explained to me, it was clear early on that most farmers did not readily welcome the modern pressurized system. They thought the lower flow (i.e., as required by its design) was insufficient and they simply preferred the more straightforward open channel that existed previously. The biggest issue was the rampant theft of water by farmers in their quest to augment the water supplied to their farms. Farmers managed this by: making illegal connections; opening their FTA when it was not their turn to do so; and/or toying with the flow limiter in their FTA to increase flow to their farms. As explained above, the consequence was disastrous in terms of water distribution efficiency. Illegal water use meant that more water was being taken out at any given time than the system was designed for. This lowered the pressure in the network which, in turn, altered the frequency and quantity of water delivered to farmers, as well as undermined their ability to properly use drip irrigation on-farm (Interviews 5;41;47;70). As the Director of the Northern Jordan Valley Directorate explained:

> When the move to the pressurized system came, the farmers were suspicious of the system and did not like it, because it was a different move. Some farmers insisted on staying with the open channel system, mainly because the pressurized system gave them 6 liters/second, which was less than what farmers had received before, and 6 liters is not enough. ... Still, some farmers preferred the open channel, and were not using the pressurized system correctly. But there is progress now in using the pressurized water, because of the water shortage. So the farmer is forced to use the pressurized system.
>
> (Interview 47)

The GTZ project leader for the water user cooperatives explained that:

> The main challenge was that farmers used to interfere in water distribution, trying to help themselves to more water. That led, gradually, to a breaking down of the regular water distribution service. It also caused a lot of physical damage to the infrastructure. The challenge was, at first, to identify suitable organizational forms that are both accepted by the farmers, and also by partner organizations like the JVA. There was quite a struggle over a year and a half, until the first group decided to become a water user council in 2001. And a little bit later, the first water user cooperatives were founded.
>
> *(Interview 5)*

The prerequisite for the pressurized network to work optimally is that all farmers must accept their allotted share of water and under no circumstance should then tamper with the system in an effort to obtain more water. Forming water user cooperatives was viewed as a way to organize and educate farmers about how to eventually achieve this outcome.

The following is intended to show that the illegal use of water by farmers has changed, and why this has occurred. The JVA was also at fault for not operating the pressurized system effectively. The root of the problem was that an increasing number of farmers adopted the more efficient drip-irrigation system which required a reduced flow, higher pressure, and increased frequency of water application—all features of a pressurized underground water distribution network. Once the pressurized network was in place, more farmers adopted drip irrigation systems; however, there was insufficient financial and technical support offered to farmers to help them adapt their new on-farm systems to the pressurized network. There were also problems related to filtration and clogging. The result was that farmers who used drip irrigation and those who still depended on surface irrigation rejected the idea of having a pressurized water distribution system predicated on a decrease in flow (25 l/s down to just 6–9 l/s). Farmers assumed a decrease in flow would mean a decrease in the volume of water reaching them. Their resistance to this change persuaded the JVA not to adhere to the proper design of the pressurized network. This meant the JVA agreed to allow a flow of up to 15 l/s. In addition, they did not implement the strict rotation schedule limiting the number of arms obtaining water simultaneously. This drastically lowered the pressure (to 1 or 1.2 bar, as opposed to the target 3 bar), and did not allow the new on-farm drip irrigation systems to operate as intended (Interviews 5; 13; 17; 75). Training farmers on the proper use of the pressurized network and drip irrigation systems was the responsibility of a French team. One of the team's consultants told me that "the water user cooperatives have a problem, which is that some farmers used to steal water. So the JVA decided to open all the secondary pipelines to the farms together—so the water flowed like it would in an open channel system—to avoid this problem. But the solution should be more training for farmers, and more control by the JVA. The JVA should monitor, but they do not want to do extra work so they take the easy way out, which is to do nothing" (Interview 75). Again, as the GTZ project leader explained, the idea is that the formation of water user cooperatives would convince ever-greater farmer numbers of the advantages in adhering to the water distribution procedure required by the pressurized network. This involves encouraging them to work with the JVA to implement the system rather than having them thwart it (Interviews 5; 17; 23; 86). These two key forces converged to make it clear to GTZ that a more effective partnership among farmers as well as between farmers and the JVA, was crucial to sustaining water resources and farmers' livelihoods. There were also other factors at play that came to the fore in 2000, including the following (GTZ, 2002):

- The JVA's lack of planning for water distribution made it nearly impossible for farmers to organize their cropping plans for the coming season. Farmers complained that the JVA did not announce its distribution plans far enough in advance, nor did it stick to the plan it announced. Clearer and more frequent communication between farmers and the JVA was needed, and water user cooperatives could help with this.
- The majority of the JVA's staff in 2008 is nearing retirement age due to a hiring freeze. This means that the bulk of knowledge about managing the irrigation system will likely be lost. Passing on to farmers the expertise gained through many years of investment in human capital formation, coupled with the substantive information base built up over the past several decades, in an organized fashion such as through a water user cooperative, would preserve the knowledge accumulated in the JVA.
- Farmers themselves have an excellent pragmatic understanding of the specific improvements the irrigation system might benefit from, and the JVA could learn a great deal from them. However, open dialogue between the two was sorely lacking. The idea being introduced was that the JVA would attend meetings of water user cooperatives periodically, and this would provide an ideal platform to enhance an information exchange between these entities.

Excerpt (pp. 208–211)

(ii) Sustainability of Supply

… the various sources of irrigation water in the Jordan Valley … include the King Abdullah Canal, wells, and side wadis. To get a better appreciation for how the various sources of water have changed over the past decade, I compiled data from the JVA Control Center in Deir Allah, which … is the central node responsible for amassing all water data. Table [6.2] lists the Valley's total inflow and outflow into the Valley between 1997 and 2006. The largest source of outflow is the transfer of water from the King Abdullah Canal to Amman. This outflow has steadily increased since 1985, reaching approximately 60 MCM/year in 2007, which is 40 percent of the water that enters the King Abdullah Canal (Interview 49; LEMA OPS, 2007). Farmers are aware that during a season when rainfall is unexpectedly low, the priority

TABLE 6.2 Jordan Valley Total Inflow and Outflow between 1997 and 2006

Inflow and Outflow (MCM)	1997	1998	1999	2000	2001	2002	2003	2004	2005	2006
Total inflow	314.5	289.9	212.5	239.7	114.7	234.0	357.3	266.0	254.7	210.5
Total outflow	204.1	232.5	175.9	170.2	124.4	148.9	208.1	222.9	206.6	169.9

Source: (from JVA 2007)

is to transfer water from the Canal to Amman. This is the expense of farmers in the north of the Jordan Valley, who rely primarily on water from the Canal for irrigation (Interviews 5; 13; 29; 75). There has been a marked reduction in inflow of almost 100 MCM per year into the King Abdullah Canal, since it opened in 1997, and this makes it imperative for farmers to do their part to ensure that the irrigation allotted to them is distributed as efficiently as possible. The best way of doing this is to take the measures necessary to the proper functioning of the pressurized network.

Compared to the open channel system, water loss in the pressurized network (conveyance system) has been significantly reduced. Efficiency in the network ranges from 75–85 percent (Van Aken et al., 2007), but it can be further improved with the cooperation of farmers. My findings suggest that water user cooperatives have indeed contributed to an improvement in the economics and operation of the pressurized network.

Water user cooperatives have also helped ensure the proper operation of the pressurized system because it requires farmers to follow a strict water rotation schedule in order to attain the required homogenous water pressure they need. Farmers taking water out-of-turn can lower the water pressure. Having farmers adhere to the water rotation schedule is achieved more easily when they are part of an organizational form that encourages information exchange about why receiving water of higher pressure is best-suited to their on-farm drip irrigation systems. In short, there is clear self-interest at stake through appropriate behavior. This also encourages trust-building, in that farmers begin counting on their neighbors not stealing extra water out-of-turn, which would foil the proper operation of the system[1] (Interviews 13; 27; 28; 46; 47; 48; 49). The latter point was emphasized by the JVA Director of the Northern and Middle Directorates of the Jordan Valley: "the water user cooperatives bring the farmers together, and gives them a stronger sense of cooperation, so they start not wanting to steal irrigation water because the farmers build feelings of closer ties to the community, and indeed closer friendships with their respective counterparts. In short, the cooperatives brought about a new culture of cooperation" (Interview 48).

The lead agricultural expert of the French-financed irrigation optimization project in the Jordan Valley offered a comment that was echoed by farmers and donor agencies alike:

Let's say, we'll improve the conditions inside the distribution network, but to have it sustainable, you need to have all the stakeholders, so all the farmers of the different branches, to follow the new rules. In a pressurized system, for example, if one farmer illegally steals water, it will affect his neighbor, who will not receive his water. So we implement, with the JVA, a system of controlling the rotation schedule, but the farmers have to follow the rules. So, through the water user cooperatives that GTZ helped create, the idea was, ok, we should have a counterpart to the JVA. The JVA implements the rules, but the farmers' cooperatives—the counterparts— will guarantee that the rules are respected by all, to keep the good level of [irrigation water distribution] service

(Interview 13).

Another way in which the presence of water user cooperatives has improved the operation of the irrigation network and promoted water conservation is that the flow of communication between officials and cooperatives about technical problems has dramatically improved. For example, problems with the water distribution infrastructure (i.e., water meters, valves, pipes) are discussed during the weekly meetings, and this has improved the reaction time for repairing and maintaining the network ... Also, excessive leaks from damaged pipes are an obvious source of water loss and the JVA has been much more receptive to responding to these issues when the complaints come from representatives of cooperatives, rather than individual farmers (Interviews 41; 42; 45; 75). The Secretary of water user cooperative PS 50, in the northern section of the Valley, made a comment echoed by other farmers I spoke with, "it is better with water user cooperatives, in the sense that if we have any problem we can tell the cooperative, and they take it to the JVA, and the JVA will come at once and fix the problem. Before the existence of the cooperative, nobody listened to our complaints" (Interview 45). In certain areas of the Jordan Valley, the number of repair and maintenance incidents has dropped markedly and the JVA staff attribute this to the increase in cooperation with the cooperatives: in PS 28 registered maintenance cases were close to 425 per year until 2002, and dropped to 115 in 2007 (GTZ 2008). Likewise, PS 50 saw a similar drop in maintenance cases from 175 to 60 cases per year in 2006 (GTZ, 2006b).

Many farmers in the Jordan Valley have been resorting to over-irrigating their fields through the storing of as much water as possible in the rooted zone of the crops because of unreliable irrigation water in terms of quantity and timing (Regner, [et al] 2006). This can lead to excessive water consumption, and also adversely affect plant development and yields. GTZ evaluation reports (GTZ, 2006c; Regner, 2005; Regner et al., 2006) suggest that because the water user cooperatives have become significantly more active in promoting more efficient water distribution, the reliability of the water supply has improved and farmers are less inclined to over-irrigate.

Note

1   According to the GTZ Project Leader, there is a trend of sharply reduced penalties issued by the JVA for violations such as illegally connecting pipes to the network,

or damaging the network, or tampering with the flow limiter in FTAs, all of which are methods aimed at obtaining extra irrigation water, in all areas of the Jordan Valley where water user cooperatives exist. But it is very difficult to prove this numerically. For example, based on the scant data that exists on penalties per year, the water user cooperative PS 28 has seen penalties decrease from 134 in 2002 to just 27 in 2007) (GTZ, 2008). Besides the reduced number of penalties, there are other indicators which point to reduced violations on the network by farmers: (i) operating pressure in the supply lines is generally high and stable, whereas before it took hours to stabilize pressure at the target level; (ii) calculated discharge for the entire pumping station is generally not exceeded. This is monitored centrally at the control center. If it is exceeding, the control center calls the head of the pumping station to reduce the discharge; and (iii) control visits of irrigation lines show little or no cases of tampering with meters or flow limiters. In Kafrein not a single illegal joint was detected since the rehabilitation of the area in 2002/2003. There are however illegal joints at a conveyance line to another area.

Commentary

The author highlights several benefits of partnerships, as well as reasons why they worked well in this case. First, the shift to a pressurized network made farmers more interdependent. Accordingly, as the author points out, the formation of water user collaboratives worked especially well because collaborating more—among themselves and with the JVA—was in the farmers' best interests. The challenge was how to shift from a pattern of antagonism to a cooperative partnership approach.

By utilizing partnerships to draw together interdependent interests, the water user collaboratives were able to improve stakeholder relationships while simultaneously making learning and communication easier. As the author notes, all of this led to clear improvements in water use and efficiency. She points to the importance of relationship building as a means of improving water management.

The excerpts also illustrate the important role international development agencies can play in establishing cooperatives and providing technical assistance. Their emphasis on facilitating knowledge transfer and collaboration among local actors was important, though even more focus on utilizing local knowledge might have been appropriate.

J. Luijendijk and W. L. Arriëns "Water Knowledge Networking: Partnering for Better Results," (2007)

Introduction

The below excerpt is taken from a Discussion Draft Paper presented by Jan Luijendijk and Wouter Lincklaen Arriëns at a 2007 UNESCO session on knowledge networking. Without focusing on a particular water body or water network, they discuss the role of knowledge networks in improving water management by building capacity and facilitating knowledge transfer across sectors and among actors. They argue that catering to priority water sector needs, focusing more on delivering results, and improving networking operations can enhance the impact of knowledge networks.

The excerpt focuses on how to define the success of knowledge networks, and on improving networking efforts to boost knowledge transfer among water managers. The authors highlight the preconditions for success, including the need for well-placed leaders and champions, the importance of connecting networks to decision-makers, and the need for interpersonal contact, storytelling and mentoring.

Excerpt (pp. 13–17)

Managing the Process—How Networks can Work Better

Key Question: How can Networks Organize for Success?

In this section we will focus on the "how to" questions of effective partnering and organizing for success. We will also refer to some recent developments that networks take advantage of, and recall some elements of earlier wisdom that continue to be relevant. A number of perspectives are offered for consideration.

Success Factors

Are there common factors that will guarantee a successful network? The answer is both yes and no. There is enough knowledge about networks to point to some factors that are necessary for success in most cases. Networking, however, remains a fast developing phenomenon, and new insights are emerging almost continuously.

"Trust is the basic lubricant for networking and sharing knowledge." (GTZ, 2006a) Successful networks tend to operate informally with few rules. For example there may be no rules for non-disclosure of important findings of a network member. Boom (2007) therefore contends that there are three important success factors: 1) trust; 2) a common goal; and 3) the need to know other members personally (not only through cyberspace). GTZ (2006a) asserts that good network management, transparency, and trust are preconditions to involve decision-makers in networks. Gloor (2006) points out that effective networks for innovation are marked by high degrees of connectivity, interactivity, and sharing.

Initiative, leadership, and vision by members are also important to network success. Networks need champions, leadership, multipliers, and standards (for example benchmarks for performance of member organizations). Networks can promote organic and incremental growth of knowledge and capacity among their members. However, in case of paradigm shifts being introduced, champions are needed, followed by leaders to internalize the changes and push through inertia against change.

Some networks, like professional associations and those focused on research, expect sustained operation over a longer term. However such sustainability may not be necessary or even possible for networks focused on innovation. Permanence is therefore an option to be selected, with success factors changing accordingly. For networks seeking sustained operation, financial contributions from members will be more important than for temporary networks that might reach their objective with time-bound support of one or more sponsors.

Wenger et al (2002) point out that the level of energy and visibility in communities of practice often increases and later decreases, in accordance with five stages of development, with corresponding developmental tensions. The first stage is marked by the discovery or imagination of potential, and is followed by coalescence with a choice of incubation or delivering immediate value. During the next stage of maturing, the community will need to choose between focusing and expanding. At its peak stage of stewarding, questions of ownership versus openness arise. And in the final stage of transformation of the community, the issue may be to let go or to live on.

Constraints to Success

When the success factors for effective networking are not achieved, problems are bound to occur, and they need addressing. However, further analysis of networking constraints is needed since not a great deal of information is available from water knowledge networks about factors that hold back their performance. In addition to the commonly raised issues of insufficient budgets and networking hardware, two factors seem particularly important.

The first is the ability of networks to reach decision-makers at both national and local levels with their products and services. This, for example, is a concern heard about networking among research organizations. The networking itself may be experienced as successful because it satisfies the professional interests of the participating researchers. However, achieving the outcome of the research depends on "clients" outside the network, and remains a challenge.

A second concern is how water organizations can be persuaded to spend more time and effort in networking. Ask a cross-section of staff in national and local governments and in development organizations including the multilateral development banks, how much time they spend as a consumer or member of networks (by regular web surfing and reading), and the result is likely to be much lower than expected. This is partly because many organizations have not yet recognized the benefits of networking for their own work, lack a corporate policy to promote networking, have yet to put in place staff incentives to spend time on networking, and have no organizational focal point to coordinate this.

Digital Divide

While ICT is driving development and networking forward with ever greater connectivity and speed, the prevailing digital divide in the world continues to impede access to such networking by many local practitioners and poor communities in the developing world, and also by the elderly who feel unable to participate. Unless water knowledge networks can find ways to bridge these divides, the risk is that the social capital of local traditional and indigenous water knowledge will be marginalized to extinction.

Also, it can be noted that modern ICT-supported networks are often premised on models, prescriptive solutions, and innovation through (perceived) paradigm shifts, while traditional low-tech communities in developing countries have a heritage of cultivation of consultative approaches and incremental improvements. Modern ICT-supported water networks could therefore be encouraged to reach out and accommodate knowledge and social capital from low-tech local communities in developing countries. To enhance knowledge and capacity development at local levels, a combination of high-tech and low-tech networks, or a cross-fertilization among such networks, might be needed, with proper interfaces that need developing.

People for Networking

Networks rely on people, however, reviews of networking experience and performance tends to overlook this obvious fact. In comparison, much more attention has been devoted to analysis of organization and ICT application for networking. The empowerment of individuals is, however, a key ingredient to capacity building of organizations and networks, together with the cultivation of vision, mission, and values. Coaching and mentoring have a high "return on investment" to enhancing networking for knowledge and capacity building.

Lank (2006) distinguishes some typical functions and roles of individuals in collaborative ventures such as networking. She identifies organizational sponsors, gatekeepers (relationship and partnership managers), partnership coordinators, advisory partnership facilitators, and project (network) managers. She also recommends organizations to consider appointing a chief relationship officer, which would help in partnering and networking. Commenting on the specific leadership qualities required for collaborative working, she quotes Doz and Hamel (1998) that "Executives do not wake up one morning with an unexplained urge to collaborate. It is not in their nature."

Networks within Networks

Networks are not homogeneous. They often comprise groups within the network. Gloor (2006) identifies three types of networks within each other, which are similar to those described by GTZ (2006a). The larger network is seen as a collaborative *insight* network which helps people with a shared interest. GTZ refers to these people as the lurkers in the network. Within that larger network, some members (people and/or organizations) take a more active role in sharing knowledge and act as a collaborative learning network that focuses on stewarding of best practices. Finally, at the core of the other two, an even smaller group works as a collaborative innovation network with total dedication on generating fundamentally new insights. These groups within the network are shown in Figure [6.7].

Gloor describes a number of conditions for collaborative innovation networks (COINS) to be successful, including 1) being a learning network, 2) having an ethical code, 3) being based on trust and self-organization, 4) making knowledge accessible to anyone, and 5) operating in internal honesty and transparency. He claims that the combination of the three networks creates a strong ripple effect.

Cutting across Boundaries

Today's ICT-supported knowledge networks can cut across boundaries and levels in hierarchies. Network members, however, are still tied to professions, position levels, and hierarchies within their organizations, and to the position of their organization in the constellations of local, regional, national, and international levels. This raises questions of who should be linked to whom in effective knowledge networks, and this question is particularly relevant when we focus on enhancing knowledge and capacity at local government level. Assuming that more networks will emerge in the coming years to link practitioners at local levels with each other, more knowledge is needed on how this can best be done. With the wide use of the internet and ICT, open access networks are expected to increase, and individuals can take part in several networks at the same time.

Language is another issue to be considered. Most local government practitioners will want to communicate in their country's language, or even in local languages. Products and services at local level will need to be disseminated in the appropriate language. Most of the ICT-supported water knowledge networks seem, however, to operate in one or few languages, and the questions arises who will network with whom in what language, and how language will be factored into a hierarchy or constellation of intersecting networks?

Another dimension of language in networking concerns the use of technical and scientific versus language for consumption of decision-makers and the general public. Networks need to consider these for effective operation and to reach their intended clients.

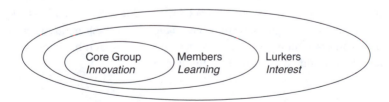

FIGURE 6.7 Groups within networks

Source: (after GTZ, 2006a and Gloor, 2006)

Assessments to Do

Many knowledge networks engage in regular surveys or assessments, and there seems to be scope for improvement by sharing knowledge and experience on how this can best be done. In a departure from past practice, more surveys and assessments will need to be focused on the needs of local practitioners and decision makers.

Gloor (2006) identifies three tools: 1) the knowledge map that focuses on "what is" and provides global or local knowledge about a topic, 2) the talent map that focuses on "who is" and shows who has the expertise on the topic, and 3) the trend map that focuses on "what will be" and presents global or local trends about the topic.

Capacity building activities, including those in the water sector, are all too often designed based on training needs assessments rather than more comprehensive diagnostic assessments taking into account the individual and institutional levels and their enabling environment. Networking that aims to enhance capacity will therefore need to engage more in diagnostic assessments with the full ownership of the organizations concerned.

Stories that Work

Why is it worth remembering that one of the most effective tools of knowledge management is to share stories and anecdotes? It is because people remember and easily identify with them. The majority of management, leadership, and self-enrichment literature published in the past decade has made extensive use of stories and anecdotes to get their messages across to their audiences. Water knowledge networks might consider adopting a similar approach in delivering some of their key messages through stories and anecdotes during networking activities and on their websites, newsletters, and publications. This may also help in reaching out to local level practitioners and decision-makers, and to externalize tacit knowledge.

Networking Push and Pull

Successful networking requires the delivery of products and services (push) that add value and thereby attract (pull) existing and new members. More research seems needed on the type of products and services that consistently produce added value and create "pull." For example, anecdotal feedback to the authors suggests that network members might value to receive a regular synthesis of good practices in their topic of interest, for example an annual overview of what is happening in countries around the world on the topic of water legislation.

Networking Everywhere

Recognizing the need to make best use of time and financial resources, the organizers of large water events are increasingly supportive of organizing side events, and these provide a cost-effective opportunity for networks to conduct face-to-face meetings of their members. With some advance organization on the part of the host organization, even more use could be made of water events by creating opportunities for country delegations to meet and forge partnerships, ranging from a straightforward exchange of information to setting up exchange visits, staff exchange programs, and developing joint programs of collaboration or twinning. Concluding such partnerships could also be recognized as one of the many objectives of water events.

Incentives for Change

Of all the products and services that networks can offer their clients, which ones are most likely to trigger change and improvements in performance? The Asian Development Bank's recent experience in helping regional water networks to introduce performance benchmarking and peer reviews has been welcomed by participating member organizations, some of which have given feedback that it has already changed the mindset of their staff. The ADB is supporting separate networks for water utilities, river basin organizations, and national water sector apex bodies, and all are now benefiting from performance benchmarking, and the latter two from peer reviews (ADB, 2007).

Commentary

The authors' early points about needing champions to initiate the creation of networks, and needing to connect networks to decision-makers, are fundamental—especially at the organizational and governmental levels. We strongly believe that champions are necessary to build momentum behind a new idea or capability within an organization, and to make sure that key internal stakeholders and decision-makers are on board (Movius and Susskind 2009). At the same time, the notion of developing water networks for societal learning puts a burden on individuals who have the most experience with water management, are interested in sharing experiences and knowledge, want to learn more, and want to take an active role in creating and sustaining networks and partnerships.

In our view, the authors are correct that learning networks require in-person contact, including mentoring. Mentoring is important because transferring knowledge and building capacity require interpersonal skills that are best taught through face-to-face observation and coaching. Networks can connect mentors and mentees, but this relationship should also be allowed to develop between colleagues and contacts in a variety of other ways.

Finally, we endorse Gloor's notion that collaborative learning networks should try to make knowledge accessible to everyone. The challenges identified, including technology gaps and language barriers, certainly pose problems. It is up to those who generate and share water knowledge to allow that knowledge to be "re-shared" in ways that encourage consumers to overcome those barriers. In other words, water networks for societal learning should adopt an "open source" ethos in which those with knowledge promote access to their ideas and products.

References

Asian Development Bank. 2007. "Information on networks and partnerships." From: http://www.adb.org/water/operations/partnerships.

Biggs, H. and Rogers, K.H. 2003. An adaptive system to link science, monitoring and management in practice, in J.T. du Toit, K.H. Rogers, and H.C. Biggs (eds.) *The Kruger Experience: Ecology and management of savanna heterogeneity.* Washington, DC: Island Press.

Biggs, H.C., Breen, C.M., and Palmer, C.G. 2008. Engaging a window of opportunity: Synchronicity between a regional river conservation initiative and broader water law reform in South Africa, *International Journal of Water Resources Development, 24*(3): 329–343.

Biggs, H.C., Westley, F.R., and Carpenter, S.R. 2010. Navigating the back loop: fostering social innovation and transformation in ecosystem management. *Ecology and Society, 15*(2): 9.

Boom, D. 2007.Unpublished draft paper on knowledge economies in Asia for a seminar at the ADB Institute. Asian Development Bank, Manila, Philippines.

Braack, L. 1997a. *A revision of parts of the management plan for the Kruger National Park, Vol. VII: An objectives hierarchy for the Kruger National Park.* Skukuza: South African National Parks.

Braack, L. 1997b. *A revision of parts of the management plan for the Kruger National Park. Vol. VIII: Policy proposals regarding issues relating to biodiversity maintenance, maintenance of wilderness qualities, and provision of human benefits.* Skukuza: South African National Parks.

Breen, C., Dent, M., Jaganyi, J., Madikizela, B., Maganbeharie, J., Ndlovu, J., et al. 2000. "*The Kruger National Park Rivers Research Programme,*" Final Report, Water Research Commission, Pretoria.

Breen, C., Quinn, N., and Deacon, A. 1994. "A description of the Kruger Park Rivers Research Programme," Second Phase: Programme description: pp. 43.

Camacho, A.E., Susskind, L., and Schenk, T. 2010. Collaborative planning and adaptive management in Glen Canyon: A cautionary tale, *Columbia Journal of Environmental Law, 35*:1

Courcier, R., Venot, J.P., and Molle, F. 2005. *Historical transformations of the lower Jordan River basin (in Jordan): changes in water use and projections (1950–2025) (comprehensive assessment research report 9).* Colombo, Sri Lanka: IWMI Comprehensive Assessment.

Davies, B.R., O'Keeffe, J.H., and Snaddon, C.D. 1995. River and stream ecosystems in Southern Africa. Predictably unpredictable, in C.E. Cushing, K.W. Cummins, and G.W. Minshall *River and Stream Ecosystems.* New York: Elsevier Press.

Doz, Y. and Hamel, G. 1998. *Alliance Advantage: The art of creating value through partnering.* Boston: Harvard Business School Press.

Dube, D. and Swatuk, L. 2002. Stakeholder participation in the new water management approach: A case study of the Save Catchment, Zimbabwe, *Physics and Chemistry of the Earth, 27*: 867–874.

Du Toit, J.T., Rogers, K.H., and Biggs, H.C. (eds.) 2003. *The Kruger Experience. Ecology and Management of Savanna Heterogeneity.* Washington DC: Island Press

Fisher, R. and Ury, W. 1991. *Getting to Yes: Negotiating Agreement Without Giving In.* New York, NY: Penguin Books.

Folke, C., Carpenter, S., Elmqvist, T., Gunderson, L., Holling, C.S., Walker, B., et al. 2002. Resilience and sustainable development building adaptive capacity in a world of transformations, *Ambio*, *31*(5): 437–440.

Freitag, S., Biggs, H., and Breen, C.M. In review. Fifteen years of the spread and maturation of adaptive management in South African National Parks: Organisational learning in systems perspective, in W. Freimund, S. McCool, and C.M. Breen (eds.) *Engaging Complexity in Protected Area Management: Challenging Occam's Razor*. Pietermaritzburg, South Africa: University of KwaZulu-Natal Press.

Fuller, B. 2006. "Trading zones: cooperating for water resource and ecosystem management when stakeholders have apparently irreconcilable differences." *Dissertation, Massachusetts Institute of Technology, Department of Urban Studies and Planning.*

Gloor, P. 2006. *Swarm Creativity: Competitive Advantage through Collaborative Innovation Networks*. Oxford: Oxford University Press.

GTZ. 2000a. *History and Lessons Learned from the Formation of the Mazowe Catchment Council*. Zimbabwe: Harare.

GTZ. 2000b. *Project Appraisal Report: Irrigation Water Management in Jordan*. Amman, Jordan: GTZ/JVA.

GTZ. 2001. *Water Resources Management for Irrigated Agriculture – Annual progress report June 2001 and May 2002*. Amman, Jordan: GTZ/JVA.

GTZ. 2002. *Water Resources Management for Irrigated Agriculture: Annual progress report June 2001 and May 2002*. Amman, Jordan: GTZ/JVA.

GTZ, 2006a. *Work the Net – A Management Guide for Formal Networks*. New Delhi: GTZ/JVA.

GTZ. 2006b. *Water Resources Management for Irrigated Agriculture Presentation – Optimizing Water Management*. Amman, Jordan: GTZ/JVA.

GTZ. 2006c. *Economic Impacts of the Introduction of Participative Irrigation Management in the Jordan Rift Valley*. Amman, Jordan: GTZ/JVA.

GTZ. 2008. *GTZ in Jordan*. Retrieved October 1, 2008, from http://www.gtz.de/en/weltweit/maghreb-naher-osten/675.htm

Holling, C.S. 2001. Understanding the complexity of economic, ecological, and social systems, *Ecosystems*, *4*(5): 390–405.

Holling, C.S. and Gunderson, L.H. 2002, Resilience and adaptive cycles, in L.H. Gunderson and C.S. Holling (eds.) *Panarchy: Understanding Transformations in Human and Natural Systems* (pp. 25–62). Washington DC: Island Press.

Jezewski, J. and Roberts C.P.R. 1986, *Estuarine and Lake Freshwater Requirements*, Technical Report TR129, Department of Water Affairs.

JVA. 2007. *Data from JVA control center in Deir Allah*. Deir Allah, Jordan: JVA.

King, J.M., Tharme, R.E., and De Villers, M.S. 2000. "Environmental flow assessments for rivers: manual for the Building Block Methodology," Water Resources Commission Report TT 131/100, Pretoria, South Africa.

Kingsford, R.T., Biggs, H.C., and Pollard, S.R. 2011. Strategic adaptive management in freshwater protected areas and their rivers, *Biological Conservation*, *144*(4), 1194–1203.

Kujinga, K. and Manzungu, E. 2004. Enduring contestations: Stakeholder Strategic action in water resource management in the save catchment area, Eastern Zimbabwe, *Eastern Africa Social Science Research Review*, *20*(1): 67–91.

Lank, E. 2006. *Collaborative Advantage: How Organizations Win by Working Together*. New York: Palgrave Macmillan.

Latham, C.J.K. 2001. Manyame Catchment Council: A review of the reform of the water sector in Zimbabwe, 2nd WARFSA – WaterNet Symposium, October 2001, Cape Town.

Lave, J. and Wenger, E. 1991. *Situated Learning. Legitimate Peripheral Participation.* Cambridge: University of Cambridge Press.

LEMA OPS. 2007. *LEMA Operations database 2000–2007.* Amman, Jordan: LEMA OPS.

Luijendijk, J. and Arriëns, W.L. 2007. *Water Knowledge Networking: Partnering for Better Results.* The Netherlands: UNESCO-IHE.

McLoughlin, C.A., Deacon, D., Sithole, H., and Gyedu-Ababio, T. 2011. History, rationale, and lessons learned: Thresholds of potential concern in Kruger National Park river adaptive management, *Koedoe, 53*(2), Art. #996.

Movius, H. and Susskind, L. 2009. *Built to Win: Creating a World-Class Negotiating Organization.* Boston, MA: Harvard Business School Publishing.

Noss, R.F. 1990. Indicators of monitoring biodiversity: a hierarchical approach, *Conservation Biology, 4*: 355–364.

Odeh, N. 2009. "Towards improved partnerships in the water sector in the Middle East: A case study of partnerships in Jordan's water sector." *Dissertation, Department of Urban Studies and Planning, Massachusetts Institute of Technology.*

O'Keeffe, J. and Coetzee, Y. 1996. Status report of the Kruger National Park Rivers Research Programme: A synthesis of results and assessment of progress to January 1996, *Pretoria, Water Research Commission*: 63.

Plumb, D., Fierman, E., and Schenk, T. 2011. "Role Play Simulations: A Useful Roadmap for Decision Makers." *Consensus Building Institute.* (available from http://cbuilding. org/publication/article/2011/roleplay-simulations-useful-roadmap-decision-makers accessed March 16, 2012).

Pollard, S. and Du Toit, D. 2005. Achieving integrated water resource management: The mismatch in boundaries between water resources management and water supply. *Association for Water and Rural Development.* (available from http://www.nri.org/projects/waterlaw/AWLworkshop/POLLARD-S.pdf accessed March 16, 2012).

Pollard, S. and Du Toit, D., 2006. "Recognizing heterogeneity and variability as key characteristics of savannah systems: The use of Strategic Adaptive Management as an approach to river management within the Kruger National Park, South Africa," Report for UNEP/ GEF Project No. GF/2713–03–4679, Ecosystems, Protected Areas and People Project.

Pollard, S. and Du Toit, D. 2007. "Guidelines for Strategic Adaptive Management: Experiences from managing the rivers of the Kruger National Park, South Africa," IUCN/UNEP/GEF Project No. GF/2713–03–4679, Ecosystems, Protected Areas and People Project, Planning and managing protected areas for global change.

Pollard, S. and Du Toit, D. 2008. "The Letaba Catchment: Contextual profile on factors that constrain or enable compliance with environmental flows," Shared River Programme, DRAFT Report, Project K5/1711.

Pollard, S.R., Biggs, H., and Du Toit, D. 2008, "Towards a Socio-Ecological Systems View of the Sand River Catchment, South Africa: An exploratory Resilience Analysis," Report to the Water Research Commission, Project K8/591, Pretoria.

Pollard, S., du Toit, D., and Biggs, H. 2011. River management under transformation: The emergence of strategic adaptive management of river systems in the Kruger National Park, *Koedoe, 53*(2). (available from http://www.koedoe.co.za/index.php/ koedoe/article/view/1011/1260 accessed August 13, 2011).

Pollard, S.R. and Cousins, T. 2008. "Towards integrating community-based governance of water resources with the statutory frameworks for Integrated Water Resources Management: A review of community–based governance of freshwater resources in four southern African countries to inform governance arrangements of communal wetlands," Water Research Commission Report TT.328/08, Pretoria, Water Research Commission.

Pollard, S.R., Weeks, D.C., et al. 1994. Effects of the 1992 drought on the aquatic biota of the Sabie and Sand rivers, in *A pre-impoundment study of the Sabie-Sand River System, Eastern Transvaal, with special reference to predicted impacts on the Kruger national Park, vol. 2* (p. 122). Pretoria: Water Research Commission Report.

Pollard, S., Riddell, E., et al. 2010. "Compliance with the Reserve: How do the Lowveld Rivers measure up?", Report prepared for the Water Research Commission: Reserve assessment of lowveld rivers (Del. 1), n.p.

Putnam, R.D. 2002. *Democracies in Flux: The Evolution of Social Capital in Contemporary Society*. New York: Oxford University Press.

Radosevich, George. 2011. "Mekong River Basin, agreement & commission." *IUCN Water Program Negotiate Toolkit: Case Studies.* (available from http://www.iucn.org/about/work/programmes/water/resources/toolkits/negotiate/, accessed August 13, 2011).

Regner, H. J. 2005. *Improvement of water distribution in the Jordan Valley through participation of water user communities and their contribution to on-farm irrigation efficiency.* Paper presented at the ESCWA seminar on enhancing agricultural productivity through on-farm water use efficiency (23–25 November 2005), Beirut, Lebanon.

Regner, H.J., Salman, A.Z., Wolff, H.-P., and Al-Karablieh, E. 2006. *Approaches and impacts of participatory irrigation management in complex, centralized irrigation systems. Experiences and results from the Jordan Valley.* Paper presented at the Conference on International Agricultural Research for Development, University of Bonn, Germany.

Roberts, C.P.R. 1983. Environmental constraints of water resources developments, *Proceedings of the South African Institution of Civil Engineers*, n.p.

Rogers, K.H. 2005. The real river management challenge: Integrating scientists, stakeholders and service agencies, *River Research and Applications*, 22(2): 269–280.

Rogers, K.M. and Bestbier, R. 1997. *Development of a Protocol for the Definition of the Desired State of Riverine Systems in South Africa.* Pretoria: Department of Environmental Affairs and Tourism.

Sithole, B. 2000. "Telling it like it is! Devolution in the water reform process in Zimbabwe." Paper presented at the biannual meeting of the International Association for the Study of "Common Property," Wisconsin.

Social Learning Group. 2011. *Learning to Manage Global Environmental Risks, Volumes 1 and 2.* Cambridge, MA: The MIT Press.

Susskind, L. and Cruikshank, J. 1987. *Breaking the Impasse. Consensual Approaches to Resolving Public Disputes.* New York: Basic Books, Inc.

Susskind, L.E. and Cruikshank, J.L. 2006. *Breaking Robert's Rules: The New Way to Run Your Meeting, Build Consensus, and Get Results.* New York: Oxford University Press.

Susskind, L., McKearnan, S., and Thomas-Larmer, J. (eds.) 1999. *Consensus Building Handbook: a comprehensive guide to reaching agreement.* Thousand Oaks, CA: Sage.

Tapela, B.N. 2002. "Institutional challenges in integrated water resources management in Zimbabwe: a case study of Pungwe sub-catchment area." Thesis (M. Phil. (Centre for Southern African Studies-School of Government))—University of the Western Cape.

Tapela, B.N. 2006. Stakeholder participation in the transboundary management of the Pungwe river basin, in A. Earle and D. Malzbender (eds.) *Stakeholder Participation in Transboundary Water Management* (pp. 10–34). South Africa: African Centre for Water Research.

U.S. Army Corps of Engineers. "Engineer Regulation 1105–2–100: Appendix D – Amendment #1, Economic and Social Considerations." *USACE Planning Guidance: Hydropower*, June 30, 2004 (available from: http://planning.usace.army.mil/toolbox/guidance.cfm?Option=BL&BL=Hydropower&Type=None&Sort=Default, accessed October 31, 2011).

Van Aarde, R., Whyte, I., and Pimm, S. 1999. Culling and the dynamics of the Kruger National Park African elephant population, *Animal Conservation*, 2(4): 287–294.

Van Aken, M., Courcier, R., Venot Jean-Philippe., and Molle, F. 2007. *Historical Trajectory of a River Basin in the Middle East: The Lower Jordan River Basin (in Jordan)*. Amman, Jordan: International Water Management Institute.

Van den Belt, M. 2004. *Mediated Modeling: A System Dynamics Approach to Environmental Consensus*. Washington, DC: Island Press.

Van Wilgen, B.W. and Biggs, H. 2010. A critical assessment of adaptive ecosystem management in a large savanna protected area in South Africa, *Biological Conservation*, n.p.

Venot, J.P. 2004. *Reclamation's History of the Jordan River Basin in Jordan, A Focus on Agriculture: Past Trends, Actual Farming Systems and Future Prospective*. Amman, Jordan: Mission Regionale Eau et Agriculture.

Venter, F.J. and Deacon, A.R. 1995. Managing rivers for conservation and ecotourism in the Kruger National Park, *Water Science and Technology*, 32: 227–233.

Venter, F.J., Gerber, F., Deacon, A.R., Viljoen, P.C., and Zambatis, N. 1995. 'Ekologiese Impakverslag: An ondersoek na die voorgestelde Maroelastuwal in die Krokodilrivier' [Ecological Impact Report: An investigation of the proposed Maroela weir in the Crocodile river], Report Nr 1/95 Scientific services, Skukuza.

Venter F.J., Naiman, R.J., Biggs, H.C., and Pienaar, D.J. 2008. The evolution of conservation management philosophy: Science, environmental change and social adjustments in Kruger National Park, *Ecosystems*, 11: 173–192.

Weeks, D.C., Pollard, S.R., et al. 1992. "Downstream effects on the aquatic biota of the Sand River following the collapse of Zoeknog dam," Report submitted to Water Research Commission.

Wenger, E., McDermott, R., and Snyder W. 2002. *Cultivating Communities of Practice: A Guide to Managing Knowledge*. Boston, MA: Harvard Business School Publishing.

Werick, B. 2007. Changing the rules for regulating Lake Ontario levels, in *Computer Aided Dispute Resolution, Proceedings from the CADRe Workshop*, Institute for Water Resources, September 2007: 119–128.

World Bank Environment Department. 2007. "Strategic Environmental Assessment and Integrated Water Resources Management and Development." *World Bank* http://www.google.com/url?sa=t&rct=j&q=&esrc=s&source=web&cd=3&ved=0CDYQ FjAC&url=http%3A%2F%2Fsiteresources.worldbank.org%2FINTRANETENVIR ONMENT%2FResources%2FESW_SEA_for_IWRM.doc&ei=h5-uTufyJ8fX0QH SxuTNDw&usg=AFQjCNF1G68HlzYI2PGgU8fnRkhDlfHhOw&sig2=tDCaZ-O6mWHAd6LSC35qRw(accessed October 31, 2011).

Wright, G. and Cairns, G. 2011. *Scenario Thinking: Practical Approaches to the Future*. New York: Palgrave/McMillan.

7

THE INDOPOTAMIA ROLE-PLAY SIMULATION

(with Catherine M. Ashcraft)

Introduction

The Indopotamia role-play simulations are designed to teach about negotiating transboundary or boundary-crossing water conflicts using the Water Diplomacy Framework (WDF) presented in this volume. There are four separate segments of the game. Each explores an important step in the negotiations required to resolve difficult disputes within a water network. The first gives game players an opportunity to apply what they have learned (by reading the relevant chapters of this book) about exploring interests and building coalitions. The second, that can be played at a later date or immediately after the first segment finishes, asks the players to share information and generate new information by applying joint fact-finding techniques. The third segment presses the group to engage in value creation through the generation of development options or packages and asks the group to prioritize these possibilities. The final segment gives the stakeholder representatives a chance to see if they can negotiate a formal agreement to manage the land and water resources shared by the three countries in the Indopotamia River Basin that includes provisions for addressing future problems or changes that may impact their agreement, such as new data, future disputes, or implementation failures.

Each segment takes about two hours to play and requires nine players. It is probably best if the participants stay in the same role for all four segments because the role-specific information they receive in advance of each segment is cumulative. Eight players are assigned stakeholder roles ranging from the representative of the Regional Development Bank (reminiscent of the World Bank), to diplomatic representatives from the countries Alpha, Beta, and Gamma (introduced in the opening fable), to representatives of various non-government actors in the water network. The ninth participant is expected to play the part of a professional mediator. Assigning this role to someone who has facilitation experience is probably a good idea, although not absolutely necessary.

It is important to note that the four segments of the Indopotamia role play simulation that appears in this chapter explore only some of the complexities presented in the water management fable presented in the opening chapter. In the real world, there are an infinite number of variations of the Alpha, Beta and Gamma fable that play out; the Indopotamia game in Chapter 7 offers one such variation. While the same basic story continues to unfold across the planet, the details in each situation change. The version of the Indopotamia game in this chapter emphasizes the dynamics and puzzles of transboundary water negotiation that we have highlighted in the book. However, they compress, leave out or alter societal, natural and political details. Indeed, the details of the fable in the opening chapter and the material in the Indopotamia game instructions diverge. Students need to pay close attention to their Confidential Instructions as they begin each segment of the Indopotamia game.

At the beginning of each segment, players are handed General Instructions summarizing technical and geo-political information about the water network that they will need to play their parts. Players are also given Confidential Instructions that circumscribe their interests as well as what outcomes they prefer and what they may not accept with regard to the water choices they face. In order to preserve the usefulness of these teaching exercises, Confidential Instructions for each player are not included in this volume. Instead, they need to be downloaded (at very modest cost) from the Clearinghouse of the Program on Negotiation at Harvard Law School (www.pon.org). Anyone wishing to use these materials must register. In this way, we hope to ensure that future game players will not have easy access to the Confidential Instructions of all the other role players when they participate in these exercises.

Each segment of the role-play simulation generates carefully structured opportunities to discuss the natural, social, and political dimensions of the water-resource management dispute confronting the parties. High degrees of uncertainty and complexity prevail. Agreements are possible in the third and fourth segments, but they are not easy to reach. For experienced water professionals, the simulations will feel realistic. While the group deliberations will be a lot shorter than what would probably be required in "real life," the tensions among the parties will not seem contrived. The simulation was developed by drawing on the details of highly relevant boundary-crossing cases.

The teaching value of these games depends on the quality of the debriefings that follow the play of each segment. A Game Manager or Instructor will need to be completely familiar with the Water Diplomacy Framework and the full contents of this book to lead an effective discussion of what "worked" and why.

Those who want to discuss their reactions to any or all segments of the simulation are encouraged to join one of the public forums on the Water Diplomacy web site (www.waterdiplomacy.org).

Teaching Notes

This is a nine-party, mediated, multi-issue negotiation game involving a dispute over the allocation of land and water resources shared by three countries in an international river basin. The complete materials for the Indopotamia game include:

PLATE 1 California Aqueduct
Credit: © trekandshoot | Dreamstime.com

PLATE 2 Chattahoochee River
Credit: niksnut, Creative Commons

PLATE 3 The Dead Sea, Jordan Bank
Credit: Jan Smith, Creative Commons

PLATE 4 Nwashitshaka River in Kruger National Park
Credit: Jeppestown, Creative Commons

PLATE 5 Piave River Basin
Credit: Paolo Tonon, Creative Commons

PLATE 6 River Pungwe
Credit: Bill Higham, Creative Commons

PLATE 7 Thousand Islands International Bridge
Credit: © Wangkun Jia | Dreamstime.com

PLATE 8 Caption: Three Gorges Dam
Credit: © Wei Wei | Dreamstime.com

PLATE 9 Sindh Province, Pakistan
Credit: UK Department of International Development, Creative Commons

PLATE 10 Siem Reap, Cambodia
Credit: Steve Nelson, Creative Commons

PLATE 11 Burkina Faso

Credit: Jeff Attaway, Creative Commons

PLATE 12 Ciampea, Indonesia
Credit: Danumurthi Mahendra, Creative Commons

- Teaching Notes
- General Instructions for each of the four segments of the game
  - ○ The map in Appendix A is best viewed in color, but can also be viewed in black and white.
- Confidential Instructions for each of the nine roles for each of the four segments of the game
  - ○ The Mediator's Instructions include forms for reporting the results of Segments 2 through 4.
- Summary form for contrasting the results of Segments 2 through 4 if multiple groups play the game at the same time.

The game is designed to be played in four separate segments. Each explores an important element of the mutual-gains approach to negotiation. The game provides opportunities to discuss the natural, societal, and political dimensions of science-intensive policy disputes in which high levels of uncertainty are involved. The game also introduces water professionals and aspiring water professionals to the WDF—see Table 7.1 for its most important elements.

TABLE 7.1 Indopotamia: Comparison of Conventional Conflict Resolution Theory with the Water Diplomacy Framework (WDF)

	The Water Diplomacy Framework (WDF)	Conventional conflict resolution theory (applied to water and other common-pool resources)
Domains and scales	Water crosses multiple domains (natural, societal, political) and boundaries at different scales (space, time, jurisdictional, institutional).	Watershed or river-basin falls within a bounded domain.
Water availability	Virtual or embedded water, blue and green water, technology sharing and negotiated problem-solving that permit re-use can "create flexibility" in water for competing demands.	Water is a scarce resource, and competing demands over fixed availability will lead to conflict.
Water systems	Water networks are made up of societal and natural elements that cross boundaries and change constantly in unpredictable ways within a political context.	Water systems are bounded by their natural components; cause–effect relationships are known and can be readily modeled.
Water management	All stakeholders need to be involved at every decision-making step including problem framing; heavy investments in experimentation and monitoring are key to adaptive management; the process of collaborative problem-solving needs to be professionally facilitated.	Decisions are usually expert-driven; scientific analysis precedes participation by stakeholders; long-range plans guide short-term decisions; the goal is usually optimization, given competing political demands.

TABLE 7.1 Continued

	The Water Diplomacy Framework (WDF)	Conventional conflict resolution theory (applied to water and other common-pool resources)
Key analytic tools	Stakeholder assessment, joint fact-finding, scenario planning and mediated problem-solving are the key tools.	Systems engineering, optimization, game theory, and negotiation support-systems are most important.
Negotiation theory	The Mutual Gains Approach (MGA) to value creation; multiparty negotiation keyed to coalitional behavior; mediation as informal problem-solving are vital to effective negotiation.	Hard bargaining informed by prisoner's dilemma-style game theory; principal–agent theory; decision-analysis (Pareto optimality); theory of two-level games all apply.

There are three key propositions critical to using the Water Diplomacy Framework (WDF).

Proposition #1: Boundaries and Representation in Water Networks should be Considered Open-ended and Continuously Changing

A common presumption in water management is that the boundaries of water "systems" are set in natural, societal, and political terms. The WDF challenges this idea, presuming that coupled natural and societal networks continue to evolve and are open, not closed. Only a careful analysis of each situation will clarify which natural and societal variables are most important in temporarily defining the nodes and dynamics of particular water networks. While states are still sovereign, the emerging global emphasis on governance (as opposed to government) means that nongovernmental actors and interests animate water management as much as states do and, thus, must be represented in any boundary-crossing water negotiations. The complexity of these coupled natural and societal systems also means that the tools required to manage water resources go beyond what water engineers and public-policy analysts have typically used.

Proposition #2: Modeling and Forecasting for Water Management should account for Variability and Uncertainty

Another common notion in water management is that the supply of water in any particular location or region can be modeled and forecasted reasonably far into the future, and that water allocations can be optimized (by experts) among multiple (competing) uses. The WDF assumes that the supply and quality of water are more unpredictable than that (and becoming even more so, for example, because of a changing climate). They can rarely be forecasted with confidence and certainly not by experts working alone. There are too many non-objective judgments that can influence the outcome. In addition, the WDF frames water as a flexible resource in terms of virtual or embedded water, blue and green water, new technologies for improving water quality, and how changing attitudes toward the role of water in sustainable development make it difficult to model and forecast what is happening to water supplies and water quality. Again, new tools are required to model emerging water concerns. Joint fact-finding and collaborative decision making—rather than expert analysis—are required to develop the necessary tools and ensure transparency and legitimacy.

Proposition #3: The Politics of Transboundary Water Management should be Adaptive and Negotiated using a Non-zero-sum Approach

Another dominant belief in conventional water management is that the allocation of common-pool resources (another name for public goods like water resources and the ecosystems that support them) is always a win–lose situation. More powerful political parties "win" and gain control of resources; less powerful parties "lose" and only have access to water if the more powerful nations or factions permit it. Over the past few decades, the emergence of non-zero-sum, or mutual-gains negotiation theory has challenged this win–lose logic by offering a value-creating alternative that allows groups with conflicting goals to achieve them simultaneously. The value-creating or mutual-gains approach to negotiation rests on the theory that joint fact-finding, the packaging of interlocking trades, contingent commitments, and an adaptive approach to handling uncertainty can maximize joint gains. "All-gain" negotiations usually require the assistance of a neutral facilitator or mediator.

Scenario

Eight stakeholder group representatives, including senior officials from three countries (Alpha, Beta, and Gamma), have gathered to discuss, with the help of a mediator, possible development strategies for the Indopotamia River. The three countries face significant water-management challenges, and there is no formal agreement governing how they are supposed to share or use the resources of the river basin. A multinational Regional Development Bank has played a key role in convening these negotiations and is prepared to offer substantial financial support if the countries and some non-national interest groups can come to an agreement on how to proceed.

The participants will have several separate opportunities to reach agreement on a number of key items. After a discussion period aimed at understanding the interests of all parties, they must agree on a way of generating reliable technical information about the basin (a joint fact-finding process). Second, they must generate a list of sustainable development priorities (or categories of sustainable development projects). Third, they must outline a regime or a treaty of some sort to guide their future interactions and resource-management efforts.

The nine participants include:

One representative from each of the three countries that share the Indopotamia River basin:

- Alpha [NOTE: Alpha's Ministry of Water Resources]
- Beta [NOTE: Beta's Ministry for Sustainability]
- Gamma [NOTE: Gamma's Ministry of Water and Energy]

One representative from an important local government in Alpha:

- Mu State's Economic Administration

Four representatives from intergovernmental and nongovernmental groups:

- The Global Water Management Organization (GWMO)
- The International Conservation Institute (ICI)
- Water Infrastructure Engineering and Design (WIED)
- The Regional Development Bank (RDB)

One neutral mediator who is assisting the negotiators.

Although the RDB would prefer unanimous support for each agreement, it has pledged to support any list of projects and any sustainable resource management efforts as long as: (1) at least seven participants reach agreement on that segment of the negotiations (and the same seven need not be part of the agreement for every segment); and (2) every agreement reached must be supported by all three Indopotamia countries and the RDB. If the parties do not reach an overall agreement—at least on the final segment, a treaty governing their future interactions—it is very likely that one or more countries will develop water projects on its own. This will surely trigger a political crisis in the region.

Logistics

The game should be played with nine players: one per negotiating role and one mediator. There can be many groups of nine (all in the same class or training event) playing the same game at the same time in separate rooms. If there are extra players, any negotiating role can be doubled up (although a pair playing the same role will need a little extra time to make sure they are ready to proceed together).

Segment 1 of the Indopotamia simulation focuses on building coalitions and learning how to probe the interests of others. Segment 2 explores the dynamics of information sharing, particularly scientific information. Segment 3 looks at the problems surrounding the generation of policy or program options and deciding in a multiparty context how to choose among them. Segment 4 addresses how groups can negotiate their future relationship, particularly the constraints that they can and will impose on each other. Each segment of the game is meant to be played separately. The results of one segment do not dictate what will happen in subsequent segments. The participants can proceed with the next segment of the negotiations even if their group was unable to reach agreement in any of the previous segments.

Segment 1: Exploring Interests and Building Coalitions

- 45 minutes: Read General Instructions (best done in advance of the full group meeting).
- 45 minutes: Read Confidential Instructions and meet with others playing the same role in other groups if more than one group is playing at the same time.
- 60 minutes: Negotiation and/or caucusing.
- 30 minutes: Debriefing.

Preparation

The key decision facing the participants in Segment 1 is whether or not they want to enter into a coalition, or strategic alliance, with one or more of the other parties. Therefore, the meetings called for in Segment 1 provide an opportunity for the participants to get to know each another, share background information, and learn about each other's interests.

This segment can be used to teach about Proposition 1 of the WDF: *Boundaries and representation in water networks should be considered open-ended and continuously changing*, as well as Proposition 3: *The politics of transboundary water management should be adaptive and negotiated using a non-zero-sum approach*. The boundaries of the water system are not fixed independently from what the participants decide to negotiate. As a result, different kinds of stakeholders may be in a position to contribute to the negotiations in valuable ways. Stakeholders often have very different priorities and therefore different ideas about what should be considered "in" the negotiations, and what should be left "out." The participants can learn how to question and verify their assumptions about one another's interests and, also, about their own assessment of the issues that will have to be addressed to reach any agreements. The mediator can play an important part in helping the participants better understand one another, and in shaping an agenda for further discussions.

Time Required

Players will need at least 45 minutes to read the General Instructions, but more time will be helpful. The General Instructions for Segment 1 are the longest of the instructions for any of the segments. These instructions provide background information about the Indopotamia River basin, each of the participants and their countries or organizations, and an update on several key development issues. Although some of the participants may think they know where the Indopotamia River is, "in real life," the instructor should point out that the name and the river are, in fact, fictional. Players should focus on the short-term and long-term issues laid out in their instructions, and not try to augment or interpret the scenario with reference to some situation with which they think they are familiar. It is best if the General Instructions are distributed several days before beginning Segment 1 of the simulation. This will give all the players a better chance to study the scenario and absorb all the technical information.

Players will need at least an additional 45 minutes to read their Confidential Instructions, which may be distributed in advance of Segment 1 if the instructor is certain that all participants will be present for the game. If not, (usually, this turns out to be a wrong assumption) Confidential Instructions should be distributed immediately before the actual play of the game to ensure that each role is appropriately assigned. Confidential Instructions spell out the key concerns, as well as the priorities, of each country or group representative.

When distributing the Confidential Instructions, the instructor should explain that there is a one-page worksheet attached at the end of each set of Confidentials, summarizing the mandates that each player must follow. These are not optional. They are not meant to be interpreted or ignored, regardless of how experienced the role-players may be. The dramatic tension in the game is a product of the interaction of these mandates. Moreover, they do represent a composite of real-life views expressed by role players in parallel transboundary water negotiations.

If there are multiple groups playing the game simultaneously, the instructor may want to give people who have been assigned the same role (who will be sitting at different tables once the game begins) a chance to confer during the preparation period. They don't have to reach agreement on how they are going to play the role, but talking with others who have the same assigned role usually makes it easier for everyone to get into their "character". If any of the roles are doubled (i.e., if there are two players per role at a single table) these pairs will need additional time to develop a joint strategy and figure out which of them will speak at which points in the game.

Segment 1 negotiations require at least 60 minutes. Before beginning, make sure that all parties understand their instructions and that the mechanics of the game are clear. Assure participants that they are free to meet one on one, in small groups, or in a large group around the table as they see fit, although the mediator has explicit procedural instructions that he or she will try to implement. However the participants at a table choose to proceed, they should interact within a defined space (ideally in a separate room) so they can easily find other players from their group throughout the negotiations.

Debriefing

Reserve at least 30 minutes to debrief the results of Segment 1. As mentioned above, this segment can be used to teach about Proposition 1 of the WDF: *Boundaries and representation in water networks should be considered open-ended and continuously changing,* as well as Proposition 3, *The politics of transboundary water management should be adaptive and negotiated using a non-zero-sum approach.*

Specifically, the debriefing is intended to highlight: (1) the difference between interests and positions; (2) the need to reveal information about one's own interests in order to get others to talk about their interests (along with the dangers of revealing too much information); and (3) key differences between multiparty negotiations and two–party negotiations, especially regarding the emergence of coalitions. Listed below are some questions that instructors may find useful for getting the discussion going and underscoring the most important take-aways:

- What happened in your group once the negotiations began? Did the group stay together around the table, or did you immediately split into caucuses?
- Did you try to create a "winning coalition?" If so, how many people did you need to talk to? How did you decide in which order to speak to people?
- Did you try to create a "blocking coalition" to avoid being left out or disadvantaged by someone else's proposed agreement? If so, who did you speak to and in what order?
- How did you try to keep your coalition together? Did you make promises or commitments to each other? Were these "iron-clad?"

Groups that spend more time caucusing tend to form coalitions of various kinds. This can be valuable for groups that are able to build a "winning" coalition, but it can be problematic for parties that are left out. Why was it that some parties stayed at the table, while others did not? If one participant pushed to stay together as a group or to break into smaller discussions, what reasons did he or she give? Often, a participant who fears being disadvantaged if a coalition forms without him or her, or the mediator, will make a compelling request that they stay together based on a principle, such as efficiency or transparency.

Did participants try to create "winning coalitions," in support of a particular proposal, or "blocking coalitions" to avoid being left out or disadvantaged by the emergence of other coalitions? Ask about the differences between the two kinds of coalitions. In particular, how many people did each participant need to talk to, to form each kind of coalition? How many people were actually needed to block an agreement from going forward? Recall the decision rule spelled out in the General Instructions: (1) at least seven out of the eight participants must agree to a final package for each segment of the negotiations coming up; and (2) the agreement must include the representatives from each of the three Indopotamia countries and the RDB. Given that a coalition of any two players can oppose any agreement, it may be easier to form a blocking coalition than a winning coalition. The decision rule also means that not all participants are equal, with some exercising veto power over an agreement.

Ask how participants decided whom to speak with first. In some cases the three countries will decide to get together and form a coalition under the assumption that if they lead, the other parties will follow. If this occurs, ask the non-country participants how they felt about it. Did they join the potential winning coalition, or did they feel isolated and compelled to form a blocking coalition of their own?

An agreement among only a few participants can make it more difficult to reach a larger consensus agreement. Parties who commit to a specific course of action (i.e., to support or oppose a particular position on a specific issue) may be overconfident in the appeal of their proposal to other participants or its potential for success. Even if the other participants put together alternative proposals that meet the interests of at least some of the coalition members very well, one or more coalition members may be inflexible and unwilling to consider any alternatives besides the one they propose. On the other hand, new proposals may fracture a nascent coalition, exposing inherent instability—a characteristic of multiparty negotiations. Participants entering into coalitions should therefore consider strategies to exit them without damaging their reputations.

- How did you try to discover the interests of the other parties? What questions were the most productive?

The Confidential Instructions provide the parties with information about their own interests, and assumptions about the positions and interests of others. Some of the information and assumptions in the Confidential Instructions are incomplete or incorrect. So ask whether any of the participants were surprised by what they learned in their one-on-one conversations. There are many opportunities to create value in negotiations, but to do so, the parties must move beyond positional statements and learn more about the interests of the others. Ask the participants to contrast positions with interests to be sure they are clear about the difference. They should also ask which questions produced the most useful information. These will often be questions that begin with "Why ..."

- How did you build your credibility with the other stakeholders?

There is a history of mistrust among some of the stakeholders, making it difficult for them to trust one another. Are there ways in which the participants tried to build their credibility and trust with the other stakeholders? The most compelling ways to build trust are: (1) say what you mean; and (2) mean what you say. If a party tries only to elicit information from others, while withholding information about his or her own interests, others will be unlikely to share very much. Instead, this tactic may cause others to feel used or taken advantage of, engendering bad feelings. Therefore, parties need to share information about their own interests to build trust, generate credibility, and create value. But to protect themselves from being taken advantage of, they can ask questions and they can try to make sure that other players reciprocate by sharing information about their interests. Parties should also "mean what they say," (i.e., only say things that are true and will subsequently turn out to be true). This means parties should resist the temptation to exaggerate, bluff, or make up information.

Another compelling way to build trust is to listen attentively and show empathy for other people. Establish interpersonal connections by demonstrating an understanding of the other person's perspectives and interests. It is difficult to build trust if you show no empathy, even if you always tell the truth.

Making others feel that they have been heard can dramatically alter the tone of a negotiation. However, showing empathy does not mean agreeing with everything the other participants want. On the contrary, participants need to identify and explain their own interests. Players must find a balance between showing empathy for others and asserting their own perspectives and interests. This is hard to do if everyone stays entirely in his or her official roles the whole time. It would be better to make some personal connections. While players do not have to be "nice" to one another, it helps to at least be civil.

- What helpful things did the mediator do?

It is important to ask how groups made use of the mediator. The mediator's role in this first segment is not well defined, but he or she can contribute to enhancing the credibility of the parties and help them move beyond positions to hear one another's interests. For example, if there is a history of bad relationships among some of the parties, the mediator can facilitate better communication and trust building between them.

Segment 2: Information Sharing and Knowledge Generation

Time Required

- 30 minutes: Read General Instructions (best done in advance).
- 30 minutes: Confidential Instructions and same-role meetings.
- 90 minutes: Negotiation.
- 30 minutes: Debriefing.

Preparation

The RDB has convened this meeting in the hope that the parties will agree on how to generate, share, and use technical and scientific data to make decisions about the management of the River Basin. Prior to this meeting, almost no data have been shared by the parties because of political concerns and lack of capacity. This segment can be used to teach about Proposition 1 of the WDF: *Boundaries and representation in water networks should be considered open-ended and continuously changing*, and Proposition 2 of the WDF: *Modeling and forecasting for water management should account for variability and uncertainty*. As in Segment 1, the participants are not likely to make much progress in their negotiations until they recognize the unique knowledge that different stakeholders can contribute. Segment 2 requires the participants to grapple with a situation in which uncertainty and complexity are paramount.

The key decision facing the participants in Segment 2 is: "What should be the design of a joint fact-finding process?" Any agreement needs to address these questions:

- Who should participate in collecting and analyzing data?
- How should the scope of any studies be set?

- What role should experts play?

If an agreement cannot be reached, the RDB will probably not provide funds for future sustainable development efforts in the River Basin.

If possible, distribute the General Instructions in advance. Players will need at least 30 minutes to read the General Instructions for this Segment, but more time would be helpful. Players will need an additional 30 minutes to read their Confidential Instructions. If multiple groups are playing the game simultaneously, give people assigned the same role a chance to sit together while they are studying their Confidential Instructions. If any roles are doubled up, those players should have an opportunity to caucus and develop a joint strategy before the full group convenes for its discussions.

Negotiations will require at least 90 minutes. Before beginning, make sure that all participants understand their instructions and the mechanics of the game. Multiple groups playing the same game should meet in separate rooms and operate completely independently from one another. Remind the players that time is limited and they should be as succinct as possible in their exchanges. Once the simulation begins, the mediator will ask each player to give a brief opening statement. Then, the mediator will review the timetable for the rest of the meeting, and remind everyone about the ground rules *to which they have all agreed*. Informal caucuses are permitted. If multiple groups are playing the game at the same time, the results from each group should be reported separately at the end of the negotiation period using the reporting form provided with the mediator's Confidential Instructions. The reporting form notes: (1) whether or not agreement was reached (and who signed the agreement); and (2) if an agreement was reached, what were the terms.

Emphasize the following points:

- The representatives of all three countries and the Regional Development Bank must be party to any agreement
- Three other parties are needed for an agreement, not including the mediator (i.e., a total of seven out of the eight negotiating roles must be party to any agreement)
- For a party to agree on a proposed plan, the agreement must not include any provisions that are identified as unacceptable in that player's Confidential Instructions.

Debriefing

Reserve at least 30 minutes for the debriefing. As mentioned above, Segment 2 can be used to teach about Proposition 1 of the WDF: *Boundaries and representation in water networks should be considered open-ended and continuously changing*, and Proposition 2 of the WDF: *Modeling and forecasting for water management should account for variability and uncertainty*. Specifically, the Segment 2 debriefing is intended to highlight the usefulness of joint fact-finding. The debriefing should begin with a posting of outcomes. Listed below are some useful questions for stimulating discussion.

- What happened in each group once the negotiations began? What agreements, if any, were reached?
- What impacts, if any, did the outcome of Segment 1 negotiations have on the Segment 2 discussion?

Within the rules of the game (i.e., the role players' mandates for each of the three issues), three seven-way agreements are possible in this Segment. Each seven-way agreement leaves out a different party: Mu State, ICI, or WIED. Unless the parties modify the options in a manner consistent with their mandates, which is certainly permitted, no eight-way agreement is possible.

Possible Outcomes

TABLE 7.2 Indopotamia: Three Possible Outcomes for Segment 2: Information Sharing and Knowledge Generation

	Issues			*Who can't agree to this outcome?*
	Participation	*Scope*	*Role of experts*	
Possible Agreement # 1	Limited to experts designated by each country	Broad: study any sustainable development questions, any issues in the basin, and use data methods according to the best current methods	Expert driven	ICI
Possible Agreement # 2	Open to experts designated by each country and official observers	Broad: study any sustainable development questions, any issues in the basin, and use data methods according to the best current methods	Co-equal decision-making by experts and non-experts	WIED
Possible Agreement # 3	Open to country experts and self-designated representatives from nongovernmental organizations and civil society	Limited: study immediate water-management issues, the international river, use data methods according to countries' current capabilities	Expert driven	Mu

Explore differences between multiple group outcomes to discover why some groups were able to create more value than others. You may ask the players whether they felt the best possible agreement was reached. Some participants may have had (financial) resources they could have contributed to a possible agreement, but were not asked to do so.

Much can be learned by exploring why a group was not able to reach agreement. It is fair to ask groups that did not reach agreement to describe the stumbling blocks they encountered. Emphasize that not reaching agreement does not mean a group "failed." The objective is not necessarily for groups to reach agreement, but rather for them to explore the dynamics of transboundary water negotiations. In the groups that did not reach agreement, coalitions may have formed that bound parties to support or oppose outcomes that were, in fact, minimally acceptable. This, in turn, could have made it impossible to get enough players on board to reach a "winning" agreement.

Ask the participants to discuss how their feelings about the outcome in Segment 2 might affect their willingness to negotiate in subsequent segments of the Indopotamia game. In two of the three possible agreements, the deals involve giving at least one party its top priority on all three issues under discussion. When this happens, one of the other parties (either ICI or Mu State) cannot agree to the package. Some of the others, who can accept the package, do not receive their top preferences on any issue. Given these differences, ask if the participants are "happy" with the outcome and what it means to be "happy" in this context. Some players may be OK with the outcome because they feel they asserted their interests very well. Others may feel they built relationships and trust, which will be important in later negotiation segments. Still others may simply be content they did better than their best alternative to a negotiated agreement (BATNA).

For others, even if the outcome is better than their BATNA, they may be unhappy if they feel other parties claimed more value than they needed to. If one player perceives another as too competitive, aggressive, inflexible, or unwilling to share information essential to the search for mutual gains, future value-creating opportunities may be restricted. Worse yet, a seven-way agreement excludes at least one party. Ask how this outcome will change the excluded party's attitude toward future negotiations. A seven-way agreement in which some participants are excluded or are unhappy can pose considerable risk to relationships that are just developing.

- How can a joint fact-finding process contribute to transboundary water negotiations?

This question is an opportunity to review which stakeholders need to be included and what elements of the water system need to be represented. Sovereign countries are obviously paramount in international decision making, but it is impossible to represent all the elements of the water network adequately without thinking of some way of involving nongovernmental interests, interests outside the geopolitical boundary of the basin, and reliable sources of scientific information.

A few key points: *Multiple kinds of expertise are valuable.* Ask participants about the different kinds of information each player can contribute and how they valued these in the course of the negotiations in Segment 2.

Experts from different disciplines must be willing and able to share data with one another and with stakeholders. Negotiations need to be informed by credible scientific and technical expertise. Although an agreement may be reached based solely on political considerations, such agreements are unlikely to be effective if they are not also based on credible forecasts about water availability and appropriate assumptions about impacts on ecological, socioeconomic, and cultural variables.

Specialized knowledge from stakeholders is important. Stakeholders have local knowledge and can contribute understanding that outside experts cannot.

Experts need to interact with knowledge users throughout the fact-finding process. At the beginning, experts can help users frame study questions. During fact-finding, experts can highlight disagreements attributable to different technical methods and assumptions. When a draft report is ready, experts should consult with users to get their reactions. When findings are set, users can help experts decide how to interpret and present the results. Final decisions may still be made by those with the formal authority to act, but the credibly of those actions will be enhanced when they are preceded by joint fact finding.

• What are the things a joint fact–finding process cannot accomplish?
Joint fact-finding cannot produce unassailable answers to scientific or technical questions. Even when experts and stakeholders share information, knowledge gaps and limits to understanding remain. Participants must acknowledge that complexity and uncertainty will make it impossible to know the future. Their task is to quantify uncertainty as best as they can, and decide how to proceed in the face of what they cannot know.

Segment 3: Option Generation

Time Required

• 30 minutes: Read General Instructions (best done in advance)
• 30 minutes: Confidential Instructions and same-role meetings
• 90 minutes: Negotiation
• 30 minutes: Debriefing

Preparation

The negotiations in Segment 3 aim to reach agreement on development priorities for the Indopotamia basin. This segment can be used to teach about Proposition 1 of the WDF: *Boundaries and representation in water networks should be considered open-ended and continuously changing,* and Proposition 3 of the WDF: *The politics of transboundary water management should be adaptive and negotiated using a non-zero-sum approach*. To reach agreement, the participants must expand the sustainable development issues they address cooperatively, to address the top priority concerns of all or most of the participants. At the same time, the participants face constraints, such as limited financial resources, which mean they cannot accomplish everything. The participants must reconcile their competing interests by creatively packaging interlocking trades.

The key decisions facing the participants in this segment are: (1) What sustainable development priorities should be funded?; and (2) What criteria should be used to decide among possible development priorities?

An initial list of ten possible sustainable development priorities is included in the General Instructions. Each specifies a category of projects, which, if agreement is reached, would be eligible to receive funding from RDB. The anticipated costs associated with each development priority are listed. The participants must discuss and rank selection criteria. Finally, the participants must use the selection criteria to make final choices. RDB has agreed to fund the top choices of the group, as long as at least seven of the eight representatives agree to the package. The seven must include the representatives from the three countries and the RDB.

If possible, distribute the General Instructions in advance, since players will need at least 30 minutes or more to read them. Allow at least an additional 30 minutes for players to study their Confidential Instructions. If multiple groups are playing the game simultaneously, give players assigned the same role a chance to sit together while they are reading their Confidential Instructions. If any roles are doubled up, these players should have an opportunity to develop a joint strategy before meeting with the rest of their group.

Negotiations will require at least 90 minutes. Before beginning, make sure that all parties understand their Instructions, as well as the mechanics of the game. Multiple groups should ideally meet in separate rooms, completely independent from one another.

Debriefing

Reserve at least 30 minutes for the debriefing. As mentioned above, this Segment can be used to teach about Proposition 1 of the WDF: *Boundaries and representation in water networks should be considered open-ended and continuously changing*, and Proposition 3 of the WDF: *The politics of transboundary water management should be adaptive and negotiated using a non-zero-sum approach*. Therefore, the debriefing of Segment 3 is intended to highlight the dynamics involved in creating and distributing value. Begin by posting the outcomes. Listed below are some useful questions for organizing the discussion.

• What happened in each group? What agreements, if any, were reached? *Please note*: given their individual mandates, the participants cannot reach agreement unless they modify the preliminary descriptions of their priorities contained in their Instructions. Two possible seven-way agreements (including sample modifications) are detailed in Table 7.3.

TABLE 7.3 Indopotamia: Two Possible Outcomes for Segment 3: Option Generation

| | Sustainable development priority (listed by estimated cost) | | | Modifications to existing priority descriptions |
	Large	Medium	Small	
Possible Agreement # 1	Land management and conservation	Monitoring and information systems	Knowledge and support for water financing	Land management has to be modified to include wetland restoration to increase storage of water in the system, augment supply, and provide near-term benefits(Mu cannot agree to this outcome)
Possible Agreement # 2	Urgent water supply and sanitation	River basin planning	Environmental compliance and enforcement	River basin planning has to be modified so studies also consider impacts on existing water uses (ICI cannot agree to this outcome)

• How did your group try to create value?

Addressing water conflicts means managing complex networks. Groups will only be able to reach agreement if they recognize that they are not in a zero–sum negotiating situation. Water resources are variable and dynamic, not fixed. Water can be used multiple times in multiple ways, and can be conserved if managed properly. So, if players can work together, invest intelligently, and draw on their problem-solving capabilities, there is likely to be more water available than they initially imagined.

In the negotiations it is critical that participants discuss how to manage interactions between societal and natural elements (water networks) of the river system. To reach agreement, groups will need to craft a negotiating agenda that ensures all participants' top-priority concerns will be addressed. It is also important that groups reserve time to create new options or modify preliminary proposals. Sometimes, value can be created by defining or redefining terms (such as "management") more clearly. In other instances, creating value may require adding a new issue not previously on the agenda. Focusing only on infrastructure investment, or just science, or just public perceptions won't work. The group's task is to formulate a "package" that guarantees to all parties that their most important concerns will be achieved.

Groups that reach agreement usually share information about their priorities and interests, and explain *why* certain outcomes are unacceptable. In this way others can see which features of a proposed agreement need to be changed. Groups that try to resolve issues one at a time will fail. While it may be necessary to make a first pass through the agenda to see which items are most controversial, the parties need to consider full packages if they want to build a consensus.

• How did your group try to distribute value?

Participants should try to use objective criteria or standards as the basis for making final decisions. This will make it easier to maintain good working relationships. If groups get into a political test of will, relationships will deteriorate quickly. Participants should respect the fact that everyone at the table will have to explain to his or her constituency or membership why they agreed to the package that they ultimately accepted. They will want to refer to reasons that justify their concurrence.

Ask the participants about the criteria they decided on and how they agreed on a ranking. To be successful, participants must explain which standards or criteria enabled them to produce an acceptable package. By contrast, it is likely that groups that did not reach agreement were unable to agree on which issues were most important. Some players may have alienated other participants through their use of hard-bargaining tactics, such as bluffing or holding out unreasonably on every issue.

• What helpful things did the mediator do?

In addition to keeping the process on track and helping the group remain aware of time limitations, a mediator can play a critical role in highlighting who cares most about which issues, and why. By helping parties share information and express their interests, a mediator can assist negotiators in identifying the objective criteria they can use to make decisions. Mediators can also help a group formulate and remain attentive to ground rules, and produce a definitive summary of whatever the group has decided.

Segment 4: Deal Making

Time Required

• 30 minutes: Read General Instructions (best done in advance)
• 30 minutes: Confidential Instructions and same-role meetings
• 90 minutes: Negotiation
• 30 minutes: Debriefing

Preparation

The key decision facing the participants in Segment 4 is what the final package should include regarding governance arrangements. This Segment is intended to challenge participants to address all three Propositions of the WDF. The stakeholders have different ideas about how to set the boundaries of the water system, in particular what should be the geographic scope of the agreement and what range of issues should be covered. The stakeholders also disagree about the best way to deal with future problems or changes that may impact their agreement, such as new data, future disputes, or implementation failures.

Please note: given the constraints imposed in the Confidential Instructions, unless the participants share information about their priorities and interests, they will be unable to create value and reach agreement.

Specifically, participants need to reach agreement on four key points: the form of the agreement; its geographic scope and the range of issues it will encompass; procedures for reviewing and revising the agreement; and data sharing. As in previous negotiating sessions, for an agreement to be reached at least seven of the eight negotiators must agree to the package, including the representatives from the three countries and the RDB.

If possible, distribute the General Instructions in advance and give players at least 30 minutes to read the General Instructions. Players will need at least an additional 30 minutes to read their Confidential Instructions. If multiple groups are playing the game simultaneously, participants assigned the same role should be given a chance to sit together while they study their Confidential Instructions. If roles are doubled, the players involved should have a chance to develop a joint strategy.

Negotiations require at least 90 minutes. Before beginning, make sure that all parties understand their Instructions and the mechanics of the game. Multiple groups should meet in separate rooms and should operate completely independently from one another.

Debriefing

Reserve at least 30 minutes for the debriefing. The debriefing of Segment 4 is intended to highlight the importance of ongoing relationships and how the trust they have built will allow them to deal with unanticipated changes in the River basin. Begin the debriefing by posting outcomes. Listed below are some useful questions for organizing the discussion.

- What happened in your group? What agreements, if any, were reached?
- What impacts did negotiations during the previous Segment(s) have on today's discussion?

Please note: the participants will not be able to reach agreement unless they modify the existing options, in this case to address some participants' concerns about the future. Two possible seven-way agreements (including sample modifications) are detailed in Table 7.4:

TABLE 7.4 Indopotamia: Two Possible Outcomes for Segment 4: Deal Making

	Form of agreement	Scope: Geographic and range of issues	Review and revision procedures	Data procedures	Special provisions	Who cannot agree to this outcome?
Possible Agreement #1	Declaration	Broad	Yes	Inform	• Need to create new dispute-resolution mechanisms • Projects have to be compatible with ten-year joint river basin plan priority	ICI
Possible Agreement #2	Convention	Narrow	Yes	Consent	• Revisit agreement in ten years • WIED provides parties (especially Gamma) with assistance for implementation • Monitoring and data-sharing through neutral platform	Mu

Water networks are open ended and constantly changing, and the Indopotamia scenario suggests enormous uncertainty about the future. The participants, therefore, need to fashion a joint-management agreement that builds on currently available data, as well as information they don't have now. They need to take into account their existing understanding of the basin and improve it through ongoing collaborative monitoring. Differences in scientific estimates about the future are at the heart of every question they try to address. Some groups figure out a way to deal with the uncertainties and some don't.

• What kinds of "predicable surprises" did the group discuss?
• What part of any agreement is most vulnerable to future perturbations?

Groups that reach agreement will usually have discussed participants' concerns about the future and uncertainties about implementing whatever agreement they reach. Some examples include: problems putting the agreement into practice due either to lack of will or lack of capacity; the identification of new priority concerns; changes in relationships among the countries or other parties; conflict over future projects; and emergencies, such as a civil war or a cholera outbreak.

Instead of arguing about whose predictions about the future are right, groups that reach agreement will have discussed strategies for what to do in the event different forecasts turn out to be incorrect. Such strategies include: (1) specifying dispute-resolution mechanisms; (2) creating provisions to monitor implementation; (3) setting ground rules and timetables for review and revision of the current framework agreement; and (4) contingent agreements. Contingent agreements take advantage of differences in the parties' perspectives, specify alternate mechanisms for dealing with various future outcomes, and include clear monitoring arrangements so the parties know what is actually occurring and, therefore, which contingent commitments should be triggered. (See recommended background reading by Bazerman and Gillespie for more information on contingent agreements.)

To generate solutions that are implementable and enforceable, transboundary water disputants need to take account of the natural, societal, and political dynamics involved. The players have to share information about their different interests and predictions of what will happen. Ask how the parties accomplished this. Typically, groups will have used some of the techniques that proved successful in previous segments. They will have asked questions about one another's interests, brainstormed opportunities to create as much value as possible before allocating gains and losses, and used objective criteria to justify how to distribute value.

Also discuss the importance of building adaptive-management mechanisms into the agreement as a way of anticipating "predictable surprises" and maintaining good working relationships. Transboundary water-resource agreements should be contingent—building in ongoing monitoring and dispute-resolution mechanisms is crucial. Continuous relationships require attention—the dynamic nature of water networks means that the parties will have to constantly update any agreement, including the terms of their governance system.

Background Readings

Bazerman, M.H. and Gillespie, J.J. 1999. Betting on the Future: The virtues of contingent contracts, *Harvard Business Review, 77*(5): 155–160.

Bazerman, M.H. and Watkins, M.D. 2004. *Predictable Surprises: The Disasters You Should Have Seen Coming, and How to Prevent Them.* Boston, MA: Harvard Business School Publishing Press.

Fisher, R., Ury, W., and Patton, B. 1998. *Getting to Yes: Negotiating Agreements Without Giving In.* New York, NY: Penguin Books.

Karl, H.A., Susskind, L.E., and Wallace, K.H. 2007. A dialogue, not a diatribe: Effective integration of science and policy through joint fact finding, *Environment: Science and Policy for Sustainable Development, 49*(1): 20–34.

Lewicki, R.J., Gray, B., and Elliott, M. (eds.). 2003. *Making Sense of Intractable Environmental Conflict: Concepts and Cases.* Washington, DC: Island Press.

Mnookin, R., Peppet, S., and Tulumello, A. 2001. *Beyond Winning: Negotiating to Create Value in Deals and Disputes.* Cambridge, MA: Harvard University Press.

Raiffa, H. 1982. *The Art and Science of Negotiation.* Cambridge, MA: Harvard University Press.

Rofougaran, N.L. and Karl, H.A. 2005. *San Francisquito Creek — The Problem of Science in Environmental Disputes: Joint Fact Finding as a Transdisciplinary Approach toward Environmental Policy Making.* U.S. Geological Survey Professional Paper 1710.

Susskind, L.E. and Ashcraft, C. 2010. How to reach fairer and more sustainable agreements, in J. Dore, J. Robinson, and M. Smith (eds.) *Negotiate: Reaching Agreements over Water.* Gland: IUCN.

Susskind, L.E., and Cruikshank, J.L. 2006. *Breaking Robert's Rule: The New Way to Run Your Meeting, Build Consensus, and Get Results.* New York, NY: Oxford University Press.

Susskind, L., Levy, P.F., and Thomas-Larmer, J. 2000. *Negotiating Environmental Agreements: How to Avoid Escalating Confrontation, Needless Costs and Unnecessary Litigation.* Washington, DC: Island Press.

Susskind, L., McKearnan, S., and Thomas-Larmer, J. 1999. *The Consensus Building Institute: A Comprehensive Guide to Reaching Agreement.* Thousand Oaks, CA: Sage Publications.

TABLE 7.5 Indopotamia Segment 2: Summary Form for Reviewing Results from all Groups

Group no.	Agreement			Participation	Scope	Role of experts	Additional details of the agreement
	Yes or no?	*No. in agreement (7 or 8)?*	*If 7, who is left out?*				
1							
2							
3							
4							

TABLE 7.6 Indopotamia Segment 3: Summary Form for Reviewing Results from all Groups

Group No.	Agreement			Sustainable development funding priorities (by cost scale)			Decision criteria (list top 5)	Additional details of the agreement
	Yes or no?	No. in agreement (7 or 8)?	If 7, who is left out?	Large	Medium	Small		
1								
2								
3								
4								

TABLE 7.7 Indopotamia Segment 4: Summary Form for Reviewing Results from all Groups

Group no.	Agreement			Form of agreement	Scope	Reporting and revision procedures	Data procedures	Additional details of the agreement
	Yes or no?	No. in agreement (7 or 8)?	If 7, who is left out?					
1								
2								
3								
4								

General Instructions for Segment 1: Interests and Building Coalitions*

The three countries sharing the Indopotamia River basin face significant water-management challenges, but lack a formal agreement governing international cooperation on these issues. The Regional Development Bank is prepared to assist in addressing these challenges through financing and implementing water-development projects, if the countries and some key interest groups can come to an agreement on how to proceed. If the parties cannot reach an agreement, it is likely

* From the Program on Negotiation at Harvard Law School: An Inter-University Consortium to Improve the Theory and Practice of Conflict Resolution by Catherine M. Ashcraft. Copyright © 2011 by the Water Diplomacy Workshop and by the President and Fellows of Harvard College.

that one or more countries will develop water projects on their own, which could cause a political crisis in the region.

The Parties

The Regional Development Bank (RDB)

If the parties can reach an agreement, the RDB is eager to provide significant support for new development projects in the region. The RDB believes its participation will provide substantial benefits in accordance with its mandates: (1) to help its member countries reduce poverty; (2) to help mobilize resources for environmentally sustainable growth; and (3) to foster regional integration.

Gamma's Ministry of Water and Energy

The least-developed country in the basin and only recently emerged from years of civil war, Gamma has a strong interest in a new agreement to develop water resources to improve the lives of its people and strengthen its economy.

Alpha's Ministry of Water Resources

The largest and most-developed country in the basin, Alpha has a long history of water use and has extensive water infrastructure in place. Historically, Alpha has opposed a regional approach to water management, preferring bilateral agreements.

Beta's Ministry for Sustainability

Beta is generally in favor of a regional water agreement that includes measures implemented by/in upstream countries to address their poor land- and water-management practices, which Beta blames for downstream flooding and environmental damage.

Mu State's Economic Administration

Situated within the delta, Mu State is home to Banaga, Alpha's most industrial city. Mu State is pleased by the prospect of new development projects, but wants to ensure that local governments share in their benefits, which has not always been the case.

Global Water Management Organization (GWMO)

GWMO's regional office in Banaga, Alpha is one of ten global research centers supported by governments, private foundations, and international and regional organizations. GWMO's mandate is to improve how water and land are managed for the environment, the livelihoods of poor communities, and good will, through an integrated approach to water-resources management.

International Conservation Institute (ICI)

ICI is an independent foundation and global organization, which acts locally through a network of country and regional offices to help societies to: (1) conserve their natural biodiversity; (2) use their natural resources in an equitable and eco-logically sustainable manner; and (3) reduce pollution. ICI is concerned about the potential negative environmental and social impacts of new development projects.

Water Infrastructure Engineering and Design (WIED)

WIED is a globally respected engineering and design consortium, whose members are active in water-development projects in more than 30 developing countries and in this region. WIED is enthusiastic about the possibility of a new agreement for more efficient development of the Indopotamia.

Mediator

All participants agreed to select a representative from Adiuto to serve as the media-tor for the negotiations. Adiuto is a not-for-profit organization of practitioners and theorists in the fields of negotiation, dispute resolution, and dispute system design, which works with governments, intergovernmental organizations, non-governmental organizations, experts, communities, and business to negotiate more effectively, build consensus, and resolve disputes.

The Development Process

Several months ago the Regional Development Bank entered into discussions with the three countries toward initiating a cooperative development strategy for the Indopotamia River. Based on these discussions, the RDB convened these negotia-tions, in cooperation with several other regional organizations and the three national governments. Civil society groups have criticized past development projects for neglecting to involve people who bear the negative consequences of the projects. Sensitive to this criticism, the RDB has invited a number of different participants to these negotiations and will not move forward with development plans unless the proposals receive significant support from them.

Although the RDB would prefer unanimous support for any final agreements, it has pledged financing and support if: (1) at least seven of the eight participants agree to a final package in each segment of negotiations; and (2) the agreements include the representatives from each of the three Indopotamia countries and the RDB. Each country has veto power over the final agreements, as the RDB requires support from all basin countries on projects like these, in international river basins likely to have transboundary impacts. As the only party with the capacity to initiate transboundary development, the RDB can also veto any final agreement.

If these negotiations fail and the participants cannot reach agreements, there will be considerable uncertainty over the future funding of any regional water projects. Countries may try to secure alternate financing to proceed unilaterally with their water-development plans, which could spark an international political crisis.

General Descriptions of the Indopotamia River Basin

Below are general descriptions of the Indopotamia River basin, its politics, and economic situation. Each party's Confidential Instructions contain additional details.

Water

The Indopotamia River flows roughly 1,500 km from high mountains located in Alpha and Gamma, through terraces, hills, and vast floodplains before discharging into Ruo Bay through an extensive delta shared by Beta and Alpha (see map and table in Appendix A). The high, mountainous areas in southern Gamma and a few areas in Alpha are prone to severe thunderstorms and landslides. Gentle hills and plains characterize much of the central basin in Alpha. About 70 percent of Beta is situated at or below one meter above sea level. The Indopotamia basin encompasses about 1,016,000 square kilometers. Sixty-nine percent of the basin is farmed, 12 percent is wetlands, 9 percent grassland, 6 percent urban and industrial areas, and 4 percent forest. Agriculture accounts for 85–95 percent of all water withdrawals in each of the countries. In the last decade, the Indopotamia experienced devastating dry season droughts and wet season floods.

The countries urgently need short-term arrangements to address projected water shortfalls for the coming year's dry season, and recurrent flooding in the wet season. Some of the River's variability is natural, but climate change and land-use practices may be exacerbating this variability. The poorest residents of the region have access to far less clean water than the World Health Organization standards set as a bare minimum. A shift to using groundwater reduced exposure to pathogens and increased the reliability of water for growing crops, but also contributed to widespread poisoning from naturally occurring arsenic in central Beta and the delta region.

Interest in new water-development projects to respond to these challenges has sparked a "race to develop," as countries, concerned about how projects in other countries will affect their own ability to use the basin's water, try to act now to develop the water.

Politics

All three countries associate water development closely with their national identities and feelings of self-determination. Gamma, only recently emerged from decades of civil war with a weak power-sharing democratic government, sees water development as a way to attract foreign investment and grow the economy. Beta voted in a new government last year on a platform to address widespread poverty by combating government inefficiencies and improving infrastructure and basic services, including flood protection. Water development is a major election issue in advance of Alpha's federal elections next year with opposition groups, including the main contender from Mu State, criticizing the ruling party for not developing available water resources more extensively for the benefit of state and local economies.

In general, Beta and Gamma consider Alpha's existing water development to be a limitation on their own future water development and are mistrustful of any plans by Alpha to expand its water projects. Alpha contends that Beta and Gamma have yet to provide credible data demonstrating transboundary harm from any of its water projects.

Other major disputes further complicate relations between the countries. Ten years ago, Alpha began diverting water at Sami, near its border with Beta, to supply water to its port city Banaga for domestic and industrial use and to improve navigability of the shipping channel. Beta requested additional information about the diversion, but Alpha delayed sharing information until it could gather sufficient data. Despite some subsequent expert-level meetings, the two countries have not been able to agree on what data are relevant to the discussion about the diversion, so it remains a major dispute between them.

A few years ago, Alpha constructed a hydropower dam on a tributary near its border with Gamma and established a security presence to protect the facility. Gamma objects to Alpha's increased security presence in this area, which the two countries fought over 20 years ago. Gamma also accuses Alpha of ignoring criminal groups that cross into Gamma for illegal logging and then transport the logs to Alpha to sell them and finance rebel activities within Gamma. Alpha vehemently denies any involvement in illegal activities or in its neighbor's political affairs.

Economy

Alpha is the regional economic powerhouse. Nevertheless, some of Alpha's rural population are very poor and vulnerable to crop failures in years with weak monsoons. Beta's economy has seen strong growth over the last decade, despite challenges, including a large population, widespread poverty, poor education, and a lack of physical infrastructure. Each year, Beta is forced to spend significant sums on flood and drainage control and infrastructure repair. Gamma is one of the least-developed countries in the world. Agriculture is the main contributor to its economy, but in recent years Gamma's economic growth has declined due to poor crop harvests.

Despite acute food-security concerns, national priorities may be shifting from using water for agricultural production to supplying urban households and businesses.

Demand for power is growing and access to reliable electricity is a concern for all of the Indopotamian countries. Gamma and Alpha have been looking into developing new hydropower facilities to meet growing demand, and there seems to be some interest in connecting and sharing power resources between the three countries.

The Upcoming Week of Negotiations

All eight negotiating parties have agreed to attend all four segments of negotiations. The first negotiation segment will last 60 minutes. The negotiation sessions for the other three segments will last 90 minutes each, and all negotiations must stop at the end of the session. A professional mediator may assist the parties. A schedule for all negotiation segments follows.

Segment 1: Interests and Building Coalitions

- *Key decision*: Do I want to enter into a coalition (strategic alliance) with any other party or parties?

Segment 2: Information Sharing and Knowledge Generation

- *Key decision*: How should a joint fact-finding process be designed?

Segment 3: Option Generation

- *Key decision*: What sustainable development priorities should we fund?
- *Key decision*: What criteria should we use to decide between possible priorities?

Segment 4: Deal Making

- *Key decision*: What are our strategies for agreeing on and an implementing a final package?

Logistics for Segment 1

- You will have 45 minutes to prepare for the upcoming negotiation with workshop participants playing the same role from other groups. Please see Appendix B for Frequently Asked Questions.
- You will go to your negotiating table and then have 60 minutes to caucus with other members of your negotiating group. Please stay at or near your table so the other members of your negotiating group can find and speak with you.
- The workshop will reconvene for 30 minutes to debrief the negotiation.

Appendix A

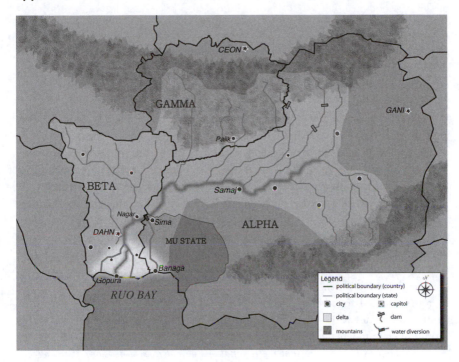

FIGURE 7.1 Indopotamia river basin map

TABLE 7.8 Indopotamia: The Basin and its Countries

	Total country area (km²)	Country's area in the basin (km²)	% Total area of basin contributed by country	% Total area of country lying in the basin	Dependency ratio*
Alpha	2,032,000	690,880	68	34	34
Beta	186,612	182,880	18	98	91
Gamma	237,066	142,240	14	60	6
		Total basin area = 1,016,000			

Note: *The water-dependency ratio compares a country's internal renewable water resources, or the annual flow of all rivers and groundwater produced from precipitation within the country's territory, to the total renewable water resources available to a country per year. Lower numbers indicate a lower dependency, meaning a country generates most of its water resources internally. A higher number indicates a higher dependency, meaning the country receives a greater percentage of its water resources from other countries.

Appendix B: Frequently Asked Questions

How will decisions be made within this negotiation? The participants do not need to reach an agreement on Segment 1. For Segments 2 through 4, if possible, decisions should be made by consensus. If a consensus cannot be reached, then a seven-out-of-eight vote is needed to approve any plan. The Ministries from each of the three countries and the representative from the Regional Development Bank must be part of any agreement.

How much information may I share with other participants? Each participant may explain his or her objectives and underlying interests to others in as much or as little detail, and with as much or as little accuracy, as he or she thinks is appropriate. However, participants may NOT show their Confidential Instructions to any other player. (There is no way in "real life" to prove that you are telling the truth!)

How closely do I have to follow my Confidential Instructions? Participants must adhere to their confidential mandates, even if in "real life" they do not share those interests or beliefs. Participants will learn the most from the simulation if they play their roles with energy, drama, and a sense of humor. Be as creative as possible within your constraints to develop constructive approaches to the issues.

Are side caucuses (meetings) allowed? The participants at each table are not required to remain seated together the entire time. That is, side caucuses are permitted among the parties at the table. If multiple groups are playing the game, each table should operate independently of all the other tables.

What happens in the event that the negotiations do not reach a decision? If no agreements are reached (i.e., if proposals do not receive at least three votes in addition to the three Ministries and the Regional Development Bank), there will be considerable uncertainty over the future funding of regional water projects. The Regional Development Bank will withhold funding for any projects with transboundary impacts until an agreement can be reached. Countries may try to secure alternate financing to proceed unilaterally with their water-development plans. If this happens, it is unclear if or when negotiations could resume in the future.

Do we have to stick with the policy options outlined in our Instructions, or can we invent other options? The group can invent other hybrid options as long as they are consistent with the information provided in the General and Confidential Instructions.

What is the best outcome possible? Several creative outcomes are possible. In general, for all parties, the best outcome is one that produces an agreement and still allows each party to feel optimistic about meeting its own interests.

General Instructions for Segment 2: Information Sharing and Knowledge Generation

As part of the overall consensus on international governance of the Indopotamia, the Regional Development Bank (RDB) requires an agreement on a joint fact-finding (JFF) process. In the past, the three countries shared very little data with one another due to their adversarial relationships, a lack of scientific capacity to carry out research, and civil unrest. Lacking mutually acceptable scientific data, each stakeholder tries to get its own data to either allege transboundary harm or bolster its defense against such attacks. Instead of informing wise decisions, use of data becomes another expression of power. Those who know more than others, or can buy science to support their positions, try to gain an advantage over others. Used in this adversarial way, scientific knowledge has brought only more confusion and distrust to the basin.

The RDB is hopeful that a joint fact-finding process can generate needed knowledge in a way in which it will actually be used to inform decision-making. Joint fact-finding is a participatory process of inquiry in which stakeholders are directly involved in knowledge production and decision-making. Once a group of stakeholders is identified, together, they design a process of scientific study—by first defining the scope of study, which will define the central questions the study will and will not examine. They also need to agree on guidelines for the roles of experts in interacting with stakeholders to implement the research strategy, deal with uncertainty, and apply findings. In this case, the RDB hopes to use the results of the joint fact-finding process to inform the preparation and financing of future priority sustainable development projects and mechanisms for attracting additional international financing.

At this meeting the participants need to agree on the design of the joint fact-finding process. Their design needs to include agreement on each of three decisions:

- Who should participate in collecting and analyzing data?
- How should the scope of any studies be set?
- What role should experts play?

To provide a starting point for this meeting, the RDB conducted interviews with the participants and prepared several alternatives for each of the decisions. Each of these options is discussed in more detail below. The RDB has pledged to provide funds for the joint fact-finding process if: (1) at least seven out of the eight participants agree to the package; and (2) the agreement includes the representatives from each of the three Indopotamia countries and the RDB. If an agreement cannot be reached, the RDB will not provide the funds and there will be no joint fact-finding process.

Design of the Joint Fact-Finding Strategy

Participation in Data Collection and Analysis

The participants disagree over who should participate in the joint fact-finding process. Some prefer to limit participation to experts designated by each country. Others prefer to open participation to experts designated by each country and official observers. Country delegates could each invite up to a certain number of official observers to participate, based on a person's or a nongovernmental organization's (NGO's) expertise and ability to fill in gaps in national capacity. Official observers could be academics, part of a civil society, a business or intergovernmental group, or any other kind of expert. The NGOs would prefer to self-designate representative experts to participate as observers. Some of those who prefer limiting participation to designated national experts feel ceding authority to NGOs could establish a dangerous precedent of voting rights for NGOs in intergovernmental agreements. They are also concerned that NGO delegates could outnumber national delegates and dominate the research agenda.

The options for participation in data collection and analysis:

- *Option 1*: Participation limited to experts designated by each country.
- *Option 2*: Participation open to experts designated by each country and official observers.
- *Option 3*: Participation open to country experts and self-designated representatives from nongovernmental organizations and civil society.

Scope of Study

There is significant disagreement over the questions that should be researched. Some say a relatively limited scope of inquiry offers the best path to producing findings within a realistic period of time to inform development-planning decisions. Others support a broader scope for the joint fact-finding process. There are three issues at stake under this decision:

1. There is debate over whether to focus on a limited number of questions of immediate water management or broader sustainable development questions, including food security, access to electricity, and alternative energy. For example, should research focus only on climate-change impacts on flooding, water quality and dry season water supply, or on broad issues of climate-change mitigation and adaptation?
2. Should the process focus on the more limited geography of the international River or the broader area of the international basin? Should studies primarily investigate issues related to the main River stem, major tributaries, and closely associated areas of the basin that directly affect the River, such as the riparian zone? Or should studies instead investigate issues throughout the basin, since the activities within it can be expected to have an impact on water resources?
3. There is disagreement over whether to collect and share data based on the current capabilities of countries or based on the most current methods, even if countries are not yet capable of complying. Under the more limited approach, in order to compare data, the countries would still need to develop harmonized methods for collecting data, analyzing it, and reporting it. The data would be coarser out of necessity, for example including monthly but not daily

precipitation data. The broader approach would include more parameters and finer data based on the best data-sharing methods possible given budgetary and other constraints.

The options for defining the scope of the study:

- *Option 1*: Limit the scope to study immediate water-management issues, the international River, and use data methods according to countries' current capabilities.
- *Option 2*: Broaden the scope to study any sustainable development questions, any issues in the basin, and use data methods according to the best current methods.

Role of Experts

Some of the stakeholders are very concerned about the role of experts in the joint fact-finding process. Some participants want to see co-equal roles for experts and non-experts throughout the process, including decisions on choice of methods, choice of data, interpretation of the data, and formation of policy prescriptions. Supporters of the co-equal approach contend that information users should explain to experts why they want to know what they do. Experts can then help knowledge users frame specific study questions. Once experts think they know how to address the study questions, non-experts can hold the experts accountable for their approach to understanding. Experts should brief information users on disagreements about methods and assumptions. In the latter stages, non-experts and experts interpret study results and make the normative leap to policy prescriptions together.

Other stakeholders would prefer to see a more expert-driven process. According to advocates of this approach, experts and non-experts still interact throughout the process, but their roles are distinct. Experts have complete control over the choice of methods, choice of data, interpretation of data, and policy prescription.

The options for the roles of experts:

- *Option 1*: Expert driven.
- *Option 2*: Co-equal control by experts and non-experts.

Logistics

- You will have 30 minutes to prepare for the upcoming negotiation with workshop participants playing the same roles at different tables.
- Go to your negotiating table.
- You will then have 90 minutes to negotiate an agreement on the three decisions to define the joint fact-finding process. Please see Appendix A for the negotiation ground rules on which all participants previously agreed.

An agreement is reached if:

- At least seven out of the eight participants at your table agree to the design; AND
- The agreement includes the representatives from each of the three Indopotamia countries and the RDB.

The workshop will reconvene for 30 minutes to debrief the negotiation.

Appendix A: Ground Rules

In preparation for these negotiations, the following ground rules were circulated and unanimously accepted by all participants:

1. All negotiators pledge to negotiate "in good faith," meaning they are willing to reach an agreement and will give their best effort to do so by participating fully in all meetings.
2. Participants agree to candidly and accurately represent their constituents' interests.
3. Participants agree to make a good-faith effort to understand one another's concerns.
4. Everyone is entitled to pursue his or her own organizational interests, but they will also seek to take account of the interests of the whole group (and look for mutually advantageous suggestions).
5. Each person may disagree with a proposal and, in so doing, accepts responsibility to offer an alternative proposal to accommodate his or her own interests and those of others.
6. All the parties have already agreed to have the mediator facilitate these negotiations. However, if the group decides the mediator is biased it can choose to dismiss the mediator.
7. The mediator will manage the meeting.
8. Only one person will speak at a time. People should speak only after the mediator recognizes them. Participants will not interrupt one another.
9. People should try to make their points as succinctly as possible and avoid asking an overwhelming number of questions in the interest of giving all participants the opportunity to speak.
10. No one will make personal attacks or disparage another's concerns.
11. Participants and the mediator may call for a caucus with one or more parties at any time. The caucus will be kept as brief as possible and last no more than five minutes. Anything said to the mediator in a caucus is confidential, meaning the mediator will not share what is said with the full group or any other parties.
12. The mediator will keep a record of all parts of an agreement and at the end of the negotiation everyone agreeing will initial the agreement, if one is reached.

General Instructions for Segment 3: Option Generation

Today you will try to reach agreement with the other negotiators on the selection of development priorities for the Indopotamia basin. The Regional Development Bank (RDB) has conducted a preliminary study, and is suggesting ten possible sustainable development priorities that appear to offer transboundary benefits. These are described in Appendix A.

Discussing Sustainable-Development Funding Priorities

The negotiations will begin with a review of the list of sustainable-development priorities. Each priority specifies a category of projects that, if an agreement is reached, will be the first to receive financing from the RDB. The group should spend about 15 minutes reviewing the list. Your objective is to make sure that whatever funding priorities are ultimately selected meet your country's or your organization's interests (including what, in your view, is best for the region as a whole). The group may also decide to further specify or modify the priorities, given their extended descriptions.

Discussing Selection Criteria

After discussing the list of options (i.e., the high-priority development categories for the region), you'll have to settle on a relatively short list. The RDB has provided a set of criteria that it thinks you should use in setting priorities. (These are included in Appendix B.) The group should spend about 20 minutes discussing the list of selection criteria (and you may also add to this list), and then rank the list of criteria in terms of their importance. By the end of the 20 minutes you should have a list of criteria for selecting sustainable development priorities, and be clear about how the group rates the criteria in terms of importance.

Agreement on Sustainable Development Priorities

In the final part of today's negotiations, the group will try to agree on a package of sustainable-development priorities everyone can live with. It will need to narrow the choices by examining the list of projects in light of the selection criteria the group has agreed upon. Be ready to use the ranked list of criteria to justify why you prefer particular options as compared to others. Explain how the projects you prefer will meet both your interests and those of others.

The RDB has agreed to fund the top choices of the group as long as at least seven out of the eight negotiators have agreed to the package. The seven must include the representatives from the three countries and the RDB.

A final note: be careful with your time. You've got to get through all three activities:

- Discussing funding priorities for sustainable development projects

- Reaching agreement on decision criteria and ranking them
- Coming up with a package of sustainable-development priorities for the region.

Logistics

- You will have 30 minutes to prepare in conjunction with workshop participants who are playing the same role as you at other tables.
- Then you will head to your negotiating table. (This will be the same table you were assigned on other segments.)
- You will have 90 minutes to reach agreement on a package of sustainable-development funding priorities.

An agreement is reached if:

- At least seven of the eight negotiators agree; AND
- The agreement includes the representatives from each of the three Indopotamia countries and the RDB.

The workshop will reconvene for 30 minutes to debrief the results of the negotiation.

Appendix A: Possible Sustainable-Development Priorities for the Indopotamia Region

Sustainable-development priorities are categorized according to the overall anticipated cost of achieving their intended objectives. Please note: In some cases these costs represent only the initial funding, which the RDB is confident will attract additional financing from other sources to cover the total costs of achieving the objectives.

- Large = $5 million or more
- Medium = Between $1 million and $5 million
- Small = $1 million or less

Extended Descriptions of the Sustainable-Development Priorities

Land Management and Conservation Initiatives

A project in this category would involve planning to restore, protect, and improve productive functions of land in the three countries within the Indopotamia basin. Estimates indicate that just over two million hectares, or about 2 percent, of the total basin would benefit from improved management and conservation initiatives. Degraded and at-risk land will be rehabilitated to enhance sustainability.

TABLE 7.9 Possible Sustainable Development Priorities for the Indopotamia Region

Priority size category	Sustainable development priority	Description	RDB cost range ($US millions)
Large	Land management and conservation initiatives	Develop a comprehensive and sustainable land-management approach to provide economic and social benefits for those who depend on the land; preserve the land's ecological integrity.	6
Large	Urgent urban water supply and sanitation rehabilitation	Rehabilitate and improve the most urgent existing water systems in major urban areas, including water supply and distribution, sanitation, and new storage systems. Public hygiene campaigns and feasibility studies for urban water supply would also be included.	5
Medium	River basin planning	Develop an integrated approach to *future* development of all of the Indopotamia basin's water supplies, including 25-year plans for sustainable water use in different sectors (e.g., agriculture, urban supply, navigation and transport, hydropower) and stress loads on water resources.	3.5
Medium	Delta flood risk and drought management	Reduce the vulnerability of *delta* communities to the impacts of floods and droughts. Develop an inventory of existing water uses in the delta and review ongoing and planned disaster risk-reduction strategies in the delta. Make recommendations for changes to increase resiliency of delta. Develop periodic progress reviews and an interactive knowledge map to allow countries to update their efforts.	3.5
Medium	Monitoring and information support system	Strengthen hydrometeorological, surface and groundwater monitoring networks, integrate historical databases, and make observations accessible through a GIS-based database.	3
Medium	Community basin management	Identify and implement community-based investments for improving community water systems (e.g., small dams, flood warning, watershed restoration, small-scale irrigation and water-supply equipment rehabilitation) to conserve natural resources, reduce poverty, and build community capacity.	3
Small	Regional energy sector strategy	Develop a comprehensive, participatory regional strategy for the energy sector to ensure efficient and affordable access to clean energy and provide energy security in the region. Formulate an action plan to address challenges in the energy sector.	1

TABLE 7.9 Continued

Priority size category	Sustainable development priority	Description	RDB cost range ($US millions)
Small	Assessing the economics of climate change	Compare the costs and benefits of unilateral and regional action to mitigate and adapt to climate change. Develop country and regional action plans to adapt development to expected future climatic conditions.	1
Small	Knowledge development for water financing	Help countries prioritize water investments, develop capacity, and implement institutional reforms. Develop sector-based case studies of water management, highlighting best practices.	1
Small	Improving environmental compliance and enforcement	Develop principles and good practices of environmental compliance and enforcement. Finance country adoption strategies and strengthen enforcement capabilities of environmental agencies.	1

Other areas will benefit as a result of more sustainable land management. By addressing land degradation, a project in this category would provide social and economic benefits for rural livelihoods, as well as protect the ecological functions (i.e., eco-services) of the land for the future. Protecting eco-services would benefit individual countries, the Indopotamia region, and the River. The total cost of the initial phase of this project is estimated at $40 million. The RDB is confident it can leverage initial funding of $6 million through a variety of other multilateral funding sources to cover the total costs of achieving the objectives.

Urgent Urban Water Supply and Sanitation Rehabilitation

Projects in this sustainable development category would seek to rehabilitate or improve the urban water systems in each of the three countries within the Indopotamia basin in most urgent need of repair. Specifically, projects would seek to rehabilitate or improve:

- Water-supply and distribution systems, including pump stations and chemical dosing, to significantly increase daily water production.
- Sanitation, including outfall sewer systems, treatment plants, and disposal of effluent and sludge, to significantly increase sewage capacity in the basin.
- Storage systems.

The proposed project would also include:

- A public hygiene campaign to inform people in cities of the importance of proper sanitation and personal hygiene, reducing infrastructure damage and vandalism.
- Feasibility studies for options to expand urban water supply and sanitation to be considered for financing with additional funds.

The projects seek to improve the health and well-being of the population through equitably providing improved water-supply and sanitation services. People living in high-density urban areas are at the greatest risk of diseases like cholera, and have little access to private or public health services. Businesses that rely on clean water supplies will be able to increase their operations. The project will also reduce pressure on the health sector, provide jobs, reduce water-treatment costs, stimulate the national economies, and improve the sustainability of the Indopotamia region and river. Much of the existing infrastructure in the basin has been poorly maintained, so we need significant investment to restore these services and meet current demand. The total cost of the initial phase of this project is estimated at $50 million. The RDB is confident it can leverage initial financing of $5 million through a variety of funding sources to cover the total costs of achieving the objectives.

River Basin Planning

This sustainable development priority seeks to design an integrated basin approach to developing the Indopotamia region's water resources in the medium-term future. The projects will culminate in 25-year projections for how water demand is expected to grow in each of the three countries in different sectors, including agriculture, urban supply, navigation and transport, and hydropower. Eligible projects will also model expected stress loads on water resources associated with the projected demand. These projections can then be used, in conjunction with social assessments, to develop a plan to sustainably develop the basin's water resources over the next 25 years. The RDB estimates the cost of achieving these objectives to be $3.5 million.

Delta Flood Risk and Drought Management

This priority aims to reduce the vulnerability of delta communities from the negative impacts of floods and droughts. Separate investment projects for Beta and Alpha will finance needed flood and drought risk-management projects in the delta region of each country. Each country will first identify existing and planned water uses in the delta, and then identify ongoing and planned national disaster risk-reduction strategies in the delta. The RDB, Alpha, and Beta, with assistance from an engineering consultant, will develop recommendations and a prioritized action plan of interventions to reduce the flood and drought risks in the delta region.

Periodic progress reviews and an interactive knowledge map would be created to allow countries to update their efforts. The RDB estimates the cost of achieving these objectives to be $3.5 million.

Monitoring and Information Support System

This sustainable development category seeks to strengthen hydrometeorological, surface and groundwater monitoring networks throughout the basin, integrate existing historical databases, and make observations accessible to national authorities (and river basin authorities that might be created in the future) through a GIS-based database. Eligible projects would seek to assess the surface and groundwater resources, analyze the existing hydrometeorological and environmental monitoring network, evaluate data collection and analysis procedures in use, and improve data dissemination. The RDB estimates the cost of achieving these objectives to be $3 million.

Community Basin Management

Projects under this priority will identify and implement community-based investments for improving community water systems to conserve natural resources, reduce poverty, and build community capacity. Eligible projects could include small-scale dams, flood warning systems, watershed restoration, irrigation, and water supply equipment rehabilitation. Communities would receive training and capacity building to improve their use of best practices in the water sector. The RDB estimates the cost of achieving these objectives to be $3 million.

Regional Energy Sector Strategy

This category will seek to develop a comprehensive, participatory regional strategy for the energy sector of the Indopotamia countries. Projects would seek to expand cooperation among the three economies to identify sector challenges and priority projects, provide efficient and affordable access to clean energy, and strengthen energy security in the region. The projects would explore transboundary supply options to reduce overall energy costs, and look at opportunities for the regional energy market, in comparison to individual national markets, to achieve efficiencies of scale from large-scale investment options. The projects would examine how a regional approach would increase energy security by diversifying countries' energy sources. They would also look at the regional impacts of energy-planning options. This funding priority would aim to significantly increase the use of clean energy in the region.

Projects would use participatory processes (regional and national workshops), to encourage dialogue about energy forecasting both at the planning stage and when outputs are ready. Participants would represent a diverse cross-section of stakeholders, including policy makers, government officials, academics, development partners, the private sector, civil society organizations, and scientific groups. A website would also be created to disseminate information. The RDB estimates the cost of achieving these objectives to be $1 million.

Assessment of the Economics of Climate Change

This sustainable development priority seeks to begin a dialogue about the costs and benefits of regional (in comparison to unilateral) actions to mitigate and adapt to climate change. Projects will strive to raise public awareness—within the Indopotamia basin, policy makers, academia, media, NGOs, the private sector, and international donors—of the urgency of climate change challenges facing the region and their likely social and economic impacts. The ultimate goal is to reach a consensus among stakeholders as to the steps needed to address climate change in the Indopotamia basin, including investment programs, policies, and development actions. The RDB could then better align its operations to support the regional consensus. The first phase of the project would involve a study on the economics of climate change in the basin, comparing regional with unilateral actions. The study results would be disseminated in different languages through print and Web-based media, and participants would also work with the media to communicate and disseminate the findings. The RDB estimates the cost of achieving these objectives to be $1 million.

Knowledge Development for Water Financing

This category would seek to work with scientific organizations to develop sector-based case studies of best water-management practices. Pilot and demonstration activities will seek to develop replicable models that can be scaled up to other situations. Sectors would include irrigation, sanitation, climate-change-oriented sustainable water-use, flood management, and urban water management. The case studies would include recommendations for applying the best water-management practices more widely, including conditions under which the practices are most and least appropriate. The case studies would be packaged into print and Web-based media and widely disseminated. Overall, this category of projects will improve water governance in the region by quickly increasing access to improved water supply and sanitation, better irrigation, and reducing flood risks. The products of these projects would also help the Indopotamia countries prioritize water investments, develop capacity, and implement institutional reforms. The RDB estimates the cost of achieving these objectives to be $1 million.

Improving Environmental Compliance and Enforcement

Projects eligible for financing under this category would seek to develop principles and good practices of environmental compliance and enforcement, with a goal of reducing pollution flowing downstream and other harm to transboundary water resources. Projects would also finance strategies for countries to adopt these principles and practices and strengthen their environmental agencies' enforcement capabilities. The RDB estimates the cost of achieving these objectives to be $1 million.

Appendix B: Selection Criteria

The RDB has used this list of criteria in the past to help stakeholders decide between different sustainable development priorities:

Precedent

According to this criterion, the Indopotamia countries should consider how similar decisions have been made in other international river basins.

Return on Investment

Cost effectiveness is another way to decide between different funding priorities. This criterion considers the predicted benefits of project options, given their expected costs.

Protection of as Many Lives as Possible

This criterion considers the impacts of different priorities on reducing vulnerability to climate change and other natural disasters. It also includes improvements in health standards, such as those that lead to lower infant mortality, fewer incidences of waterborne diseases, and reduced exposure to arsenic.

Improved Food Security

Improving food security is a common goal of the Indopotamia countries. Sustainable development options may be prioritized based on how they are expected to affect food security throughout the basin and the three countries, for example, by supporting farmers (which leads to improved agricultural services), or by increasing annual crop yields.

Enhanced Equity in the Allocation of Regional Water Benefits

This criterion encourages more equitable sharing in the many benefits of the use of the Indopotamia's water. Equity considerations include benefit sharing between countries, as well as between rural and urban areas, and between households.

Improved Political Relationships

Sustainable development options may be prioritized because they support peace building, regional cooperation, and integration.

General Instructions for Segment 4: Final Packages and Implementation Strategies

Congratulations! You have already made progress in these negotiations toward regional cooperation. In this final segment of negotiations, you will try to reach agreement on a new regime for long-term development of the Indopotamia River. The Regional Development Bank (RDB) wants an agreement on governance arrangements to ensure that participants will follow through on their commitments. The RDB also wants a legal basis for follow-up investments and loans. The RDB hired a consultant to produce a Draft Agreement in advance of today's meeting, which you and the other participants received in advance of the meeting (see Appendix A for the Draft Agreement).

According to the consultant, the parties have made significant progress toward an agreement, but four issues remain unresolved. These form the agenda for today's negotiation: the form of the agreement, its geographic scope and the range of issues it will encompass, procedures for reviewing and revising the agreement, and data procedures.

Unresolved Issues

Form of the Agreement

Some parties would prefer that the Agreement take the form of a Declaration as a useful first step toward regional cooperation. Others prefer a Convention that is a legally binding set of agreements. Those who want a Convention are worried that a Declaration will not ensure concrete actions follow.

Scope: Geography and Range of Issues

There are two critical questions remaining about the scope of the Agreement. First, should the participants limit the Agreement to surface and ground water-management issues on which they can make substantial progress in a relatively short time? These would include the question of allocating water resources during the dry season and dealing with flooding issues in the wet season. Others would rather have the Agreement address the full range of natural resource-management issues that relate in any way to surface, ground, and green waters of the Indopotamia basin. This broader scope would include land-use practices, food security, sustainable development, energy supplies, and urban migration within the watershed.

Second, should all aspects of the Agreement apply to the whole River basin, or for some issues, like the definition of sustainable development, should the Agreement scope be defined separately by the three countries and cover areas outside the basin?

Review and Revision Procedures

Some parties want to make certain that the Agreement (and specific elements of it, such as funding priorities), will be reviewed according to a set schedule and revised when new information becomes available. Others are concerned about the negative impacts such a changeable Agreement might have on long-term investments.

Data Procedures

The participants disagree about sharing data with one another, especially about proposed projects that will consume water. The most contentious issue is whether sharing data and safeguarding one another's interests require only informing one another of such projects, or whether they will need to receive consent from the other countries.

Appendix A: Agreement to Cooperate on Questions Concerning the Water Management of the Indopotamia (Cambridge Agreement)

At the Cambridge Conference dealing with the issues of Indopotamia water-management, representatives of the governments of the Indopotamia countries Gamma, Alpha, and Beta, Mu State in Alpha, the Regional Development Bank, the Global Water Management Organization, the International Conservation Institute, and Water Infrastructure Engineering and Design,

- Being aware of the need for and serious challenges to comprehensive management of the Indopotamia for the benefit of the peoples of the Indopotamia countries and for the economic and social development thereof;
- Being aware of the risks posed by climate change to the Indopotamia River and the need to develop adaptation strategies to help alleviate the impacts of climate change on water management;
- Being convinced that the effectiveness of water-management actions could be increased through the multilateral cooperation of all Indopotamia countries; and
- Guided by generally accepted principles and rules of international law in accordance with the interests and the sovereignty of all Indopotamia countries,

Agree to the following:

1. Establish a Conference of the Parties (COP).

The COP will hold annual meetings to implement this Agreement. The first COP will be held within six months of this Agreement.

2. Participation

- Gamma, Beta, and Alpha are Parties to the Agreement;
- The other signatories of the Agreement, intergovernmental financial institutions and bilateral donors providing financing for implementation of the Agreement, including RDB, may attend the COP as Observers; and

- Other international and national organizations, both governmental and non-governmental, may attend the COP as Observers, as long as a majority of the Parties and Observers consider the organizations' competencies relevant to the Agreement and do not object to their participation.

3. Rules of Procedure Governing Decision-Making

Decision-making rules of the COP and of any subsidiary bodies, such as scientific working groups, will be established at the first meeting of the COP. Each Party to the Agreement will have one vote.

4. Financial Mechanisms

The Regional Development Bank (RDB) will establish a Trust Fund for the Indopotamia region.

- Through the Trust Fund, the RDB will coordinate, harmonize, and administer contributions from international financing organizations and any donors; and
- Funds from the Trust Fund will support investment programs, activities, and bodies agreed upon by the COP.

The COP will decide on a reasonable annual payment schedule for the Parties, to support the agreed-upon activities and bodies created by the COP.

- The COP will establish procedures for reviewing the amount of the agreed-upon annual payment, including conditions under which the Parties can rely on bilateral or multilateral financing to pay their dues.

5. Data Procedures

The Parties will strive to generate knowledge products, best practices, and guidelines to inform investment and promote a coordinated approach to addressing their water-management issues. Toward these objectives and to safeguard all countries' interests, the Parties will share data about planned projects that might affect the volume of water resources.

These will include:

- Improving and harmonizing data collection and analysis methodologies for monitoring the state of the basin;
- Developing protocols for sharing comparable data with the COP;
- Developing protocols for evaluating implementation of the Agreement and its overall impacts; and
- Establishing a subsidiary scientific body to implement agreed-upon data-sharing activities, coordinating joint fact-finding initiatives, and advising and informing the COP of the results of its work.

6. Review

To promote effective implementation, the COP will review the obligations of the Parties and institutional arrangements of the Agreement, including funding priorities.

7. Implementing Agency

The Parties agree to establish an agency for the purpose of implementing this Agreement. The functions of the agency will include:

- Facilitating and coordinating bi- or multilateral actions toward implementing the Agreement;
- Coordinating the annual meetings of the COP and meetings of subsidiary bodies as requested;
- Coordinating with the RDB and other international bodies;
- Entering into administrative and contractual arrangements under the guidance of the COP; and
- Compiling and disseminating reports it receives.

8. Dispute Resolution

The Parties will seek to settle disputes by peaceful means.

If the Parties cannot reach agreement by negotiation they may seek settlement according to the following provisions:

- Mediation or arbitration.
- Submit the dispute to the International Court of Justice.
- A consultative process for resolving disputes related to the Agreement and its implementation, which the COP will consider designing.

GLOSSARY

adaptive management: a decision-making process ideally suited for natural resource management given the complexity of the systems involved. Adaptive management is a flexible approach to making and implementing decisions in the face of uncertainty that requires experimentation, careful monitoring of results or impacts and subsequent rounds of purposeful adjustment.

Apalachicola-Chattahoochee-Flint (ACF): a river basin that drains 19,800 square miles of western Georgia, northern Florida, and eastern Alabama in the United States. Nearly 2.6 million people depend on the ACF for their water.

AquaPedia: a managed wiki that gathers case studies of boundary-crossing water management and water conflicts around the world. It provides reliable, relevant, and readily available water information and wisdom from users and producers of explicit and tacit water knowledge.

Arbitration: a form of alternative dispute resolution that allows a neutral party to hear a case and make a binding determination regarding the outcome.

Best alternative to a negotiated agreement (BATNA): refers to a negotiator's walk-away option: what a party is most likely to be left with if no agreement is reached.

boundary-crossing: refers to crossing domains (natural, societal, and political), scales (e.g., spatial, temporal, jurisdictional, national, institutional, knowledge) and levels (e.g., seconds, minutes, days, years, etc. for temporal scales).

CALFED Bay Delta Program: a collaboration between 25 state and federal agencies to improve California's water supply and the ecological health of the San Francisco Bay/Sacramento–San Joaquin River Delta.

capacity-building: see *social learning*

climate change: any change in climate over time, whether due to natural variability or as a result of human activity. This usage differs from that in the Framework Convention on Climate Change, where climate change refers to a change of climate that is attributed directly or indirectly to human activity that alters the composition of the global atmosphere and that is in addition to natural climate variability observed over comparable time periods. Adaptive capacity is the ability of a system to adjust to climate change (including climate variability and extremes) to moderate potential damages, to take advantage of opportunities, or to cope with the consequences. Vulnerability is the degree to which a system is susceptible to, and unable to cope with, adverse effects of climate change, including climate variability and extremes. Vulnerability is a function of the character, magnitude, and rate of climate change and variation to which a system is exposed, its sensitivity, and its adaptive capacity.

coalitions: formed when stakeholder groups join together for the sake of promoting/ protecting a common interest in negotiations. See *multiparty negotiation.*

co-evolution: a dynamic and continuous process by which water networks change over time. Water networks exist within their own environment, and they are also part of that environment. As their environment changes, they evolve to ensure their survival. Because they are part of their environment, evolution of the water network changes their environment, as well.

collaborative adaptive management (CAM): a flexible approach to designing and implementing negotiated agreements in situations with high levels of uncertainty, complexity, and controversy. CAM allows for experimentation, learning, and adjustments to earlier decisions as necessary. CAM requires a clear understanding of how to measure proposed efforts, followed by careful monitoring, usually in cooperation with all relevant stakeholders.

complex systems: the field of complex systems challenges the notion that by understanding a system at a component level we can understand the system as a whole. To put it in simple terms and borrowing the words of Miller and Page (2007): "one and one may well make two, but to really understand *two* we must know both about the nature of *one* and the meaning of *and*". Unless we understand the nature of different components (i.e., nature of *one*) of a system and their evolving and continuously changing interdependent relationships (i.e., meaning of *and*) and feedbacks, we cannot understand the behavior of a complex system.

complex water management problems: see *simple and complex water problems.*

complexity theory: attempts to explain the behavior of networks and systems (especially natural systems) that cannot be understood solely in terms of their constituent parts or mechanistic accounts of how those parts interact. Systems amenable to description using complexity theory are characterized by non-linearity, uncertainty, interactions, and emergence.

complicated water management problems: those that involve water problems that operate in somewhat unpredictable ways, and about which the means and ends of action are highly contested (regardless of how much scientific information is available because of differences in competing values).

concession trading: see *hard bargaining*

consensus building: procedures used to generate unanimity or near-total agreement among stakeholders involved in a particular conflict or decision-making situation. A five-step collaborative decision-making process: (1) Stakeholder Assessment; (2) Clarifying Responsibilities; (3) Deliberation; (4) Written Agreement; and (5) Implementation.

convening: bringing parties together in "informal" efforts to animate decision-making by government agencies or officials. Informal negotiations ought to be convened by one or more agencies or organizations with formal decision-making authority in order to make a "governance connection" between ad hoc assemblies and formal public decision-making structures. The convening agency should be prepared to make an explicit commitment to support proposals that emerge from informal problem-solving as long as representation is handled properly.

desalination: converting salt water to fresh water. Desalination techniques offer a way to cope with water scarcity and are applicable in a range of developing countries.

diplomacy: describes the practice of interaction among nations aimed at avoiding hostility among the parties. Because of the complexity of the interactions among stakeholders, resource availability, and additional confounding factors, specialized techniques have been developed and are evolving to better solve specific types of problems, leading to views of "environmental or water diplomacy" that are both more specific, focusing on allocation or management of a specific resource or group of resources, and simultaneously more broad, examining meeting the needs of diverse stakeholder groups that may not be well-represented by sovereign states or legislative entities (see *water diplomacy*).

domain: a domain is a sphere of knowledge, influence, or activity. The *Water Diplomacy Framework* focuses on three domains: Natural, Societal, and Political (see *natural domain, societal domain, political domain*).

Dublin Statement and Guiding Principles: over 500 participants from more than 100 countries, international, intergovernmental, and non-governmental organizations attended the International Conference on Water and the Environment (ICWE) in Dublin, Ireland, in January 1992. Then, world leaders assembled at the United Nations Conference on Environment and Development (UNCED) in Rio de Janeiro in June 1992 and prepared the Dublin Statement on Water and Sustainable Development. This statement and associated conference report provides recommendations for action at local, national, and international levels, based on four guiding principles: (1) water is a finite and vulnerable resource, essential to sustain life, development and the environment; (2) development and management should be based on a participatory approach, involving users, planners, and policy-makers at all levels; (3) women play a central part in the provision, management, and safeguarding of water; and (4) water has an economic value in all its competing uses and should be recognized as an economic good.

embedded water: water used to produce other products (e.g., food, clothes, cars, etc.) that are traded internationally. In terms of water diplomacy, water-intensive products should be made and exported by water-rich countries, not the other way around. Embedded water is also referred to as "virtual water."

emergent property: the emergent property of a water network is derived from the interactions and feedback of nodes and links, and is not observable or inherent in any node or link of the network considered separately.

Everglades: a vast area of wetlands in the central and southern regions of Florida. This book includes a case example that describes how a long-standing disagreement over the impact of phosphorous on downstream ecosystems was resolved with the help of a mediator.

feedback loop: mechanism or process that channels information or other reactions to past influences or activities back to various network nodes. They can be part of a chain of interacting relationships.

flexible resource: traditionally, water has been viewed and managed as a finite resource. Any water set aside for one group is presumably unavailable to other groups. Ongoing and fundamental change in the creating and sharing of knowledge and new technologies now allow the same resource to be used in multiple ways by multiple users. The *Water Diplomacy Framework* views water as a flexible resource.

Ganges River Water Supply Treaty: a 1996 international treaty between India and Nepal that regulates the sharing of river waters at Farakka for flow augmentation during the rainy season. The treaty sought to provide mutual benefits for the people of both countries, but due to the lack of public involvement in its formulation, it was not perceived as equitable.

Global Water Partnership (GWP): founded in 1996 to foster integrated water resources management (IWRM). The GWP envisions a water secure world with a mission to support the sustainable development and management of water resources at all levels. GWP takes its guiding principles from the Dublin and Rio statements (1992). More information is at www.gwp.org/en/About-GWP/

governance: the act of governing. In a water diplomacy context, it is important to consider governance in the context of formal government institutions, as well as networked, collaborative arrangements involving unofficial stakeholders.

ground rules: the procedural mandates that groups agree to live by during the deliberation and implementation of collaborative efforts.

group think: psychological phenomenon that occurs when the desire to minimize conflict in a decision-making group overrides a realistic appraisal of alternative ideas or critical evaluation of viewpoints.

hard bargaining: negotiation style that pits parties against each other, frames the outcome in terms of "win/lose" (see *zero-sum thinking*), and encourages parties to withhold information rather than share it in an effort to solve problems collaboratively. Hard bargaining leads to "concession trading," which drives parties to their lowest acceptable agreements and may produce a "winner" in the short term, but undermines trust, sabotaging future negotiations and relationships. Hard bargaining is informed by prisoner's dilemma-style game theory, principal-agent theory, and decision-analysis (Pareto Optimality).

Indopotamia: a water management role-play simulation that takes place in the fictitious Indopotamia River basin. Designed to teach about managing boundary-crossing water conflicts using the *Water Diplomacy Framework*.

Indus Water Treaty: a 1960 water-sharing treaty between India and Pakistan that provides Pakistan with assurances that the source waters of the Indus basin, which are located in India, will not be constrained so as to create drought or famine in Pakistan.

informal problem solving: forums for collaborative decision-making that allow many stakeholders to participate in solving problems, and provide policy advice with coalitional support to public officials with decision-making authority.

Integrated Water Resources Management (IWRM): a process which promotes the coordinated development and management of water, land, and related resources in order to maximize economic and social welfare in an equitable manner without compromising the sustainability of vital ecosystems and the environment.

Joint Fact Finding (JFF) or mediated modeling: a collaborative method for gathering and analyzing scientific or other technical data. The parties determine together what information would be most useful to gather, and how to gather it. Parties may agree on a particular computer-assisted modeling approach to help them work through the information. Joint fact-finding does not eliminate disagreement, it only makes clearer what the parties are likely to accept as common information and where and why they disagree about interpreting this material.

joint gains: see *mutual gains*

knowledge transfer: the concept that "lessons learned" in any given water network should be shared with other water networks in order to improve water management capacity on a global scale. See *social learning*.

Komadugu-Yobe Basin: a water basin in Nigeria. A Water Charter was developed using a participatory process that included plans for "ongoing communication, cooperation and coordination." The charter specifically addressed the responsibilities of the different stakeholders in implementing the agreement, as well as future mechanisms for cooperation among them.

Komati River Basin: a shared water-course between South Africa, Swaziland and Mozambique. In this book, we include a case-study example of a successful collaborative process that asked stakeholders to produce initiatives that would be both sustainable and politically plausible for the Komati River Basin.

levels: see *scales and levels*

mediator/professional neutral: a skilled intermediary who maintains a nonpartisan posture while working with groups seeking to reach agreement. Mediators can play a significant role in identifying and organizing participants, specifying "ground rules" for negotiation, and assisting the parties in generating a creative resolution of their differences. Mediators or other neutral professionals may be asked to monitor the implementation of agreements they help to generate, or initiate reassembly of the parties to review progress or deal with perceived violations of what was promised. Mediators have specialized training and in some contexts are formally accredited or recognized by government agencies.

Mekong River Commission (MRC): as part of the United Nations Development Program, the MRC participated in a facilitated negotiation to produce the 1995 Mekong Agreement, a framework for cooperation in the Mekong River Basin. The Mekong River is a principal water-source for several communities in China, Laos, Thailand, Myanmar, Cambodia, and Viet Nam. The Mekong Agreement was signed by the four lower basin countries: Cambodia, Laos, Thailand, and Viet Nam.

multiparty negotiation: any negotiation with more than two parties. The key difference between multiparty and two-party negotiation is the opportunity for "coalition building," or for like-minded stakeholders to gain leverage over other parties. Both blocking and winning coalitions are inherently unstable.

mutual gains: almost all parties can get better results from a value-creating negotiation than from hard bargaining. A mutual gains approach to negotiation rests on the assumption that joint fact-finding, the discovery of interlocking trades, contingent commitments and an adaptive approach to handling uncertainty can "maximize joint gains." "All-gain" negotiations usually require the assistance of a neutral facilitator or mediator to manage the problem-solving process.

natural domain: in the context of water diplomacy, refers to three variables: water quantity (Q); water quality (P); and ecosystems (E) and the interdependancies and feedbacks between them. This domain interacts with the societal and political domains.

nearly self-enforcing agreements: many negotiations result in an agreement that is little more than a statement of goals or objectives. These tend to fall apart during implementation. Nearly self-enforcing agreements include contingent commitments to deal with uncertainty. They also include dispute-resolution provisions that state how parties will proceed if and when commitments are not being met. These agreement packages include provisions such that the constituent groups would rather see the agreement succeed than fail.

non-stationarity: see *stationarity*

non-zero-sum negotiation: a joint problem-solving framework that assumes outcomes favorable to one party are not necessarily detrimental to another. Requires a commitment to search for value-creating trades.

organizational learning: refers to adaptive behavior in a public or private organization in response to changes in its environment. Usually requires modifications in basic operating rules in light of experience.

Piave River Basin: river basin in northeastern Italy. Game-theory has been used to analyze water allocation negotiation in this basin following a traditional zero-sum approach to negotiation.

policy-shed: A policy-shed is a geographic area over which a governmental entity has legislative authority, such as a nation, state, province, county, or municipality.

political domain: competition, interconnection, and feedback among natural and societal processes occur within this domain (see: *natural domain, societal domain*).

pre-negotiation: steps that take place prior to negotiations in an effort to ensure the best "groundwork" is in place for effective negotiation. Includes identifying the right stakeholders, developing an agenda and timetable for the negotiation, and establishing joint problem-solving procedures. Usually refers to the work that individual parties do on their own to prepare for negotiations as well as the work done on behalf of all the participants by a convener or a *professional neutral*.

problem-shed: a "geographic area that is large enough to encompass management issues, but small enough to make implementation feasible."

professional neutral: An individual or team who helps parties navigate the negotiation process. A professional neutral is usually trained as a mediator or facilitator and does not have a personal stake in the outcome of the negotiation. See *mediator*.

Program on Negotiation (PON): is an inter-university consortium dedicated to improving the theory and practice of negotiation and dispute resolution. Includes faculty, students, and staff from Harvard University, Massachusetts Institute of Technology, and Tufts University.

regime: refers to either the behavior of a natural process (such as annual rainfall) or a set of rules circumscribing the way governments or agencies will relate to each other.

representation: in most negotiations, the individuals at the table are not simply advocates for their own personal interests, but represent larger groups of stakeholders who count on them to speak on their behalf. Engaging a full range of stakeholder interests is vital to the credibility of water diplomacy efforts.

riparian: refers to geographic areas and habitats that exist at the interface between land and rivers or streams. Riparian zones have important ecological contributions to species biodiversity, soil conservation, and overall ecosystem health.

scales and levels: we define scales as the spatial, temporal, quantitative, or analytical dimensions used to measure and study any phenomenon. We denote levels as the units of analysis located at different positions on each scale (e.g., levels refer to seconds, days, seasons, decades, etc. for temporal scales). An understanding of the scope of a particular problem in terms of domain, scale, and level is not easily transferrable to other domains, scales, or levels.

scenario planning: a tool to deal with uncertainty and complexity in a contingent fashion. Scenario planning involves simulating alternative futures and generating a "portfolio" of potential actions to meet the conditions presented in the full range of alternative futures.

sea-level rise: the primary mechanisms for measurable increases in mean global sea-level that occur within a human timescale are contributions from melting continental glaciers and thermal expansion of seawater.

simple and complex water management problems: simple problems are characterized as easily knowable, while complicated problems are not simple, but are knowable and somewhat predictable. Complex problems, on the other hand, are not easily knowable, and are usually unpredictable.

single text procedure: when a negotiation agreement is crafted by a single working group or individual with the resulting document being edited by involved parties to craft an agreement that is acceptable to all.

social learning: the learning that occurs through observations of others or interactions with others. Social learning is important for enhancing the knowledge and capability of individuals, organizations, and networks over time. Reflection on experience in negotiations should be used to strengthen the underlying capacity of the agencies and actors involved. Also called societal learning, or (at the organizational level) capacity building.

societal domain: in the context of water diplomacy, refers to three variables: social values and cultural norms (V); assets, including economic and human resources (C); and governance and institutions (G), and the interdependencies and feedbacks between them. This domain interacts with the natural and political domains.

societal learning: see *social learning*

sovereignty: the quality of having independent authority over an area (such as a geographic territory). The issue of sovereignty, particularly at a national level, is a critical dynamic in water diplomacy because approaches to water management may fall into two categories: decisions that a higher authority can enforce, and those that the parties themselves must negotiate and mutually enforce, as with neighboring sovereign nations.

stakeholder assessment: an early step in the consensus-building process used to identify appropriate stakeholder groups and their representatives. Often undertaken by a professional neutral. Typically involves confidential interviews with a wide range of groups, organizations, or individuals that might be concerned about decisions or policies about to made or changed. The neutral summarizes his or her findings in a draft document that protects participants' anonymity, while providing transparency regarding the issues at hand.

stakeholders: individuals, groups, or organizations with an interest in a project or decision. In water management, stakeholders usually include a large and diverse set of organizations and individuals including governments, corporations, special-interest groups (environmental, business, recreation, etc.), residents (local, regional, etc.). Stakeholders often have conflicting interests, biases, and pre-existing relationships with other stakeholders.

stationarity: an attribute of a situation in which statistical properties are unlikely to change over time. The properties of a non-stationary process will change over time.

systems: a system is usually defined as an interconnected set of components organized in a bounded domain to achieve an objective. Systems are often more than the sum of their parts. "Systems engineering" works beautifully when systems are bounded and the cause–effect dynamics involved are well understood. Sending Apollo to the moon or optimizing water distribution are brilliant success stories of the systems engineering approach to solving complicated, but well-defined, scientific problems. When system boundaries are ill defined and cause–effect relationships are not well understood, a systems engineering approach may not provide much insight.

systems engineering: according to the International Council on System Engineering Fellows: "System engineering is an engineering discipline whose responsibility is creating and executing an interdisciplinary process to ensure that the customer and stakeholder needs are satisfied in a high quality, trustworthy, cost efficient, and schedule compliant manner."

uncertainty: a state of having limited knowledge where it is difficult to exactly describe the current state or a future outcome.

value creation: in negotiation theory refers to the generation of options, packages or deals that respond to the interests of all the parties. This is usually accomplished through trades that take advantage of differences in the priorities or interests of each "side." Value creation is most likely to be successful if the parties consider ways of meeting their own interests while simultaneously meeting the interests of others.

virtual water: see *embedded water*

water diplomacy: rooted in the ideas of complexity theory and negotiation, water diplomacy is a theory and practice of adaptive water management being developed at Tufts, MIT, and Harvard. Conflicts over water occur when natural, societal, and political forces interact, creating complex water-networks. Management of them becomes critically important as population growth, economic development, and climate change create pressure on finite water-resources. Science or policymaking alone is not sufficient. Sustainable solutions can only come from diplomacy that takes science, policy, and politics into account.

Water Diplomacy Framework (WDF): The WDF is rooted in complexity theory and non-zero-sum negotiation, and seeks to bridge objective knowledge and contextual understanding of water. The WDF posits that complex water-problems might be more effectively managed by thinking about water as a flexible resource and invoking three key assumptions about water networks: (1) water networks are open and continuously changing as a function of the interactions among natural, societal, and political forces; (2) water network characterization and management must account for uncertainty, nonlinearity, and feedback; and (3) management and evolution of water networks ought to be adaptive and negotiated using a "non-zero-sum" approach.

Water Diplomacy Graduate Program at Tufts University: educates doctoral students who will become the next generation of teachers and scholars of water diplomacy. Supported by the Integrative Graduate Education and Research Traineeship (IGERT) of the US National Science Foundation, this program will educate interdisciplinary water professionals to think across boundaries, integrate explicit and tacit knowledge, and link knowledge and action from multiple perspectives to help resolve water issues through mutual-gains negotiations.

Water Diplomacy Research Coordination Network (Water Diplomacy RCN): a National Science Foundation-funded partnership among research centers and reflective water professionals from around the world involved in a range of water issues.

Water Diplomacy Workshop (WDW): a one-week "train-the-trainer" event that seeks to build the capacity of mid- to senior-level water professionals to use the Water Diplomacy Framework for managing simple, complicated, and complex water problems. Through highly interactive presentations and exercises, it helps participants master important water-network management tools, and gain the skills needed to teach these tools to others.

water network: a network (or graph) made up of a collection of nodes (vertices) and links (edges) between the nodes. These links can be directed or undirected, and weighted or un-weighted. A water network can be described as an interconnected set of nodes representing natural, societal, and political variables and processes. The flow of information among these nodes to update their status makes them dynamic.

wicked problems: a class of social problems that cannot be successfully tackled using traditional analytical approaches that work successfully on simple problems. They use this term to describe social planning problems that are often impossible to characterize or solve because of incomplete, often contradictory and changing requirements. Wicked policy problems cannot be definitively described because in a pluralistic society there are no objective definitions of public good or equity. Consequently, policies that attempt to address social problems in general cannot be objectively categorized as correct or incorrect and are unlikely to yield "optimal solutions." We use this term to describe complex water-management problems that cross multiple boundaries and involve multiple stakeholders with competing objectives.

World Commission on Dams: an independent commission formed in 1997 that released ten key guidelines for dam building. In this book, we highlight a Nepal case where pro- and anti-dam stakeholders undertook a series of joint studies to assess Nepal's hydropower experience as an initial step to developing a set of country-specific guidelines based on the World Commission on Dam's report.

zero-sum negotiation: this is a game-theoretic formulation that assumes a favorable outcome for one party in a negotiation or a conflict must always be matched by non-favorable outcomes for another party in the same situation. Thus, the sum of all outcomes is zero.

zero-sum thinking: a win/lose perspective on negotiations that assumes gains to one party result in matching losses to other parties. In water negotiations, this thinking manifests itself in the notion that water can only be used once, and by only one party.

Zone of Complexity: a setting in which water management problems involve intricate linkages between natural and societal forces. Complexity theory—drawing on assumptions about nonlinearity, emergence, interaction, feedback, mutation, and adaptation—can be used to identify the parameters of management problems that fall into the Zone of Complexity. Problem-solving in this zone is distinct from what works in simple and complicated settings.

Zone of Possible Agreement (ZOPA): the overlapping negotiating space between all parties' *BATNAs* (on the low-end) and the maximum amount of value that each party can imagine and create (the high end).

INDEX

Note: Page numbers in **bold** are for figures, those in *italics* are for tables.